THE MANY SCALES IN THE UNIVERSE

The Many Scales in the Universe

JENAM 2004 Astrophysics Reviews

Edited by

J.C. DEL TORO INIESTA
Instituto de Astrofísica de Andalucía, CSIC, Spain

E.J. ALFARO
Instituto de Astrofísica de Andalucía, CSIC, Spain

J.G. GORGAS
Universidad Complutense de Madrid, Spain

E. SALVADOR-SOLÉ
Universitat de Barcelona, Spain

and

H. BUTCHER
ASTRON, Dwingeloo, The Netherlands

Springer

A C.I.P. Catalogue record for this book is available from the Library of Congress.

Additional material to this book can be downloaded from http://extras.springer.com.

ISBN-10 94-017-8146-X
ISBN-13 978-94-017-8146-6
ISBN-10 1-4020-4526-3 (eBook)
ISBN-13 978-1-4020-4526-4 (eBook)

Published by Springer,
P.O. Box 17, 3300 AA Dordrecht, The Netherlands.

www.springer.com

Printed on acid-free paper

Contents

Preface

This book gathers the invited conferences presented at the Thirteenth Joint European and National Astronomical Meeting (JENAM) organized by the European Astronomical Society (EAS) and the Spanish Astronomical Society (Sociedad Española de Astronomía, SEA) and hosted by the Instituto de Astrofísica de Andalucía (IAA - CSIC). The event, held from September 13 to 17, 2004 in Granada, was at the same time the Sixth Scientific Meeting of the SEA. The proceedings of such national meetings are traditionally collected in a series of books generically entitled *Highlights of Spanish Astrophysics*, which will be momentarily interrupted in this occasion as the contributions of the Spanish astronomers to the meeting share the present publication with those of their colleagues from other European countries or, to be more exact, from countries all over the world. The meeting brought together, indeed, more than 450 participants from 30 different countries, making it one of the most successful JENAMs ever celebrated. This success was undoubtedly due to the fact that, as readily seen from the titles of the parallel sessions, the scientific scope of this JENAM reached, for the fist time, all fields of astronomy and astrophysics. In fact, there was still another parallel session opened to teachers and professionals of popularisation in astronomy whose proceedings are published elsewhere.

As members of the SOC we were aware about the risk that, in willing to cover all fields of astronomy, no particular topic would be treated duly in depth, which might make the meeting loose interest for potential attendants. For this reason, a dozen of renowned specialists in different fields were invited to review central aspects in the different domains of astrophysics. We want to thank here all of them for the superb *mise á jour* they offered to the audience. These main talks constitute the main body –the printed part– of the present book, becoming in this way an excellent tool for astronomers to have a general up-to-date look on the most exiting items of current astronomical research. The papers are organized per sessions, per scales of the Universe indeed, according to the structure of the meeting; the last session is devoted to instrumentation. It would certainly be a pity not to preserve at the same time the remaining 150 short oral contributions and 324 posters presented at the meeting with interesting results on a variety of specific items. For this reason, attached to the book is a CD including all these contributions in full format. We hope they will be very useful for astronomers working on the corresponding domains.

The Editors
Granada, July 2005

JENAM 2004
SEA
eas
http://www.iaa.es/jenam2004/
The many scales in the Universe
Roads to cosmology
The life of galaxies
Your favourite stars and their environments
The Sun and planetary systems
Real and virtual instruments
Teaching and communicating astronomy
GRANADA
Joint European and National Astronomical Meeting
13-17 September, 2004
CSIC
esa

Committees

Honour Committee

H.R.H. Felipe de Borbón y Grecia
Prince of Asturias

H.E. Mª Jesús Sansegundo
Minister of Education and Science

H.E. Manuel Chaves González
President of the Autonomous Region of Andalusia

H.E. Cándida Martínez López
Counsellor of Education and Science of the Autonomous Region of Andalusia

H.E. Francisco Vallejo Serrano
Counsellor of Innovation, Science, and Business of the Autonomous Region of Andalusia

H.E. Antonio Martínez Caler
President of the Province of Granada Government

H.E. José Torres Hurtado
Mayor of Granada

H.E. Carlos Martínez Alonso
President of the National Council for Scientific Research

H.E. David Aguilar Peña
President of the University of Granada

Scientific Organizing Committee

E.J. Alfaro (Granada, executive co-ordinator)
G. Bernabéu (Alicante)
H. Butcher (Dwingeloo, co-chair)
M. Castellanos (Madrid)
R. Domínguez–Tenreiro (Madrid)
F. Figueras (Barcelona)
M.A. Gómez-Flechoso (Madrid)
J. Gorgas (Madrid)
B. Gustaffson (Upsala)
A. Herrero Davó (La Laguna)
V. Martínez (Valencia)
V. Martínez Pillet (La Laguna)
P. Papaderos (Göttingen)
J.M. Rodríguez–Espinosa (La Laguna)
E. Salvador–Solé (Barcelona, co-chair)
A. Sánchez–Lavega (Bilbao)
J. Silk (Oxford)
M. Stavinschi (Bucharest)
N. Thomas (Bern)
J.C. del Toro Iniesta (Granada)

Convenors of the Parallel Sessions

Roads to Cosmology
R. Domínguez Tenreiro (Madrid)
E. Gaztañaga (Barcelona)
J. Silk (Oxford)

The Life of Galaxies
J.M. Vílchez (Granada)
F. Figueras (Barcelona)
R. Bender (Munich)

Your Favourite Stars and their Environments
A. Herrero (La Laguna)
A. Alberdi (Granada)
B. Gustaffson (Uppsala)

The Sun and Planetary Systems
V. Martínez–Pillet (La Laguna)

A. Sánchez–Lavega (Bilbao)
N. Thomas (Bern)

Real and Virtual Instruments
J.M. Rodríguez–Espinosa (La Laguna)
A.I. Gómez de Castro (Madrid)
M. Dennefeld (Paris)

Teaching and Communicating Astronomy
V. Martínez (Valencia)
G. Bernabéu (Alicante)
M. Stavinschi (Bucharest)

Local Organizing Committee

Antxon Alberdi
Emilio J. Alfaro
Begoña Ascaso
Daniel Cabrera
Benigno Cantero
Ricardo Casas
Antonio J. Delgado
Daniel Espada
Omaira González–Martín
Carolina Kehrig
Lucas Lara
Silbia López de Lacalle
M. Ángeles Martínez–Carballo
David Orozco
Mayra Osorio
Daniel Reverte
José Ruedas
Jose Carlos del Toro Iniesta (Chair)
José Sabater
M. Carmen Sánchez–Gil
M. Jesús Vidal

THE COSMIC MICROWAVE BACKGROUND ANISOTROPIES: OPEN PROBLEMS

Enrique Martínez-González and Patricio Vielva
Instituto de Física de Cantabria, CSIC–Universidad de Cantabria
martinez@ifca.unican.es,vielva@ifca.unican.es

Abstract We present a review of some interesting theoretical and observational aspects of the Cosmic Microwave Background anisotropies. The standard inflationary model presents a simple scenario within which the homogeneity, isotropy and flatness of the universe appear as natural outcomes and, in addition, fluctuations in the energy density are originated during the inflationary phase. These seminal density fluctuations give rise to fluctuations in the temperature of the Cosmic Microwave Background (CMB) at the decoupling surface. Afterward, the CMB photons propagate almost freely, with slight gravitational interactions with the evolving gravitational field present in the large scale structure (LSS) of the matter distribution and a low scattering rate with free electrons after the universe becomes reionized by the first stars and QSOs. These secondary effects slightly change the shape of the intensity and polarization angular power spectra of the radiation, the so called C_ℓ. The C_ℓ contain very valuable information on the parameters characterizing the background model of the universe and those parametrising the power spectra of both matter density perturbations and gravitational waves. The extraction of this richness of information from the C_ℓ is complicated by the superposition of the radiation coming from other Galactic and extragalactic emissions at microwave frequencies. In spite of this, in the last few years data from sensitive experiments have allowed a good determination of the shape of the C_ℓ, providing for the first time a model of the universe very close to spatially flat. In particular the WMAP first year data, together with other CMB data at higher resolution and other cosmological data sets, have made possible to determine the cosmological parameters with a precision of a few percent. The most striking aspect of the derived model of the universe is the unknown nature of most of its energy contents. This and other open problems in cosmology represent exciting challenges for the CMB community. The future ESA Planck mission will undoubtely shed some light on these remaining questions.

Keywords: Cosmic Microwave Background, Data Analysis, Cosmological Parameters

J. C. del Toro Iniesta (eds.), The Many Scales in the Universe, 1–23.

1. Introduction

In recent years there has been an explosion of cosmological data allowing a strong progress in the characterization of the cosmological model of the universe. In particular, recent data of the temperature of the Cosmic Microwave Background (CMB) have played a crucial role in the determination of the cosmological parameters. Many experiments aimed to map the temperature of the CMB have been and are being carried out, and many others are now being planned to extend present capabilities in resolution and sensitivity. From the cosmological data already collected we know that the spatial geometry of the universe is close to Euclidean, with most of the energy density in the form of the so called "dark energy (DE)" (with an equation of state close to a cosmological constant) and most of the matter density in the form of the so called "cold dark matter (CDM)" (weakly interactive matter with negligible velocity). About the primeval density perturbations we know that they were close to the adiabatic type (constant ratio of the matter number density to photon density for each matter component) and Gaussianly distributed with a nearly scale-invariant power spectrum (i.e. the gravitational potential fluctuations are the same at all scales). The generic predictions of inflation —a very brief episode of drastic expansion in the very early history of the universe close to the Planck time— are consistent with the characteristics of the model of the universe and of the primeval density fluctuations just mentioned, the so called "concordance model". Thus, inflation provides us with a plausible scenario within which we can understand the horizon and flatness problems (i.e. why causally disconnected regions appear statistically similar in the CMB sky and why the density of the universe is so close to the critical one) and also the origin of the density perturbations. Although a physical model of inflation is still lacking, however, it provides us with a phenomenological scenario within which we can conceptually understand some fundamental problems related to the origin of the "special" properties of the universe —i.e. its homogeneity and critical density— and its large scale structure (LSS) matter distribution which otherwise will have to rely on ad hoc initial conditions. In the past decade two of the experiments on-board of the NASA COBE satellite, FIRAS and DMR, established unambiguously the black-body electromagnetic spectrum of this radiation (Mather et al. 1994, 1999) and the level and approximate scale-invariant shape of the spectrum of density fluctuations (Smoot et al. 1992). The former data rebated some previous results which indicated possible distortions from the black-body spectrum, establishing the thermal origin of the CMB with a high precision. The latter confirmed the gravitational instability theory for the formation of the LSS and determined an initial spectrum of fluctuations characterized by density fluctuations with approximately equal amplitude when entering the horizon. These fundamental results set the basis for the later developments in our understanding of the universe and opened the era of precision cosmology. At the end of the decade several experiments determined that the universe is close to spatially flat (BOOMERANG, De Bernardis et al. 2000; MAXIMA, Hanany et al. 2000). More recently the NASA WMAP satellite (Bennett et al. 2003) nicely confirmed this result and, together with other higher resolution CMB experiments as well as the galaxy survey 2dFGRS (Percival et al. 2001), determined the cosmological parameters with a few percent errors. The combination of different cosmological data sets not only helps to improve the precision of the cosmological parameters but, what is more important, shows the compatibility of the different data sets in the context of the concordance model. The most important features of this model is the distribution of the energy content of the universe with about 70% of dark energy, 25% of cold dark matter (CDM) and only 5% of baryons. The detection of the polarization fluctuations by DASI (Kovac et al. 2002) and the later determination of the temperature-polarization angular cross-power spectrum by WMAP (Kogut et al. 2003) has strongly confirmed the concordance model. Very recently, there have been several works finding evidences of positive cross-correlations between the CMB temperature map and the LSS distribution of galaxies (see e.g. Boughn and Crittenden 2004). These results represent an independent piece of evidence of the existence and dominance of dark energy in the recent history of the evolution of the universe (assuming that the universe is close to flat). Although the concordance ΛCDM model represents a good fit to the CMB data as well as to other cosmological data sets —namely LSS galaxy surveys, primordial Big-Bang nucleosynthesis, measurements of the Hubble constant, SN Ia magnitude-redshift diagram— there are however some problems related to the observations and also to their interpretation.

The aim of this paper is to review some of the most relevant theoretical and observational results on the CMB field and to point out some open problems associated to them. For more detailed discussions the reader is refereed to the many good reviews written in the literature, some of them more oriented to the theory (e.g. Hu and Dodelson 2002, Challinor 2004, 2005, Cabella and Kamionkowski 2004) and others to the experiments (e.g. Barreiro 2000, Bersanelli, Maino and Mennella 2002, Mennella et al. 2004).

The rest of this review is as follows. In Section 2 we describe the main properties of the CMB temperature anisotropies and the physical effects that originate them. In Section 3 the most relevant properties of the CMB polarization anisotropies are considered. The cosmological parameters are defined in Section 4 and their values determined from recent CMB data and in combination with other cosmological data sets are presented in Section 5. Finally, open problems in the CMB field are discussed in Section 6.

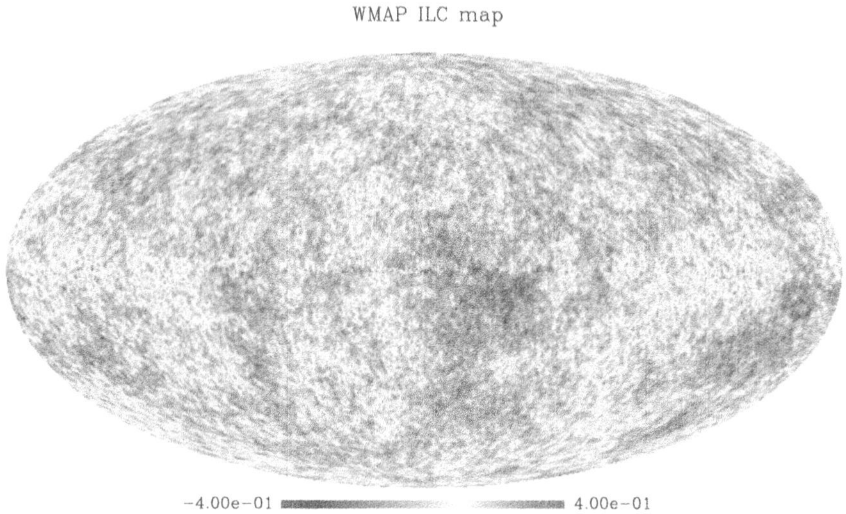

Figure 1. The CMB as seen by WMAP (data obtained from the LAMBDA wave page of NASA).

2. CMB temperature anisotropies

Description

The anisotropies of the CMB —the fluctuations in the intensity of this radiation as a function of the direction in the sky— are interpreted as a realization of a random field on the sphere. Behind this interpretation is the idea that we aim to understand and possibly explain the anisotropies in a statistical manner. Because of the blackbody spectrum of the CMB, the anisotropies in the intensity are generally given as temperature ones. Since the temperature fluctuations are a function of the spherical coordinates it is convenient to expand them in spherical harmonics,

$$\frac{\Delta T}{T}(\vec{n}) = \sum_{\ell,m} a_{\ell m} Y_{\ell m}(\vec{n}), \tag{1}$$

where $a_{\ell m}$ are the spherical harmonic coefficients. A very important quantity is the CMB anisotropy angular power spectrum, C_ℓ, the second order moment of the $a_{\ell m}$ defined as

$$< a_{\ell m} a^*_{\ell' m'} >= C_\ell \delta_{\ell\ell'} \delta_{mm'}. \tag{2}$$

The null correlation of the harmonic coefficients for different ℓ or m is due to the homogeneity and isotropy of the universe —i.e. the Cosmological Principle assumed as a fundamental pillar in cosmology and consistent with all of our observations up to date. Here it is important to notice, however, that the previous equation does not hold if the universe possesses a nontrivial topology. An important property of the C_ℓ is that if the anisotropies are Gaussian —as predicted by Inflation— then all the statistical information is contained in it (reported deviations from Gaussianity will be discussed in section 6). The correlation function of the temperature fluctuations, $C(\theta) \equiv < \Delta T/T(\vec{n}_1)\Delta T/T(\vec{n}_2) >$ with $\vec{n}_1\vec{n}_2 = \cos(\theta)$, is related to the C_ℓ through the Legendre transform

$$C(\theta) = \sum_\ell \frac{2\ell+1}{4\pi} C_\ell P_\ell(\cos\theta). \tag{3}$$

The isotropy of the field is now reflected in the independence of the correlation from the direction. Although the two quantities $C(\theta)$ and C_ℓ contain the same information, however the null correlation of the $C_{\ell's}$ for different values of ℓ makes the latter quantity preferable for cosmological studies. The quantity which is usually displayed is $\ell(\ell+1)C_\ell/2\pi$, i.e. the power per logarithmic interval in ℓ for large ℓ. Since the $\ell = 1$ moment is dominated by our motion only moments with $\ell \geq 2$ are considered.

There is a fundamental limitation in the accuracy with which the angular power spectrum C_ℓ can be determined due to the fact that we can only observe one last-scattering surface —the error associated to it is called "cosmic variance". Since there are $2\ell + 1$ $a_{\ell m}$ coefficients for a given ℓ then for Gaussian temperature fluctuations the cosmic variance is easily calculated (from the dispersion of a chi-squared distribution with $2\ell + 1$ degrees of freedom)

$$\Delta C_\ell = \frac{1}{\sqrt{\ell + 0.5}} C_\ell. \tag{4}$$

There are other sources of error which should be added to the cosmic variance. One is the fraction of the sky f_{sky} covered by an experiment, which increases the error by a factor $f_{sky}^{-1/2}$ (Scott et al. 1994). Another one is the sensitivity of the experiment whose noise power spectrum adds to the C_ℓ of the cosmic signal in equation 4. Finally, there is a source of error coming from the process to separate the cosmic signal from the other foregrounds, namely Galactic emissions[1] (synchrotron, free-free and thermal dust), extragalactic sources, galaxy clusters and the lensing effect from the LSS. This is usually a complex task to perform which requires multifrequency observations and whose error is difficult to estimate a priori. Estimate of the errors assuming different situations

[1]See e.g. Watson et al. (2005) for evidence of possible anomalous Galactic emission

and methodology can be found in, e.g., Hobson et al. (1998), Tegmark et al. (2000), Bouchet and Gispert (1999), Vielva et al. (2001), Maino et al. (2002), Herranz et al. (2002), Martínez-González et al. (2003), Delabrouille et al. (2003).

In figure 1 the map of temperature anisotropies as measured by WMAP is shown (Bennett et al. 2003). Although a large effort has been made to eliminate all foreground contributions however some residuals from Galactic foregrounds can still be seen in the central horizontal band of this map. In order to avoid the introduction of this contamination in the data analysis a mask covering that and other regions of the map is usually applied.

Physics

The CMB anisotropies are usually divided in primary and secondary, depending if they are produced before or after the last-scattering surface. The primary anisotropies are the most interesting ones to study the cosmological parameters characterising the universe as well as its basic matter and energy constituents. The secondary ones are produced by scattering of the CMB photons from ionized matter —either generated after the reionization of the universe or present as hot gas in the central regions of galaxy clusters and whose interaction with the CMB photons is known as the Sunyaev-Zeldovich effect (SZ, Sunyaev and Zeldovich 1972)—, and by gravitational interactions acting on them during their travel from the last-scattering surface to the observer —causing both redshift and lensing effects. They are expected to be sub-dominant, at least in relation to the C_ℓ, up to $\ell \approx 2500$.

The main physical effects producing the primary anisotropies and observed in the synchronous-comoving gauge (the observer's gauge) can be summarized in the following equation (Martínez-González, Sanz and Silk 1990, Sanz 1997):

$$\frac{\Delta T}{T}(\vec{n}) \approx \frac{\phi_e(\vec{n})}{3} + \int_e^o \frac{\partial \phi}{\partial t} dt + \vec{n} \cdot (\vec{v}_o - \vec{v}_e) + \left(\frac{\Delta T}{T}(\vec{n})\right)_e \tag{5}$$

The first two terms in the r.h.s. account for the gravitational redshift suffered by the CMB photons in their travel toward us along the direction $\vec{n}$. They are called the Sachs-Wolfe (SW) and Integrated Sachs-Wolfe (ISW) effects, respectively (Sachs and Wolfe 1967). The third term is a Doppler effect due to the motion of the emitters at the last-scattering surface. Finally the fourth term is the temperature fluctuation at that surface.

Since the gravitational field has a large scale of interaction the SW and ISW effects dominate the angular power spectrum at the largest angular scales (ℓ smaller than a few tens) producing an approximate plateau (see figure 2). Notice also that these scales are the most affected by the cosmic variance (see equation 4). The ISW effect has two contributions: an early one before recom-

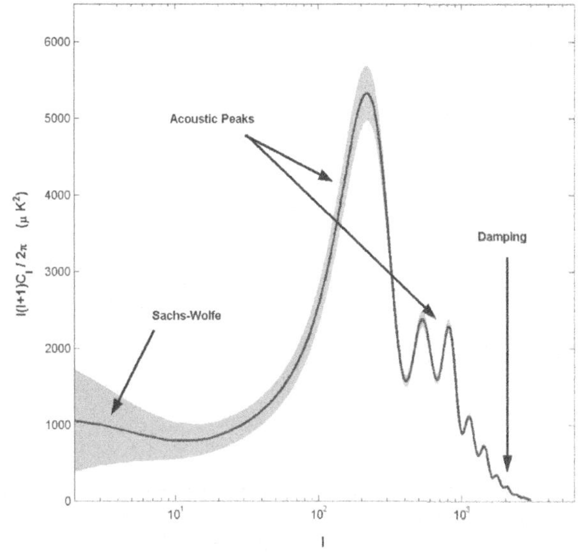

Figure 2. C_ℓ for the best fit model given in Bennett et al. (2003) (see also table 2). The range of ℓ where the different physical effects dominate is indicated. The gray band envolving the C_ℓ is the unavoidable error due to the cosmic variance (see text). The spectrum has been computed using the CMBFAST code (Seljak and Zaldarriaga 1996).

bination produced by the imperfect coupling of photons and baryons causing variations in the gravitational potential with time, and a late one after recombination due to changes in the gravitational potential with time. The late ISW effect is produced by the linear evolution of the large-scale matter distribution at later times if the universe is different form Einstein-de Sitter. As we will see at the end of Section 5 this is an independent test from the standard one (involving usually CMB, SN Ia or LSS data sets) about the existence of a dark energy component dominating the dynamics of the universe at recent cosmic times.

The Doppler effect plus the temperature fluctuations at the last-scattering surface dominate the shape of the angular power spectrum at ℓ larger than a few tens. At multipoles above a hundred, the angular power spectrum exhibits a sequence of oscillations called acoustic peaks. They are driven by the balance between the gravitational force pulling to compress over-dense regions and radiation pressure pushing in the opposite direction. The position and amplitude of these oscillations are very much determined by the total energy content of the universe (or equivalently its geometry) and the nature and amount of the different components.

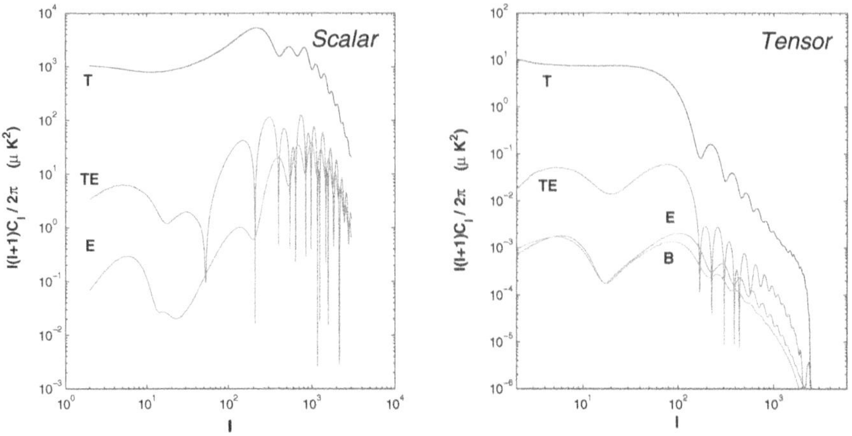

Figure 3. Temperature and polarization angular power spectra for scalar and tensor perturbations. The ratio tensor/scalar has been chosen to be $r = 0.01$. Apart from this parameter, the vlues for the rest of the parameters have been fixed to the values of the best fit model given in Bennett et al. 2003 (see also table 2). All the spectra have been computed using the CMBFAST code.

Finally, at large ℓ ($\gtrsim$ 1000) the C_ℓ power starts to decrease due to the width of the last-scattering surface and the imperfections of the coupling of the photon-baryon fluid (Silk effect, Silk 1968).

In Figure 2 one can see the range of ℓ where the different effects dominate the angular power spectrum.

3. Polarization

Thomson scattering of the radiation generates linear polarization at the end of recombination when the growth of the mean free path of the photons allow anisotropies to grow (for a detailed description of the physics of polarization see, e.g., Challinor 2005). The expected level of polarization is only of $\approx 5\%$. From the Stokes parameters Q, U two rotationally invariant quantities can be constructed E, B (often referred to as the E-mode and B-mode, see Zaldarriaga and Seljak 1997, Kamionkowski et al. 1997). Under parity transformations E remains unchanged and B changes sign, and therefore the cross-power $< a^B_{\ell m} a^{T*}_{\ell m} >=< a^E_{\ell m} a^{B*}_{\ell m} >= 0$. It is for these parity properties that only three angular power spectra are required to characterise CMB polarization: $C^E_\ell, C^B_\ell, C^{TE}_\ell$.

An important property of the E,B decomposition is that whereas the E-mode polarization can be generated by both density perturbations and gravitational waves the B-mode can only be generated by the second ones. Therefore the detection of B-mode polarization is a unique proof of the existence of primordial gravitational waves, opening the door to new areas of physics. In particular, it can also be used to impose strong constraints on the energy scale of inflation and on the shape of the inflaton potential. A complication comes from the lensing effect due to the gravitational potential of the LSS which converts part of the E-mode polarization in B-mode. Fortunately, this effect dominates over the primary one only at $\ell \gtrsim 100$ leaving the rest of the spectrum unaltered. All the temperature and polarization angular power spectra for scalar and tensor perturbations are plotted in figure 3.

As for the temperature angular power spectrum, the accuracy in the determination of the polarization angular power spectra has also a fundamental limit imposed by the cosmic variance. For Gaussian distributed anisotropies the errors in $C_\ell^E, C_\ell^B, C_\ell^{TE}$ are given by:

$$\Delta C_\ell^{E,B} = \frac{1}{\sqrt{\ell + 0.5}} C_\ell^{E,B}, \quad \Delta C_\ell^{TE} = \frac{1}{\sqrt{\ell + 0.5}} \left(C_\ell^T C_\ell^E + C_\ell^{TE^2} \right)^{1/2}. \quad (6)$$

There are a number of Galactic and extragalactic foregrounds which complicate the observation of the CMB polarization. Although their relevance depends very much on the frequency they are expected to be very harmful, specially for the B-mode due to its relatively small amplitude. For a recent estimate of the effect of foregrounds on the polarization observations see Tucci et al. (2005).

Only recently the DASI experiment has been able, for the first time, to detect anisotropies in polarization (Kovac et al. 2002). Afterward, the WMAP experiment measured the TE angular cross-power spectrum with more precision and covering a much wider range of scales (Kogut et al. 2003). More recently the CBI experiment has measured the E angular power spectrum with more resolution than DASI allowing the detection of the second, third and fourth acoustic peaks (Redhead et al. 2004b).

In addition to the temperature anisotropies, the anisotropies in polarization contain very relevant and independent information. In particular, in the standard model the maxima in the E spectrum are out of phase with those in the T spectrum due to the fact that polarized radiation is sensitive to the velocity of the fluid, and the velocity and density are out of phase in an acoustic wave. This shift is precisely what has been recently reported by Readhead et al. (2004b) based on CBI data (see figure 7 below). As we will see later, anisotropies in the polarization are very relevant to confirm the best fit model given by the temperature data and to constrain specific parameters as the optical depth to which they are very sensitive.

4. Cosmological parameters

There are a number of cosmological parameters that account for very different fundamental physical properties of the universe and that influence the radiation angular power spectrum in many ways. These parameters characterize the background model of the universe (assumed to be the homogeneous and isotropic Friedmann-Robertson-Walker model), the primordial scalar and tensor fluctuations and the reionisation history. At present around 12 parameters are considered for the data analysis.

The parameters characterizing the background model of the universe are the following:

- *Physical baryonic density,* w_b: $w_b = \Omega_b h^2$.
- *Physical matter density,* w_m: $w_m = \Omega_m h^2$. Ω_m is given by the sum of the baryonic density Ω_b, the CDM density Ω_{CDM} and the neutrino density Ω_ν.
- *Physical neutrino density,* w_ν: $w_\nu = \Omega_\nu h^2$ (up to now only upper limits are found for this parameter).
- *Dark energy equation of state parameter,* w: $w \equiv p_{DE}/\rho_{DE}$.
- *Dark energy density,* Ω_{DE}: In case w were constant and took the value -1 then the dark energy takes the form of a cosmological constant and its energy contribution is represented by Ω_Λ.
- *Hubble constant,* h: $h \equiv H_0/100\text{km s}^{-1}\text{ Mpc}^{-1}$.

The reionisation history of the universe influences the CMB by a single parameter:

- *Optical depth,* τ: $\tau = \sigma_T \int_{t_r}^{t_0} n_e(t)dt$, where σ_T is the Thompson cross-section and $n_e(t)$ is the electron number density as a function of time.

The cosmological parameters that characterize the matter and gravitational waves primordial power spectra are:

- *Amplitude of the primordial scalar power spectrum,* A_s:
  $P_s(k) = A_s(k/k_0)^{n_s}$, where $k_0 = 0.05\text{Mpc}^{-1}$.
- *Scalar spectral index,* n_s.
- *Running index,* α: $\alpha = dn_s/d\ln k$. It accounts for the deviations from a pure power-law. Its value is normally determined at the scale $k_0 = 0.05\text{Mpc}^{-1}$.
- *Tensor-to-scalar ratio,* r: $r = A_t/A_s$.
- *Tensor spectral index,* n_t: from the consistency relation of inflation it is normally assumed that $n_t = -r/8$.

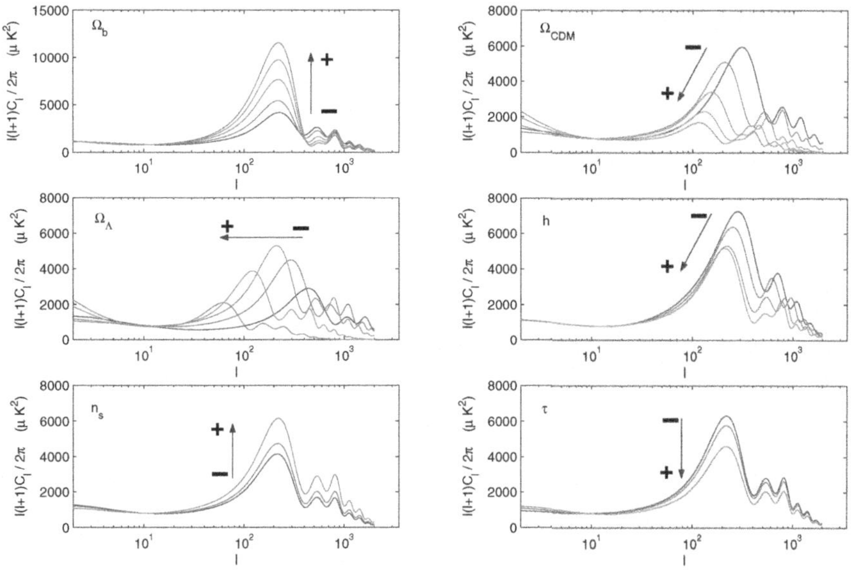

Figure 4. Dependence of C_ℓ on some relevant cosmological parameters (The C_ℓ has been produced with the CMBFAST code.

Besides those parameters there are two possible types of primordial matter density fluctuations: adiabatic (the entropy per particle is preserved) and isocurvature (matter fluctuations compensate those of the radiation conserving total energy density). As commented in the Introduction the standard model of inflation predicts fluctuations of the adiabatic type.

Analyses that combine CMB with LSS require an additional parameter accounting for the bias b of the galaxy density respect to the matter density.

In figure 4 we show the changes in C_ℓ produced by the variation of some of the most relevant cosmological parameters. The many different ways in which the C_ℓ can vary with the parameters produces degeneracies complicating their accurate determination. In particular, there is a well known geometric degeneracy involving the matter and dark energy densities. This is shown in figure 5 where almost identical angular power spectra are obtained for three different values of the curvature ($\Omega_k = 1 - \Omega_m - \Omega_{DE}$). This example illustrates the need of including in the analysis additional cosmological data sets (like SN Ia, LSS, cluster density, CMB polarization, ...) to break the degeneracies.

5. Cosmological constraints

In recent years there has been an explosion of cosmological data which have made possible a strong advance in the determination of the cosmological para-

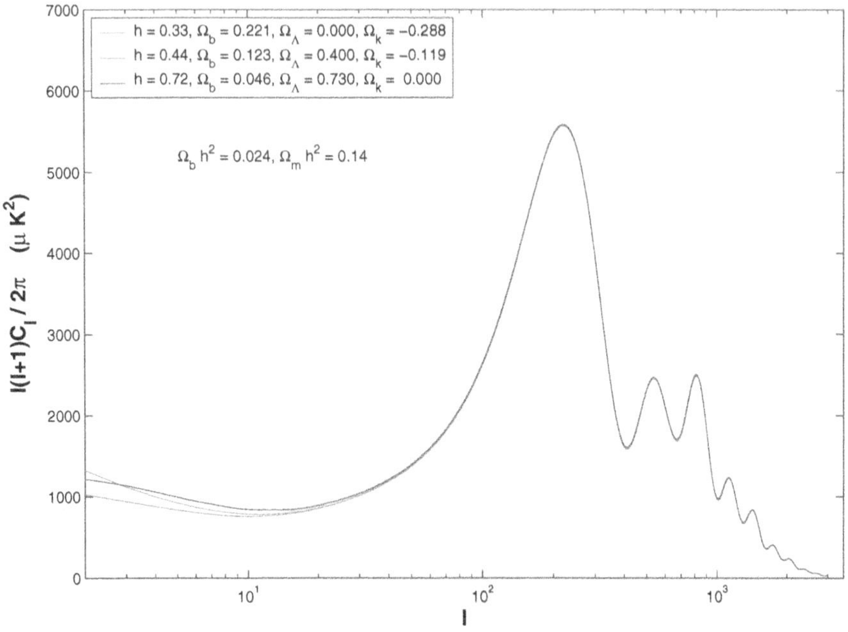

Figure 5. Geometric degeneracy for three values of the curvature Ω_k. Apart from the parameters given in the figure the rest of the parameters have been fixed to the best fit model given in Bennett et al. (2003). The three spectra have been computed using the CMBFAST code.

meters. Below we summarize the main results for CMB data alone and when combined with other cosmological data sets.

WMAP and higher resolution CMB experiments

The WMAP data alone is able to put strong constraints on some cosmological parameters when some priors are assumed in the analysis (Spergel et al. 2003). Table 1 summarizes the results when a flat universe is assumed, the prior $\tau < 0.3$ is imposed on the optical depth and only 6 parameters are considered. It is interesting to notice that the Einstein-de Sitter universe (i.e. a spatially flat universe with null dark energy) is rejected at a very high confidence level. Besides, the value of the optical depth parameter τ is essentially determined by the TE angular cross-power spectrum.

The WMAP data also test the type of the primordial fluctuations. The clear detection of the first acoustic peaks as well as the detection of the TE cross-correlation imply that the fluctuations were primarily adiabatic, in agreement with the standard inflationary model.

Table 1. Cosmological parameters using only WMAP data. In the fit the universe is assumed to be spatially flat and the value of the optical depth is constrained to $\tau < 0.3$ (from Spergel et al. 2003)

Parameter	*Values (68% CL)*
w_b	0.024±0.001
w_m	0.14±0.02
h	0.72±0.05
A_s	0.9±0.1
τ	$0.166^{+0.076}_{-0.071}$
n_s	0.99±0.04

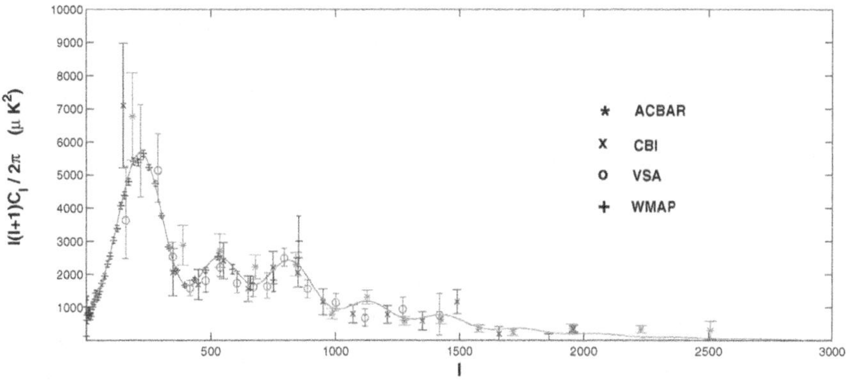

Figure 6. C_ℓ measured by WMAP, ACBAR, CBI and VSA, together with the best fit model given by Bennett et al. (2003). The values of the parameters are listed in table 2.

The situation is improved if temperature data from high resolution CMB experiments is included in the analysis (Spergel et al. 2003, Dickinson et al. 2004). In figure 6 the C_ℓ obtained from the experiments WMAP (Hinshaw et al. 2003), ACBAR (Kuo et al. 2004), CBI (Readhead et al. 2004a) and VSA (Dickinson et al. 2004) is shown. Although the polarization measurements are not yet sensitive enough to significantly improve the constraints already derived from the temperature one (except for τ), however the measured peaks in the E-mode C_ℓ, which are out of phase with the temperature C_ℓ ones, suppose an independent evidence of the standard model and, more specifically of the adiabatic type of the primordial matter density fluctuations. The TE angular cross-power spectrum from DASI (Leitch et al. 2004), WMAP (Kogut et al. 2003) and CBI (Readhead et al. 2004b) as well as the E-mode polarization angular power spectrum from DASI and CBI are shown in figure 7.

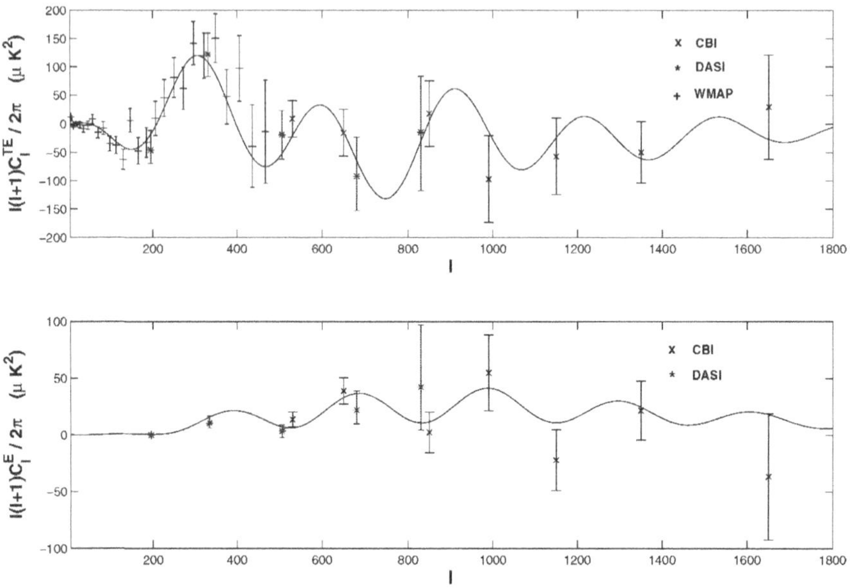

Figure 7. TE and E angular power spectra measured by DASI, WMAP and CBI (TE) and DASI and CBI (E). Also plotted is the best fit model given by Bennett et al. (2003). The values of the parameters are listed in 2.

Combination with other cosmological data sets

The present richness of data is not exclusive of the CMB. Also large galaxy surveys covering a large fraction of the sky and measuring the three-dimensional power spectrum using $\approx$ 200000 redshifts, like the 2dFGRS (Percival et al. 2001) or SDSS (Tegmark et al. 2004a), have been recently achieved. If the initial fluctuations are Gaussian, all the information is included in the power spectrum $P(k)$ and thus this is the quantity normally used for the analysis. The cosmological parameters determine the matter power spectrum at present through the initial amplitude and the spectral index A_s, n_s and the transfer function connecting linearly the present and initial $P(k)$. An additional bias parameter b, assumed to be scale-independent and linear, is needed to relate galaxy and matter density. Moreover, non-linear evolution of the matter density power spectrum should be also considered in the comparison with observations. The present matter density power spectrum as well as the luminosity distance-redshift diagram determined with SN Ia depend on the cosmological parameters in a very different way from the C_ℓ, and so the combined data sets can potentially impose much more severe constraints. Results of combining CMB data with LSS galaxy surveys (Percival et al. 2001, Tegmark et al. 2004a), HST key project value of H_0 (Freedman et al. 2001), SN Ia magnitude-redshift data (Riess et al. 2001, Tonry et al. 2003), Lyα forest power spectrum (Croft et al. 2002, McDonald et al. 2004) are given in Bennett et al. (2003), Spergel et al. 2003, Tegmark et al. (2004b) and Seljak et al. (2004). As it is shown in these papers, the constraints on the cosmological parameters are greatly improved when combining all those cosmological data sets and, what is even more important, the best fit model is an acceptable fit for all of them. Table 2 shows the best fit cosmological parameters given in Bennett et al. (2003) for the combination WMAP+CBI+ACBAR+2dFGRS. Alternative combinations include WMAP, high redshift SN Ia and abundances of rich clusters of galaxies (see e.g. Rapetti, Steven and Weller 2005 and Jassal, Bagla and Padmanabhan 2005 for a detailed analysis of the constraints imposed by these data sets on the dark energy equation of state). In figure 8 confidence contours are given for the (Ω_m, Ω_{DE}) plane combining WMAP, SN Ia (Knop et al. 2003) and galaxy cluster abundance (Allen et al. 2002). The complementarity of the three data sets is clearly noticed by the reduction of the contours of the combined data set compared to the individual ones.

Table 2. Cosmological parameters from WMAP, CBI, ACBAR and 2dFGRS combined data (from Bennett et al. 2003)

Parameter	*Values (68% CL)*
w_b	0.0224±0.0009
w_m	$0.135^{+0.008}_{-0.009}$
w_ν	<0.0076 (95% CL)
w	<-0.78 (95% CL)
Ω_{DE}	0.73±0.04
h	$0.71^{+0.04}_{-0.03}$
τ	0.17±0.04
A_s	$0.833^{+0.086}_{-0.083}$
n_s	0.93±0.03
α	$-0.031^{+0.016}_{-0.018}$
r	<0.90 (95% CL)

Integrated Sachs-Wolfe effect

There are strong evidences that the universe today is dominated by the dark energy density Ω_{DE}. The evidences came first from measurements of the luminosity curve and redshift of distance SN Ia, and later from surveys of the CMB anisotropy and LSS distribution of galaxies. More recently there have been independent tests that confirm this result. They come from the cross-correlation of the CMB map with LSS surveys which span a wide range in redshift. From the late ISW effect a non null signal is expected due to the fact

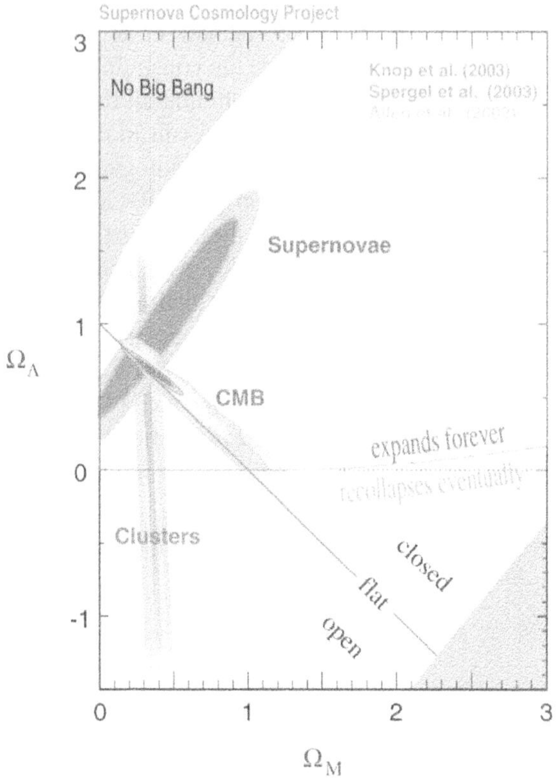

Figure 8. Confidence contours for the plane $(\Omega_m, \Omega_\Lambda)$ using SN Ia, CMB and cluster density data (taken from the Supernova Cosmology Project).

that the same gravitational potential created by the LSS at late times is also leaving an imprint on the CMB anisotropies. The amplitude and sign of the effect are determined by Ω_{DE}, Ω_k and h. If the universe is close to spatially flat, as found in a consistent way by very different cosmological data sets, then a positive LSS-CMB cross-correlation would be an unambiguous indication of the presence of dark energy.

This is what has been recently found by a number of authors using different surveys and methods (Boughn and Crittenden 2004; Fosalba, Gaztañaga and Castander 2004; Nolta et al. 2004; Afshordi, Loh and Strauss 2004; Vielva, Martínez-González and Tucci 2004). The maximum amplitude for the signal has been obtained in the latter paper by cross-correlating the radio galaxy sur-

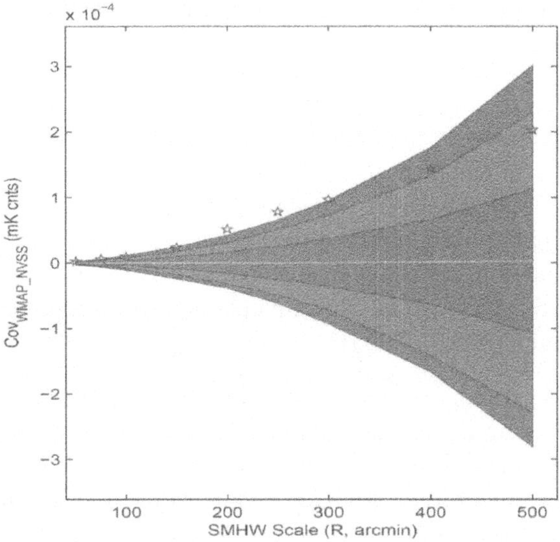

Figure 9. Detection of the ISW effect, above 3σ level, applying spherical wavelets to the NVSS and WMAP maps and computing the covariance of the wavelet coefficients as a function of the wavelet scale. The contours are 1, 2 and 3 σ (from Vielva, Martínez-González and Tucci 2004).

vey NRAO VLA Sky Survey (NVSS, Condon et al. 1998) with the WMAP data. It represents a detection $> 3\sigma$ (see figure 9).

6. Open problems

Looking at table 2 and considering the situation only ten years ago, the progress in the determination of the cosmological parameters has been spectacular. However, what is even more impressive is the concordance among all cosmological data sets available at present. This agreement in the results has made possible to build a well specified model of the universe, the so called concordance model. Nevertheless there are still pending problems related to the data fitting as well as the interpretation of the values of some of the parameters, whose solutions will need to wait for future more sensitive experiments (at least for some of them, see below).

In spite of the very precise overall fit of the theoretical C_ℓ to the WMAP data (Spergel et el. 2003), there are some significant discrepancies. There are also significant deviations from the assumed isotropic Gaussian random field hypothesis, property which is also assumed in the parameter estimation. Moreover, some parameters take uncomfortably high values and the nature of others is completely unknown. More specifically, some of the most relevant open problems are:

- *Anomalies at the lowest multipoles ($\ell = 2, 3$)*: deficit of power and the coincidence of orientations of the quadrupole and octopole in the WMAP data (Bennett et al. 2003, de Oliveira-Costa et al. 2004, Efstathiou et al. 2004). A better removal of the Galactic contamination narrowing the mask used for the analysis (as expected from the future Planck mission, see below) can help to clarify this situation. An attempt to solve this problem invokes a global non-trivial topology for the universe (Luminet et al. 2003).

- *Anomalies at intermediate multipoles ($\ell = 22, 40, 210$)*: glitches have been found around those multipoles in the WMAP data (Bennett et al. 2003). It is not clear if they are just statistical fluctuations, foreground contamination or represent real features. Some of them might be related to the asymmetries discussed below (Hansen et al. 2004).

- *Asymmetries and non-Gaussianity*: a number of authors have found significant deviations from isotropy and Gaussianity in the WMAP data. In particular asymmetries along the sky and/or non-Gaussianity have been found in the genus (Park 2004), low-order moments (Eriksen et al. 2004, Hansen et al. 2004), wavelets (Vielva et al. 2004, Mukherjee and Wang 2004), phases (Chiang et al. 2003), extrema (Larson and Wandelt 2004), directional wavelets (McEwen et al. 2005) and local curvature (Cabella et al. 2005). In addition a very cold spot has been detected and localized (Cruz et al. 2005). Although some of these results could be due to foregrounds others seem to be very difficult to explain in that way. Anisotropic models for the universe of the type Bianchi VII_h with very small vorticity and shear seem to provide a good fit to all those features (Jaffe et al. 2005); however they require a total $\Omega = 0.5$, which deviates significantly from the value of the concordance model.

- *Power excess at high multipoles ($\ell > 2000$)*: several high resolution experiments have found excess of power at multipoles $2000 < \ell < 3000$ (CBI, DASI, ACBAR). This excess can be associated, at least partially, to SZ emission from galaxy clusters (Bond et al. 2005) or to extragalactic sources (Toffolatti et al. 2005). Diego et al. (2004) suggested a sensitive way to separate between those possibilities using the skewness and kurtosis of the Mexican Hat wavelet coefficients.

- *High value of the optical depth*: from the WMAP data a high value of the Thompson optical depth is derived ($\tau \simeq 0.17$, see tables 1 and 2), implying a high redshift for the reionization of the inter-galactic medium of the universe $z \sim 20$ and, thus, that non-linear structures were already formed at these high redshifts. This result heavily relies on the high value of the TE angular cross-power spectrum at low multipoles measured by WMAP, and so a confirmation with the second and other years of data —particularly with the E-mode angular power spectrum— would be crucial. Again, sensitive polarization data covering a large fraction of the sky (as expected from Planck) would represent an important step forward to solve this problem.

- *Non-zero running index* α: WMAP data combined with other cosmological data sets slightly favours a running index different from zero (see table 2. This result is heavily supported by the data at small scales provided by the Lyα forest power spectrum (Croft et al. 2002). However, new analyses of the matter power spectrum from the Lyα forest find values of α perfectly consistent with zero (Seljak et al. 2005).

- *Isocurvature fluctuations*: Although the WMAP temperature and polarization data imply that the primordial density fluctuations were primarily adiabatic (also confirmed by the recent polarization measurements from CBI), however at the 2σ CL up to $\approx 50\%$ of the C_ℓ power can still be due to isocurvature fluctuations (Crotty, Lesbourgues and Pastor 2003). In addition, considering all possible combinations of adiabatic and isocurvature modes for each matter component will significantly lower the constraints in the parameters (Bucher, Moodley and Turok 2001). Future sensitive experiments like Planck will be able to further constraint the type of the fluctuations and their relative amounts.

- *Topological defects*: the acoustic oscillations measured in the C_ℓ by WMAP and other CMB experiments strongly reduce the possible role of topological defects in the universe. Recently, Bevis, Hindmarsh and Kunz (2004) and Fraisse (2005) have limited the contribution of topological defects (textures and cosmic strings) to the C_ℓ to be $\lesssim 20\%$. However, a fine determination of their possible sub-dominant contribution will probably have to wait for future more precise data.

- *Nature of dark matter and dark energy*: probably the most challenging problem in cosmology is to understand the nature of the dark energy and the dark matter which completely dominates the dynamics of the universe. For the former, a cosmological constant (with equation of state $w = -1$) seems to be in acceptable agreement with many different data sets. This situation is particularly embarrassing since a natural explanation in terms of the vacuum energy, given by quantum zero-point fluctuations of fundamental fields, implies a value around 100 orders of magnitude above the observed one. Regarding the dark matter, although several candidates have been proposed within particle physics theories, neither its nature nor its evolutionary history are known.

- *B-mode polarization*: detection of this mode would unambiguously indicate the existence of a primordial background of gravitational waves and thus would allow to constrain theories of the early universe. Due to its expected weak signal respect to the temperature, the E-mode anisotropies (see figure 3) and to the emission from the Galactic and extragalactic foregrounds (Tucci et al. 2005), its detection and precise determination will necessarily require the development of new very sensitive experiments.

On top of these problems there are others, more fundamental ones, related to the origin of inflation and to its dynamics: which is the particle physics framework for inflation?, how the universe started to inflate in the early universe?, which is the specific model of inflation within, e.g., the simple classification by Kinney (2002)? The ESA Planck mission, the third space mission after COBE and WMAP, scheduled for launch in 2007, will measure the temperature of the CMB over the whole sky with high sensitivity (few μK), resolution (down to 5 arcmin) and wide frequency coverage ($30 - 850$ GHz), allowing a good separation between CMB and foregrounds and thus providing an angular power spectrum with unprecedented precision limited only by cosmic variance. It will also provide a map of polarization with sensitivity beyond the one reached by previous experiments. These new measurements will certainly have a profound impact on our understanding of the origin and evolution of our universe (for more details on the Planck mission see the web page address given in the references). An open issue which may remain even after Planck is the detection of the B-mode polarisation. As stated above, the determination of this mode would have a tremendous impact on the theories of the early universe. There are now many plans to build experiments capable to reach the demanding sensitivities needed to search for the B-mode including space missions from ESA and NASA.

Finally, from the previous discussion we can state that the role played by the CMB on our understanding of the universe has been very relevant in the past and is expected to continue being so in the future.

Acknowledgments

We thank R.B. Barreiro for useful comments on the manuscript. We acknowledge financial support from the Spanish MEC project ESP2004-07067-C03-01. We also acknowledge the use of LAMBDA, support for which is provided by by the NASA Office of Space Science. We acknowledge the use of the software package CMBFAST (http://www.cmbfast.org) developed by Seljak and Zaldarriaga. The work has also used the software package HEALPix (http://www.eso.org/science/healpix) developed by K.M. Gòrski, E.F. Hivon, B.D. Wandelt, J. Banday, F.K. Hansen and M. Barthelmann.

References

Afshordi N., Loh Y.-S. & Strauss M. A., 2004, Phys. Rev. D, 69, 3524

Allen A.W., Schmidt R.W. & Fabian A.C., 2002, MNRAS, 334, L11

Barreiro R.B., 2000, New Astronomy Reviews, 44, 179

Bartolo N., Komatsu E., Matarrese S. & Riotto A., 2004, Phys. Rep., 402, 103

Bennett C. L. et al., 2003, ApJS, 148, 1

Bersanelli M., Maino, D. & Mennella A., 2002, Nuovo Cimento, 25, 1

Bevis N., Hindmarsh M. & Kunz M., 2004, Phys. Rev. D, 70, 043508

Bond et al., 2005, ApJ, 626, 12

Bouchet F.R. & Gispert R., 1999, New Astronomy Reviews, 4, 443

Boughn S. P. & Crittenden R. G., 2004, Nature, 427, 45

Bucher M., Moodley K. & Turok N., 2001, Phys. Rev. Lett., 87, 191301

Cabella, P. & Kamionkowski, M., 2004, astro-ph/0403392

Cabella P., Liguori M., Hansen F.K., Marinucci D., Matarrese S., Moscardini L. & Vittorio N., 2005, MNRAS, 358, 648

Cruz M., Martínez-González E., Vielva P. & Cayón L., 2005, MNRAS, 356, 29

Challinor A., 2004, astro-ph/0403344

Challinor A., 2005, astro-ph/0502093

Chiang L.-Y., Naselsky P.D., Verkhodanov O.V. & Way M.J., 2003, ApJL, 590, 65

Condon J.J. et al., 1998, AJ, 115, 1693

Croft R.A.C. et al., 2002, ApJ, 581, 20

Crotty P., Lesbourgues J. & Pastor S., 2003, Phys. Rev. D, 67, 123005

De Bernardis P. et al., 2000, Nature, 404, 955

de Oliveira-Costa A., Tegmark M., Zaldarriaga M. & Hamilton A., 2004, Phys. Rev. D, 69, 063516

Delabrouille J., Cardoso J.-F. & Patanchon G., 2003, MNRAS, 346, 1089

Dickinson C. et al., 2004, MNRAS, 353, 732

Diego J.M., Martínez-González E., Vielva P. & Silk J., 2004, astro-ph/0403561

Efstathiou G., 2004, MNRAS, 348, 885

Eriksen H.K., Hansen F.K., Banday A.J., Gorski K.M. & Lilje P.B., 2004, ApJ, 605, 14

Fosalba P., Gaztañaga E. & Castander F., 2004, ApJL, 597, 89

Fraisse A.A., 2005, Phys. Rev. Lett., submitted, astro-ph/0505402

Freedman W.L. et al., 2001, ApJ, 553, 47

Hanany S. et al., 2000, ApJ, 545, L5

Hansen F.K., Banday A.J. & Gòrski K.M., 2004, MNRAS, 354, 641

Herranz D., Sanz J.L., Hobson M.P., Barreiro R.B., Diego J.M., Martínez-González E. & Lasenby A.N., 2002, MNRAS, 336, 1057

Hinshaw G. et al., 2003, ApJ, 148, 135

Hu W. and Dodelson S., 2002, ARAA, 40, 171

Hobson M.P., Jones A.W., Lasenby A.N. & Bouchet F., 1998, MNRAS, 300

Jaffe T.R., Banday A.J., Eriksen H.K., Gorski K.M. & Hansen F.K.,2005, ApJL, submitted, astro-ph/0503213

Jassal H.K., Bagla J.S. & Padmanabhan T., 2005, astro-ph/0506748
Kamionkowski M., Kosowsky A. & Sttebins A., 1997, Phys. Rev. D, 55, 7368
Kinney W.H., 2002, Phys. Rev. D, 66, 083508
Knop R.A. et al., 2003, ApJ, 598, 102
Kogut A. et al., 2003, ApJS, 148, 161
Kovac J.M. et al. 2002, Nature, 420, 772
Komatsu E. et al., 2004, ApJS, 148, 119
Kuo C.L. et al., 2004, ApJ, 600, 32
Larson D.L. & Wandelt B.D., 2004, ApJ, 613, 85
Leitch et al., 2004, ApJ, submitted, astro-ph/0409357
Luminet J.-P., Weeks J.R., Riazuelo A., Lehoucq R. & Uzan J.-P., 2003, Nature, 425, 593
Maino D. et al., 2002, MNRAS, 334, 53
Martínez-González E., Sanz J.L. & Silk, 1990, ApJ, 335, 5
Martínez-González E., Diego J.M., Vielva P. & Silk J., 2003, MNRAS, 345, 1101
Mather J.C. et al. 1994, ApJ, 420, 439
Mather J.C. et al. 1999, ApJ, 512, 511
McDonald P. et al., 2004, ApJ, submitted (astro-ph/0407377)
McEwen J. D., Hobson M. P., Lasenby A. N. & Mortlock D. J., 2005, MNRAS, 359, 1583
Mennella A. et al., 2004, astro-ph/0402528
Mukherjee P. & Wang Y., 2004, ApJ, 613, 51
Nolta M. R. et al., 2004, ApJ, 608, 10
Park C.-G., 2004, MNRAS, 349, 313
Percival W.J. et al., 2001, MNRAS, 327, 1297
Planck mission, http://www.rssd.esa.int/Planck
Rapetti D., Steven S.W. & Weller J., 2005, MNRAS, 360, 555
Readhead A.C.S. et al., 2004a, ApJ, 609, 498
Readhead A.C.S. et al., 2004b, Science, 306, 836
Riess et al., 2001, ApJ, 560, 49
Sachs R. K. & Wolfe A. M., 1967, ApJ, 147, 73
Sanz J.L., 1997, In The Cosmic Microwave Background, eds. C.H. Lineweaver et al., Kluwer Academic Publishers, p. 33
Scott D.H., Srednicki M. & White M., 1994, ApJ, 421, 5
Seljak U. & Zaldarriaga M., 1996, ApJ, 469, 437
Seljak U. et al., 2005, Phys. Rev. D, 71, 103515
Silk J., 1968, ApJ, 151, 459
Smoot G. F. et al., 1992, ApJL, 396, L1
Spergel D. N. et al., 2003, ApJS, 148, 175
Sunyaev R.A. & Zeldovich Y.B., 1972, Comm. Astrophys. Space Phys., 4, 173
Tegmark M., Eisenstein D.J., Hu W. & de Oliveira-Costa A., 2000, ApJ, 530, 133
Tegmark M. et al., 2004a, ApJ, 606, 702
Tegmark M. et al., 2004b, Phys. Rev. D, 69, 103501
Tonry J.L. et al., 2003, ApJ, 594, 1
Tucci M., Martínez-González E., Vielva P. & Delabrouille J., 2005, MNRAS, 360, 935
Toffolatti L., Negrello M., González-Nuevo J., de Zotti G., Silva L., Granato G.L. & Argüeso F., 2004, A&A, in press, astro-ph/0410605

Vielva P., Barreiro R. B., Hobson M. P., Martínez-González E., Lasenby A. N., Sanz J. L. & Toffolatti L., 2001, MNRAS, 328, 1

Vielva P., Martínez-González E., Barreiro R.B., Sanz J.L. & Cayón L., 2004, ApJ, 609, 22

Vielva P., Martínez-González E. & Tucci M., 2004, MNRAS, submitted, astro-ph/0408252

Watson R.A., Rebolo R., Rubiño-Martín J.A., Hildebrant S., Gutiérrez C.M., Fernádez-Cerezo S., Hoyland R.J. & Battistelli E.S., 2005, ApJ, submitted, astro-ph/0503714

Zaldarriaga M. & Seljak U., 1997, Phys. Rev. D, 55, 1830

THE GRAVITATIONAL LENSING EFFECT IN COSMOLOGY

Genevieve Soucail
Observatoire Midi-Pyrenees
Toulouse, France
soucail@ast.obs-mip.fr

Abstract In this paper, I review some recent developments in the framework of gravitational lensing, mainly devoted to clusters of galaxies and large scale structures. Used as a powerfull tool in observational cosmology, gravitational lensing allows to assess both the questions of mass determination in clusters of galaxies and the study of background sources thanks to the magnification effect. Some cosmological implications will be reviewed.

Keywords: Cosmology, Observations, Gravitational lensing, Clusters of galaxies

1. Basics of gravitational lensing

General formalism

The general formalism of gravitational lensing has been presented in many review papers (Schneider, Ehler & Falco 1995; Mellier 1999) and is only summarized here. First, gravitational lensing is sensitive to gravity, which means all kind of matter, whatever its nature. In addition, under general assumptions (listed below), gravity is traced by the newtonian potential and there is no need for a full knowledge of General Relativity to address astrophysical issues with gravitational lensing.

So I remind the main assumptions used in this review, valid in most lensing configurations, especially those related to clusters of galaxies: the lens is "thin" compared to the cosmological distances between the source and the observer, it is stationary so the gravitational potentiel does not vary while the light travels along it. Morevoer lensing is acromatic and the distances in use are the angular-diameter distances.

Following these preliminaries, the basic lens equation, which relates the position of the source on the sky plane to its corresponding image is simply

J. C. del Toro Iniesta (eds.), The Many Scales in the Universe, 25–34.

written as

$$\left\{ \begin{array}{l} \vec{\beta} = \vec{\theta} - \vec{\alpha} = \vec{\theta} - \vec{\nabla}\psi(\vec{\theta}) \\ \psi(\vec{\theta}) = \frac{2}{c^2} \frac{D_{LS}}{D_{OS} D_{OL}} \vec{\nabla}\varphi(\vec{\theta}) \end{array} \right.$$

where $\varphi(\vec{\theta})$ is the 2D newtonian gravitational potential of the lens projected on the sky plane. The reduced potential $\psi(\vec{\theta})$ is also written in terms of the convergence κ as

$$\Delta\psi(\vec{\theta}) = 2\,\kappa(\vec{\theta}) = 2\,\Sigma(\vec{\theta})/\Sigma_{cr}$$

with $\Sigma_{cr} = \frac{c^2}{4\pi G} \frac{D_{OS}}{D_{OL} D_{LS}}$ which depends only on the geometrical lens configuration.

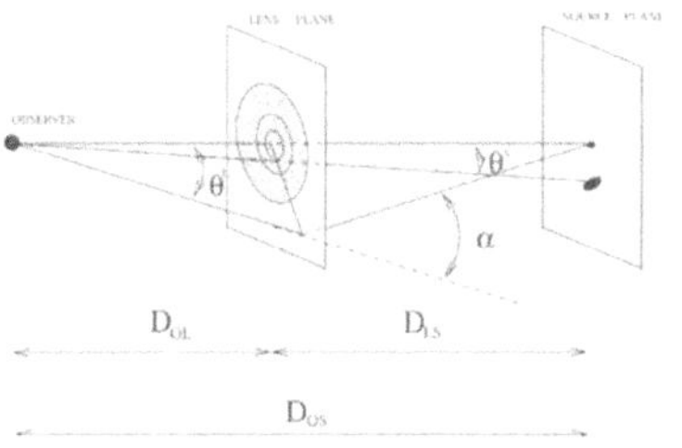

Figure 1. Usual gravitational lens configuration and the relation between the source and image positions.

The lens equation is a mapping of the source plane on the image plane, and the relation between the source size and the image size defines the magnification matrix, under the property of surface brightness conservation:

$$\mathcal{A} = \left(\frac{\partial \vec{\beta}}{\partial \vec{\theta}} \right) = \left(\delta_{ij} - \frac{\partial \alpha_i(\vec{\theta})}{\partial \theta_j} \right) = \left(\delta_{ij} - \frac{\partial^2 \psi(\vec{\theta})}{\partial \theta_i \partial \theta_j} \right)$$

This magnification matrix can be written in terms of the complex shear ($\gamma = \gamma_1 + i\gamma_2$) and convergence κ as

$$\mathcal{A} = \begin{pmatrix} 1 - \kappa - \gamma_1 & -\gamma_2 \\ -\gamma_2 & 1 - \kappa + \gamma_1 \end{pmatrix}$$

$$\left\{ \begin{array}{lcl} \kappa & = & \frac{1}{2}\mathrm{tr}\psi_{,ij} = \frac{1}{2}(\psi_{,11} + \psi_{,22}) \\ \gamma_1 & = & \frac{1}{2}(\psi_{,11} - \psi_{,22}) \\ \gamma_2 & = & \psi_{,12} = \psi_{,21} \end{array} \right.$$

The effect of the convergence on a circular source is an isotropic magnification while the anisotropic distorsion is directly represented by the shear term.

Image formation and lensing regimes

There are 3 lensing regimes which act differently on the image formation:

- The strong lensing regime corresponds to the regime of large deformations and possible image splittings. The images appear like giant arcs with very distorted shapes. They lie close to the critical lines of the lens.
- The intermediate regime corresponds to the regions of formation of the so-called arclets, i.e. distorted but single images, located outside the Einstein radius. The relations between source and image properties are still non-linear. In these regions, one can also observe the magnification bias, with usually a depletion of background sources, related to the intensity of the local magnification.
- The weak lensing regime acts when individual distorsion of the sources is not detectable because it is smaller than the intrinsic ellipticity of the galaxies. Only statistical shear can be detected by averaging the shape properties of background sources. This also corresponds to the linear regime of the source-image relation.

2. Strong lensing in clusters of galaxies

In this section we focus on some important scientific results issued from the analysis of the strong lensing effect in clusters of galaxies. This effect was highlighted by some spectacular images obtained with the HST. However it is out of the scope of this review to present a complete and up-to-date listing of all the observations and lensing models performed in clusters of galaxies. Only a few illustrative examples are presented below, trying to be representative of what can be done in a cosmological context.

The major use of gravitational lensing in clusters of galaxies is the modeling of the lensing configurations in order to constrain the mass distribution. The general scheme of any lens model is to first identify one or several families of multiple images, and build a model of the gravitational potential (or the equivalently the mass distribution) by minimizing the source distances between different positions corresponding to the multiple images. This minimisation is done by adjusting the lens potential. For illustration I present below the cluster of galaxies Abell 2218 ($z = 0.17$) observed first with the HST/WFPC2 camera (Kneib et al. 1996) and more recently with the ACS. This is a rich cluster of galaxies and a strong X-ray emitter. All these properties point to a massive cluster and a very favorable configuration for gravitational lensing. Presently at least 4 families of multiple images are clearly identified, with measured redshifts ranging from 0.702 to 5.576 (Ebbels et al. 1996; Ellis et al. 2001).

The mass modeling was built with 2 main cluster components centered on the two brightest galaxies of the cluster and adding the individual mass of the

17 brightest cluster galaxies located in the center of the cluster. For numerical simplicity, the potential was represented by a pseudo-isothermal elliptical mass distribution (PIEMD) (Kassiola & Kovner 1993) and caracterized by 3 main parameters: the central velocity dispersion σ_0, a core radius θ_c and a truncature radius θ_t, plus geometrical parameters like the position of the center or the ellipticity and position angle of the potential.

Recovering the mass distribution

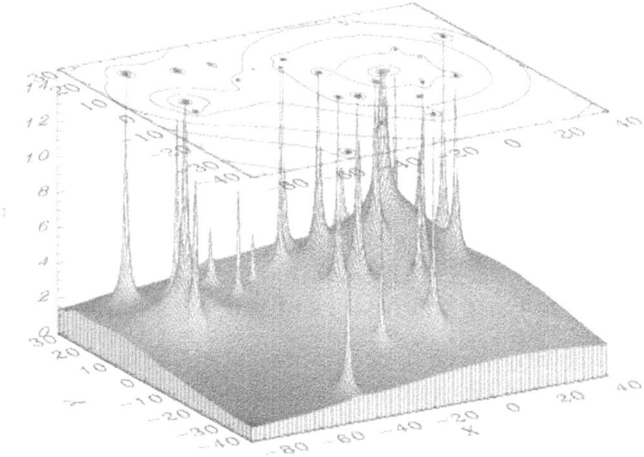

Figure 2. Model of the gravitational potential in Abell 2218, from the constraints of 4 multiple-images configuration.

The best fit results of this modeling show several significant tendancies, common to most lens models of clusters: the main massive components follow quite closely the light distribution. But in the case of A2218, the ellipticity of the main component is much smaller than that of the cD light distribution while for the secondary massive component, it is larger its optical counter-part. This may be interpreted in terms of a possible signature of a recent merger between the two components, with a dark matter distribution not yet re-adjusted to the luminous one. This conclusion is confirmed by recent X-ray observations with Chandra, also interpreted in terms of a sub-cluster merging with the main cluster. In addition to the constraints derived on the two main components, adding the individual contribution of the galaxies improves significantly the results of the models. The best fit parameters give standard values for the galaxy masses, with a slight increase of the M/L ratio with L (Soucail, Kneib & Golse 2004).

Constraining the cosmological parameters

The mass distribution models of some well studied cluster lenses are so well constrained that gravitational lenses start to be considered as potential tools to constrain the geometrical cosmological parameters, independantly from more classical tests. Indeed, lensing depends on the ratio D_{LS}/D_{OS} for each source (or each set of multiple images) multiplied by a term which depnds only on the gravitational potential. Provided the redshifts of the sources are measured accurately with spectroscopy, the ratio D_{LS}/D_{OS} will depend only on the geometrical cosmological parameters $(\Omega_M, \Omega_\lambda)$ or (Ω_M, w) for a flat universe with dark energy. A typical degeneracy in the $(\Omega_M, \Omega_\lambda)$ plane is expected (Golse, Kneib & Soucail 2002) with a caracteristic shape different from the degeneracies obtained by CMB anisotropies measurements or Supernovae fittings. A first and preliminary attempt was proposed recently on the well constrained cluster-lens Abell 2218 (Soucail, Kneib & Golse 2004), where the expected degeneracy was observed (Fig. 3). However, recent simulations (Dalal, Hennawi & Bode 2004) show that the uncertainties in the lens modeling, mainly due to density fluctuations from large-scale structures, preclude the use of individual clusters to infer cosmological parameters. More work is needed in order to fully explore the potentialities of this method. The combination of a large dataset of cluster lenses will be crucial.

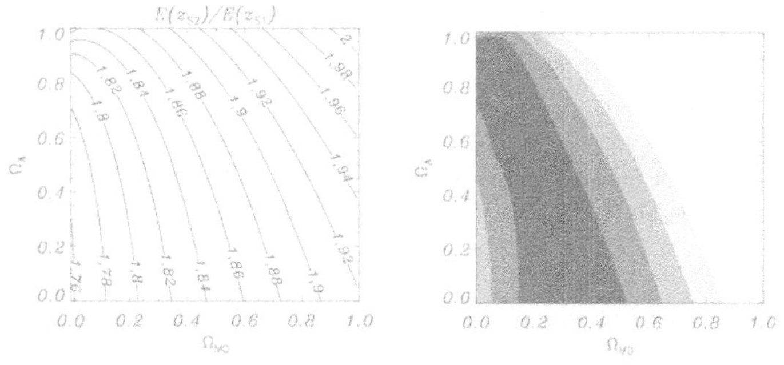

Figure 3. (left) Evolution of the ratio D_{LS}/D_{OS} for two sources at different redshifts in the $(\Omega_M, \Omega_\lambda)$ plane. (right) Results of the χ^2 analysis after the modeling of the cluster A2218. The expected degeneracy is nicely recovered.

Looking at magnified galaxies

Clusters of galaxies act as giant gravitational telescopes. This idea has been extensively used to benefit from the magnification effect and to detect very faint objects which would be un-observable out of the lens.

A recent and spectacular breakthrough was the detection of very high redshift galaxies behind cluster lenses. Presently most of the highest redshift galaxies are magnified by cluster lenses. As an illustration, we can mention two pairs of high redshift galaxies identified behind the cluster Abell 2218: $z = 5.67$ (Ellis et al. 2001) and $z \simeq 7$ (Kneib et al. 2004). Searched on purpose in the area close to the critical lines at high redshift, they are magnified by large factors, from 10 to 25! Although still controversial, the highest redshift source identified at a redshift of 10 is also magnified by 25 to 100 (Pello et al. 2004). At the limit of detectability of all the existing facilities, these sources would have been un-observable without magnification. The observing window behind cluster lenses is a unique opportunity to increase our understanding of the very early stages of galaxy formation up to the epoch of reionisation.

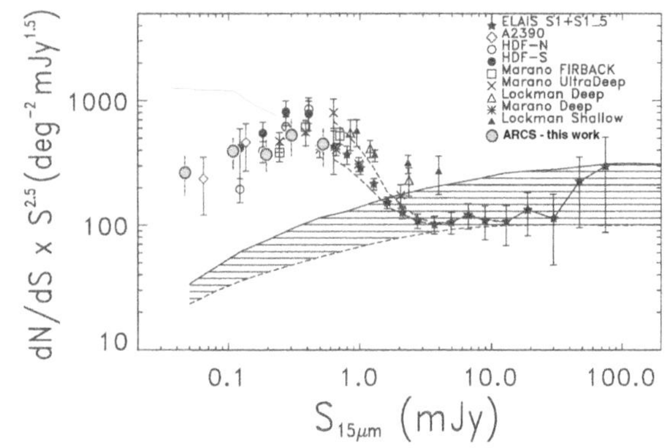

Figure 4. Extension of the faint differential number counts of mid-IR sources observed at 15 μm behind the cluster-lens Abell 2390 (from Metcalfe et al. 2003).

For illustration I also mention an interesting project developped with ISO, in which ultra-deep imaging was performed on the cluster-lens Abell 2390 with ISOCAM. The correction from the lens magnification was applied and allowed to derive the deepest counts obtained at 7 and 15 μm with ISO (Metcalfe et al. 2003). The average gain in flux was about 2 in an area of a few square arcmintues, decreasing the intrinsic flux limit to less than 0.05 mJy at 15 μm.

3. From weak lensing to masses

Outside the cluster centers, where the gravitational lensing effects are the strongest and the most spectacular, lensing is still present only slightly distorts the images of background sources. Due to the instrinsic shape of these galaxies, it is impossible to measure directly and individually the small additional shear and only statistical lensing is measured. The modeling of the image ellipticities is obtained from the 2nd order moments of the light distribution. Then it is related to the gravitational reduced shear, a complex quantity defined as

$$g(\vec{\theta}) = \frac{\gamma(\vec{\theta})}{1 - \kappa(\vec{\theta})}$$

The source and image ellipticities are related by Seitz & Schneider (1995)

$$\varepsilon^{(s)} = \begin{cases} \dfrac{\varepsilon - g}{1 - g^{\star}\varepsilon} & |g| \leq 1 \\ \\ \dfrac{1 - g\varepsilon^{\star}}{\varepsilon^{\star} - g^{\star}} & |g| \geq 1 \end{cases}$$

In the weak lensing approximation ($i.e.$ $\kappa \ll 1$), $\varepsilon \sim \varepsilon^{(s)} + g$. Assuming a random distribution of the intrinsic orientation of the galaxies, we have the basic relation of weak lensing: $< \varepsilon > \simeq < \gamma >$.

Applying this formalism to real observations is more complex because of the PSF smearing on the shape of the galaxies. To correct from this smearing, acting like a convolution of the intrinsic light distribution, several methods were developped. The local modeling of the PSF is obtained by interpolation of the shapes of nearby stars. Then the PSF correction is applied following several prescriptions: the most popular one, described by Kaiser & Squires (1993) consists in a decomposition of the total distorsion in an anisotropic PSF term + an anisotropic one + the shear term. Another approach more recently developped compute the real galaxy shape parameters after the deconvolution by the PSF (Bridle et al. 2001).

From the corrected shape parameters of the background galaxies, one then needs to reconstruct the mass profile of clusters of galaxies. Generally a χ^2 minimization on the shear profile is applied to contrain the mass distribution up to large radii (a maximum radius of 1 to 3 Mpc can easily be acheived with present-day reconstruction techniques). Another tool is to use the so-called ζ-statistics or the aperture mass, i.e. the integration of the total mass up to a given radius (Fahlman et al. 1994). This gives an underestimate of the total mass, but avoids the difficulty induced by the mass-sheet degeneracy effect (Schneider & Seitz 1995): any shear pattern obtained from a given lensing distribution (κ, γ) is equally reproduced by any distribution (κ', γ') such as

$$\kappa'(\vec{\theta}) = \lambda\kappa(\vec{\theta}) + (1 - \lambda) \qquad \gamma'(\vec{\theta}) = \lambda\gamma(\vec{\theta})$$

where λ is an arbitrary constant. This degeneracy is difficult to break and requires either more than one set of multiple images with known redshift (strong + weak lensing) or redshift information on all the background sources, although with lower accuracy (Bradac, Lombardi & Schneider 2004).

In conclusion, recovering the mass distribution from weak lensing is very promising although additional work is still necessary to fully understand the residual systematics which may bias the mass measurements in clusters of galaxies. The comparison between these mass distributions and the light distribution of the galaxies is a direct estimate of the cosmological bias. It is still not clear wether the mass-to-light ratio in clusters varies from the densest central regions towards the external ones. In principle weak lensing techniques are able to address this question although quantitative estimates require further developments.

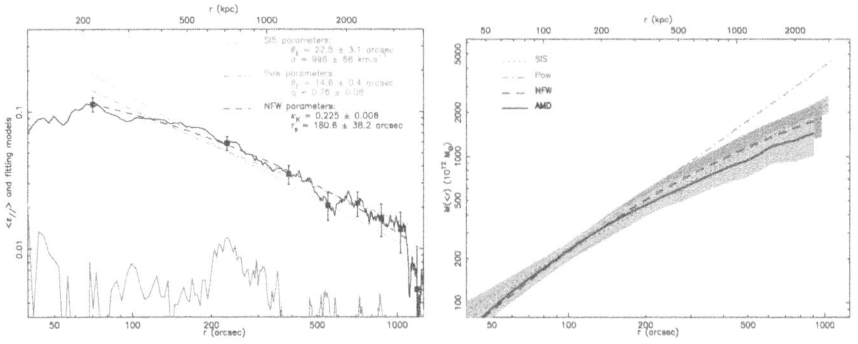

Figure 5. Example of a measured shear profile at large radius, and the reconstructed mass profile, in the cluster Abell 1689 (from Bardeau et al. 2004).

4. The cosmic shear or lensing by large scale structures

Extending the weak lensing phenomenon over the outskirts of clusters of galaxies has revealed a very powerfull tool to study the mass distribution of large scale structures. The formalism used in that case is slightly different, and one usually considers the global power spectrum of the mass distribution instead of individual mass over-densities. The usefull quantity is the 2-point shear correlation function which is related to the mass fluctuations power spectrum $P_\kappa(k)$. Recent results from several teams lead to concordant results and significant constraints on the degenrate product $\sigma_8\ \Omega_M^{0.6}$ (Van Waerbeke et al. 2002 and references therein). Breaking this degeneracy with additional constraints seem to point to a high value of σ_8 and $\Omega_M < 0.5$ (Pen et al. 2003).

The future of such measurements is expected from large dedicated imaging surveys. The CFHT Legacy Survey is already on-going and preliminary results will be available soon. It covers an area larger than 150 square degrees, with high quality imaging and multi-color photometry (allowing photometric redshift information on the background sources). With such a wide area, the expected accuracy on the cosmological parameters should be better than

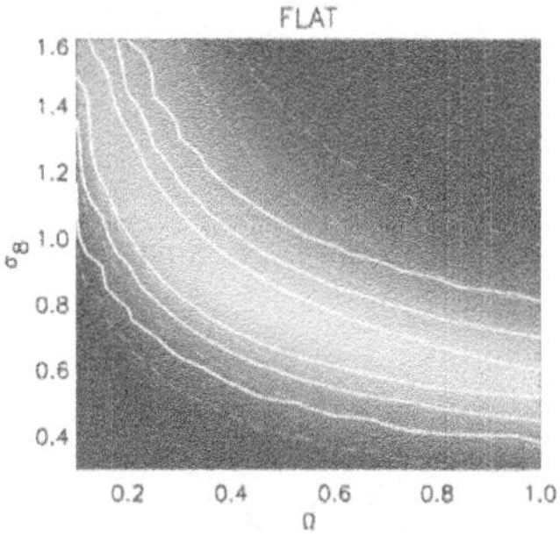

Figure 6. Degeneracy on the constraints on Ω_M and σ_8 from cosmic shear measurements in the VIRMOS-DESCART survey (Van Waerbeke et al. 2002).

10%. Other ambitious projects are under development, like the PAN-STARRS project or the LSST on the ground or space projects like SNAP-JDEM. The next decade will be very challenging, indeed!

5. Conclusions

I hope to have convinced the reader that gravitational lensing offers a lot of potentialities for Cosmology. It allows a direct determination of the mass distribution in clusters at different scales and is a complementary approach to the study of the X-ray gas distribution or the dynamics of the galaxies. The detailed modeling of some well-defined lenses confirm that in most clusters the dark matter is centrally concentrated, with a distribution characterized by a small core radius. Moreover, the magnification gain in strong lensing is used to extend the limits of available instruments and to explore the very distant Universe: high amplification events led to the discovery of some spectacular galaxies at the frontier our visible universe. Finally there are some good hopes to use gravitational lensing to get better geometrical constraints on the structure of the Universe, although this is clearly a difficult task and the expected performances are well below those achieved by other means. But it remains totally independant from other methods and is an alternative to be considered.

All these topics will be developped in the near future, thanks to new observing facilities. Extending the wavelength domain where gravitational lensing is used to the mid-IR or the sub-mm domains is possible with Spitzer and soon with ALMA. The very weak lensing regime will also be explored, thanks to the panoramic camera which start to be available on some large telescopes. Note that the study of the cosmic shear is one of the 3 core programes of the CFHT Legacy Survey. Finally one can expect that imaging from space will allow to extend the range of masses addressed by the effect of lensing (either in the very weak lensing regime by large scale structures or in the strong regime by low mass galaxies at very small angular scale).

References

Bardeau, S., Kneib, J.P., Czoske, O., Soucail, G., I., Smail, & Ebeling, H. 2004, A&A, submitted, astro-ph/0407255

Bradac, M., Lombardi, M., & Schneider, P. 2004, A&A, 424, 13

Bridle, S., Gull, S., Bardeau, S., & Kneib, J.-P. 2001, in *Proceedings of the Yale Cosmology Workshop, "The Shapes of Galaxies & their Dark Halos"*, Ed. P., Natarajan, World Scientific

Dalal, N., Hennawi, J., & Bode, P. 2004, ApJ, submitted, astro-ph/0409028

Ebbels, T. M. D., Le Borgne, J.-F., Pello, R., Ellis, R. S., Kneib, J.-P., Smail, I., & Sanahuja, B. 1996, MNRAS, 281, L75

Ellis, R., Santos, M. R., Kneib, J., & Kuijken, K. 2001, ApJL, 560, L119

Fahlman, G., Kaiser, N., Squires, G., & Woods, D. 1994, ApJ, 437, 56

Golse, G., Kneib, J.-P., & Soucail, G. 2002, A&A, 387, 788

Kaiser, N. & Squires, G. 1993, ApJ, 404, 441

Kassiola, A. & Kovner, I. 1993, ApJ, 417, 450

Kneib, J., Ellis, R. S., Santos, M. R., & Richard, J. 2004, ApJ, 607, 697

Kneib, J.-P., Ellis, R. S., Smail, I., Couch, W. J., & Sharples, R. M. 1996, ApJ, 471, 643

Mellier, Y. 1999, ARA&A, 37, 127

Metcalfe, L., Kneib, J.-P., McBreen, B., et al. 2003, A&A, 407, 791

Pello, R., Schaerer, D., Richard, J., Le Borgne, J.-F., & Kneib, J.-P. 2004, A&A, 416, L35

Pen, U., Zhang, T., van Waerbeke, L., Mellier, Y., Zhang, P., & Dubinski, J. 2003, ApJ, 592, 664

Schneider, P., Ehlers, J., & Falco, E. E. 1995, Space Science Reviews, 72, 682

Schneider, P. & Seitz, C. 1995, A&A, 294, 411

Seitz, C. & Schneider, P. 1995, A&A, 297, 287

Soucail, G., Kneib, J.-P., & Golse, G. 2004, A&A, 417, L33

Van Waerbeke, L., Mellier, Y., Pelló, R., Pen, U.-L., McCracken, H. J., & Jain, B. 2002, A&A, 393, 369

THE PLANCK MISSION

J.A. Tauber*
Astrophysics Division, Research and Scientific Support Department of the European Space Agency
ESTEC, P.O. Box 299, 2200AG Noordwijk, The Netherlands

Abstract We present an overview of the European Space Agency's Planck mission, its scientific objectives and the main elements of its technical design. The current programmatic status of Planck within ESA's Scientific Programme, implementation plans, and near-term milestones are also outlined.

1. Introduction

Planck (*http://www.rssd.esa.int/Planck*) is a space observatory designed to image the temperature anisotropies of the Cosmic Microwave Background (CMB) over the whole sky, with unprecedented sensitivity ($\Delta T/T \sim 2\times10^{-6}$) and angular resolution (better than 10 arcminutes). Planck will provide a major source of information relevant to several cosmological and astrophysical issues, such as testing theories of the early universe and the origin of cosmic structure.

The ability to measure to high accuracy the angular power spectrum of the CMB fluctuations will allow the determination of fundamental cosmological parameters such as the density parameter (Ω_0), and the Hubble constant H_0, with an uncertainty of order a few percent (e.g. Bersanelli et al. 1996, Bond et al. 1997, Efstathiou & Bond, 1999). Planck will not only measure the temperature fluctuations of the CMB, but also its polarisation state. This measurement (e.g. Seljak, 1997) will not only yield new scientific results, but will also help to analyze the CMB temperature anisotropies (in particular by resolving degeneracies in the estimation of some cosmological parameters; see Zaldarriaga et al. 1997; Kamionkowski & Kosowsky, 1998).

In addition to the main cosmological goals of the mission, the Planck sky survey will be used to study in detail the very sources of emission which "con-

*On behalf of ESA and the Planck Scientific Collaboration.

J. C. del Toro Iniesta (eds.), The Many Scales in the Universe, 35–43.

taminate" the signal due to the CMB, and will result in a wealth of information on the properties of extragalactic sources, and on the dust and gas in our own galaxy (e.g. De Zotti et al. 1999). One specific notable result will be the measurement of the Sunyaev-Zeldovich effect in many thousands of galaxy clusters (Aghanim et al. 1997; Pointecouteau et al. 1998).

2. Payload

The scientific prescription which will allow Planck to meet its ambitious objectives calls for:

- an offset telescope with a physical aperture of size $\sim$1.5 m. to achieve the angular resolution
- state-of-the-art broadband detectors covering the range $\sim$25 to $\sim$1000 GHz, to achieve the required sensitivity and the ability to remove foreground sources of emission
- a survey with all-sky coverage carried out from a far-Earth orbit
- extreme attention to rejection of unwanted systematic effects.

To achieve this prescription a payload was conceived for Planck (Bersanelli et al. 1996) consisting of three basic components: (1) a telescope and baffling system, providing the angular resolution and rejection of straylight; (2) a Low Frequency Instrument (or LFI) – an array of tuned radio receivers, based on HEMT amplifiers, covering the frequency range 25 – 80 GHz, and operated at a temperature of 20 K; and (3) a High Frequency Instrument (or HFI), consisting of an array of bolometers operated at 0.1 K and covering the frequency range 90 – 1000 GHz.

The major elements of the Planck payload and their disposition on the spacecraft can be seen in Fig. 1.

3. Programmatic aspects

Planck was selected as the third Medium-Sized Mission (M3) of ESA's Horizon 2000 Scientific Programme and is now part of its Cosmic Vision programme. Planck was formerly called COBRAS/SAMBA. After the mission was selected and approved (in late 1996), it was renamed in honor of the German scientist Max Planck (1858–1947), Nobel Prize for Physics in 1918. Planck will be launched together with ESA's Herschel Far-Infrared and Submillimetre Space Observatory in August of 2007.

Starting in 1993, a number of technical studies laid the basis for the issue in September 2000 of an Invitation to Tender (ITT) to European industry for the procurement of the Herschel and Planck spacecraft. From the submitted

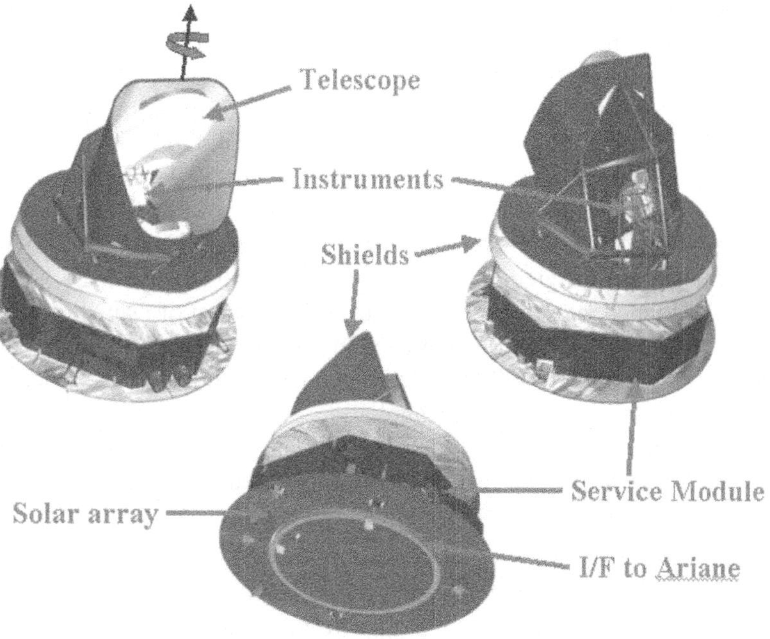

Figure 1. A conceptual view of the arrangement of the main elements of the Planck payload. The instrument focal plane unit contains both low and high frequency detectors. The function of the large shield surrounding the telescope is to control the far sidelobe level of the radiation pattern as seen from the detectors. Specular cones thermally decouple the Service Module (located below the payload) from the Payload Module, and allow the payload environment to reach its nominal temperature of $\sim$50 K. The satellite's spin axis is indicated; the Sun is always kept in the anti-spin direction (perpendicular to the solar array). The satellite views are courtesy of Alcatel Space (Cannes).

proposals, a single prime contractor, Alcatel Space (France), was selected in early 2001 to develop both Herschel and Planck spacecrafts. Alcatel Space is supported by two major subcontractors: Alenia Spazio (Torino) for the Service Module, and Astrium GmbH (Friedrichshafen) for the Herschel Payload Module; and by many other industrial sub-contractors from all ESA member states. The detailed definition work began in June 2001, and at the time of writing a full qualification model of the satellite is being integrated for test. A view of the current design of Planck is shown in Fig. 1.

Planck is a survey-type project which is being developed and operated as a Principal Investigator mission. In early 1999, ESA selected two Consortia of scientific institutes to provide the two Planck instruments: the Low Frequency Instrument will be developed and delivered to ESA by a Consortium led by Reno Mandolesi of the Istituto di Astrofisica Spaziale e Fisica Cosmica (CNR) in Bologna (Italy); similarly, the High Frequency Instrument will be provided to ESA by a Consortium led by Jean-Loup Puget of the Institut d'Astrophysique Spatiale (CNRS) in Orsay (France). More than forty European institutes, and some from the USA, are collaborating on the development, testing, and operation of these instruments, as well as the ensuing data analysis and exploitation. The capabilities of the instruments are described in Table 1.

In early 2000, ESA and the Danish Space Research Institute (DSRI, Copenhagen) signed an Agreement for the provision of the two reflectors that constitute the Planck telescope. DSRI leads a Consortium of Danish institutes, and together with ESA has subcontracted the development of the Planck reflectors to Astrium GmbH (Friedrichshafen), who will manufacture the reflectors using state-of-the-art carbon fibre technology.

In parallel, the instrument development has proceeded largely according to schedule. Qualification models are manufactured and being tested. The deliveries to ESA of instrument flight models are expected in mid-2005.

4. Scientific performance

The principal objective of Planck is to produce maps of the whole sky in ten frequency channels. The currently foreseen characteristics of the two Planck instruments (see Fig. 2) are summarized in Table 1; these characteristics largely drive the quality of the final maps.

These maps will not only include the CMB itself, but also all other astrophysical foregrounds, whether galactic (free-free, synchrotron or dust) or extragalactic in origin. All ten Planck sky maps will be used to produce a single map of the Cosmic Microwave Background anisotropies. The key that allows to reach this objective is the wide spectral coverage achieved by Planck. Each astrophysical foreground has a distinct (albeit at present poorly known) spectral characteristic. Specialized data processing algorithms (e.g. Tegmark 1997; Hobson et al. 1998; Bouchet & Gispert, 1999) will use this information to iteratively extract the signal due to each foreground component, until only the CMB signal remains. Instrumental systematic effects (e.g. Delabrouille 1998; Burigana et al. 1998; Maino et al. 1999), as well as uncertainties in the recovery of parameters characterizing the foregrounds will degrade the final noise level in the CMB maps.

Therefore the final scientific performance of the mission depends not only on the instrumental behavior, but also on the detailed nature of the various astrophysical foregrounds, the behavior of many systematic effects which produce spurious signals (such as straylight), and the ability to remove these signals from the measured data by means of data processing algorithms. Current estimates of the performance of Planck are based on simulations of the mea-

Table 1. Characteristics of the Planck Payload

Telescope	1.5 m. (proj. apert.); Aplanatic;Temp. ~50 K; $\epsilon_{system} \sim 1\%$ Shared focal plane, viewing direction offset 85° from spin axis								
Instrument	LFI			HFI					
Center Frequency (GHz)	30	44	70	100	143	217	353	545	857
Detector Technology	HEMT receiver arrays			Bolometer arrays					
Detector Temperature	~20 K			0.1 K					
Cooling Requirements	H_2 sorption cooler			H_2 sorption + 4K J-T stage + Dilution					
Number of unpolarized Detectors	–	–	–	–	4	4	4	4	4
Number of Lin. Polarised Detectors	4	6	12	8	8	8	8	–	–
Bandwidth ($\Delta\nu/\nu$)	0.2	0.2	0.2	0.33	0.33	0.33	0.33	0.33	0.33
Angular Resolution (arcmin)	33	24	14	9.2	7.1	5.0	5.0	5.0	0.5
Average $\Delta T/T$ per pixel (Stokes I) (14 mos., 1σ, 10^{-6} units)	2.0	2.7	4.7	2.0	2.2	4.8	14.7	147.0	6700
Average $\Delta T/T$ per pixel (Stokes U, Q) (14 mos., 1σ, 10^{-6} units)	2.8	3.9	6.7	4.0	4.2	9.8	29.8	–	–

surement process (e.g. Bersanelli et al. 1996; Bouchet and Gispert, 1999; Knox, 1999; Tegmark et al. 2000), which include such effects to the best of available knowledge, as well as of the signal extraction process. These simulations suggest that the ability to extract the CMB signal from the measurements will be limited mainly by the background of noise originating in unresolved structure in the various foregrounds.

The Planck instruments are also designed to provide information on the polarisation state of the CMB. Although simulations of the extraction of polarisation from Planck are at a less sophisticated level than those dealing with temperature anisotropies, it is expected that Planck will be able to measure with good accuracy the angular power spectrum of the "E-component" of CMB polarisation (Hu & White, 1997), with the consequent improved estimation of cosmological parameters (Bouchet et al. 1999; Prunet et al. 2000).

Naturally, as a by-product of the extraction of the CMB, Planck will also yield all-sky maps of all the major sources of microwave to far-infrared emission, opening a broad expanse of astrophysical topics to scrutiny. In particular, the physics of dust at long wavelengths and the relative distribution of interstellar matter (neutral and ionized) and magnetic fields will be investigated using dust, free-free and synchrotron maps. In the field of star formation, Planck will provide a systematic search of the sky for dense, cold condensations which are the first stage in the star formation process. One specific and local distortion of the CMB which will be mapped by Planck is the Sunyaev-Zeldovich (SZ) effect arising from the Compton interaction of CMB photons with the hot gas of clusters of galaxies. The very well defined spectral shape of the SZ effect allows it to be cleanly separated from the primordial anisotropy. The physics of gas condensation in cluster-size potential wells is an important element in the quest to understand the physics of structure formation and ultimately of galaxy formation.

Therefore, even though that is not its primary scientific objective, Planck will deliver high quality all-sky maps of all extended foreground emission components between cm and submm wavelengths. These maps will constitute a scientific product which is comparable to the IRAS and COBE-DIRBE maps at shorter wavelengths. The high-resolution all-sky nature of the maps, coupled with the broad spectral coverage and highly accurate calibration, will present a unique opportunity to explore in a global sense all the extended emission components of our own as well as external galaxies.

5. Mission profile

Planck will be launched together with Herschel in August of 2007 by an Ariane 5 rocket from the European spaceport in Kourou (French Guiana). Planck and Herschel will separate immediately after launch, and each will proceed independently to different orbits around the L2 point of the Earth-Sun system. At this location, the payload can be continuously pointed in the anti-Sun direction, thus minimizing potentially confusing signals due to thermal fluctuations and straylight entering the detectors through the far sidelobes.

The transit time for Planck will be between 3 and 4 months; this period will be used for commissioning of the spacecraft and instruments. The spacecraft (S/C) will be placed into a Lissajous orbit around L2 characterised by a $\sim$6 month period and a maximum elongation from L2 of about 380000 km, such that the Sun-S/C-Earth angle will not exceed 15°. From this orbit, Planck will carry out two complete surveys of the full sky, for which it requires between 12 and 14 months of observing time.

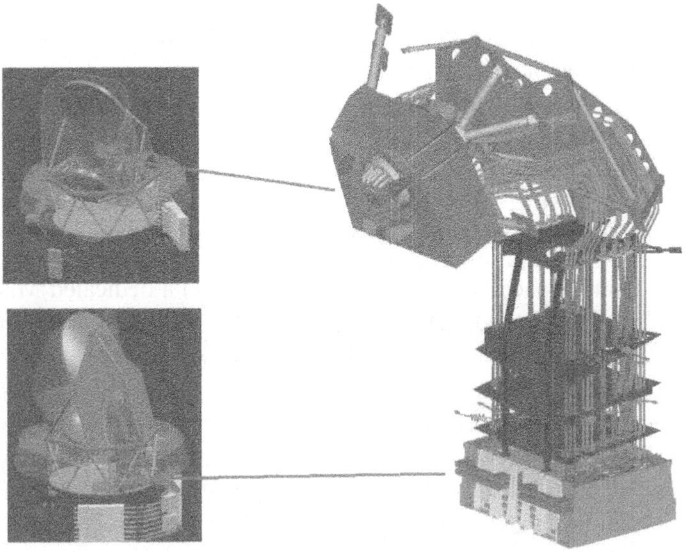

Figure 2. On the right (top) is shown the combined focal plane unit formed by the LFI and HFI (see also Note 1 on pg 2); the colors on each horn correspond to frequency channels. The HFI focal plane is inserted into the ring formed by the LFI horns, and includes thermal stages at 20 K, 4 K, 2 K and 0.1 K.The LFI horns and detectors are cooled to 20 K, and are attached by waveguides to a lower unit containing back-end amplifiers; a complex mechanical structure stiffens the assembly. On the left is shown the manner in which the various instrument units are distributed throughout the Payload and Service Modules. In general all cryogenic units are located in the Payload Module and all warm units (electronics, coolers) are located in the Service Module. The figures are courtesy of Alcatel Space (Cannes) and Laben (Milano).

The satellite will rotate at 1 rpm around a spin axis pointed within 10° of the Sun. The payload will always remain in the shadow of the Sun. The solar array ensures this as long as it is inclined with respect to the Sun-S/C line by less than 10°. The Planck telescope and focal plane define a sparsely sampled field of view (FOV) approximately 8° in diametre around a reference line-of-sight which is inclined by 85° with respect to the spin axis. As the satellite rotates, the FOV will thus trace a circle of diametre 170° on the sky.

In order to carry out its two consecutive full-sky surveys and maintain the payload in the solar shadow, the spin axis of Planck must be displaced on the average by 1° per day in the direction defined by the orbital motion of the Earth around the Sun. This is achieved by spin axis depointing manoeuvers at regular intervals. As the spin axis is displaced, the observed circle also moves and gradually covers a large fraction of the sky.

Planck will dump each day to Earth within a period of 3 hours the data acquired during 24 hours. Observations will not be interrupted during the downlink period, and the S/C will not be reoriented towards the Earth. The telemetry antenna is designed to have adequate gain within a 15° half-cone from the spin axis, ensuring that even at the extremes of its orbit the Planck telemetry can achieve full bandwidth.

6. Operations and data processing

The Planck spacecraft will be controlled from a dedicated Mission Operations Centre developed and operated by ESOC in Darmstadt (Germany). From there, the scientific data produced by Planck will be piped daily to two Data Processing Centres (DPCs), which will be developed and operated by the two Consortia selected to provide the Planck instruments.

In particular, the two DPCs will be responsible for: (a) daily and long term analysis of instrument health and performance; (b) daily analysis of science data; (c) all levels of processing of Planck data, from raw telemetry to deliverable scientific products.

The two DPCs will share a basic information management infrastructure, the Planck Integrated Data and Information System (IDIS, Bennett et al. 2000). IDIS is being conceived from an object-oriented point of view, and is planned to contain five different components: (a) a Document Management Component, containing all relevant documentation; (b) a Software Management Component, encompassing the software in common between the two Consortia; (c) a Process Coordinator Component, providing a single software environment for data processing (e.g. a data pipeline manager); (d) a Data Management Component, allowing the ingestion, efficient management and extraction of the data (or subsets thereof) produced by Planck activities; (e) a Federation layer, providing inter-connection among IDIS components (e.g. relating objects controlled by each component).

The main scientific products of the mission will be produced by the two DPCs jointly, and will consist of all-sky maps in ten frequency bands, which will be made publicly available one year after completion of the mission (i.e. in mid-2011), together with a first generation set of maps of the CMB, Sunyaev-Zeldovich effect, galactic emission (dust, free-free, and synchrotron), and point-source catalogs. The time series of observations (after calibration and position reconstruction) will also eventually be made available as an on-line archive.

Acknowledgments

Planck (*http://astro.esa.int/Planck*) is an ESA project with instruments funded by ESA member states (in particular the PI countries: France and Italy), and with special contributions from Denmark and NASA (USA). This paper draws widely from the work carried out in the past years by the Planck Instrument and Reflector Provider Consortia, ESA, ESOC, and industry (principally the former Matra Marconi Toulouse, and Alcatel Space Cannes).

References

Aghanim, N., de Luca, A., Bouchet, F. R. et al. 1997, A&A, 325, 9

Bennett, K., Pasian, F., Sygnet, J. F. et al. 2000, SPIE, 4011, 2

Bersanelli, M., Bouchet, F. R., Efstathiou, G., et. al. 1996, ESA D/SCI(96)3,

Bond, J. R., Efstathiou, G., & Tegmark, M. 1997, MNRAS, 291, L33

Bouchet, F., & Gispert, R. 1999, New Astronomy, 4, 443

Bouchet, F. R., Prunet, S., & Sethi, S. K. 1999, MNRAS, 302, 663

Burigana, C., Maino, D., Mandolesi, N., et al. 1998, A&AS, 130, 551

Delabrouille, J. 1998, A&AS, 127, 555

de Zotti, G., Toffolatti L., Argüeso, F., et al. 1999, in *Proceedings of the EC-TMR Conference on 3K cosmology* AIP Woodbury, N.Y., Eds. L. Maiani, F. Melchiorri, & N. Vittorio, 476, 204

Efstathiou, G. & Bond, J. R. 1999, MNRAS, 304, 75

Hobson, M. P., Jones, A. W., Lasenby, A. N., et al. 1998, MNRAS, 300, 1

Hu, W. & White, M. 1997, New Astronomy, 2, 323

Kamionkowski, M. & Kosowsky, A. 1998, Phys. Rev. D, 57, 685

Knox, L. 1999, MNRAS, 307, 977

Maino, D., Burigana, C., Maltoni, M., et al. 1999, A&AS, 140, 383

Pointecouteau, E., Giard, N. & Barret, D. 1998, A&A, 336, 44

Prunet, S., Sethi, S. K. & Bouchet, F. R. 2000, MNRAS, 314, 348

Revenu, B., Kim, A., Ansari, R. et al. 2000, A&AS, 142, 499

Seljak, U. 1997, ApJ, 482, 6

Tegmark, M., 1997, ApJ, 480, L87

Tegmark, M., Eisenstein, D. J., Hu, W. & De Oliveira-Costa, A. 2000, ApJ, 530, 133

Zaldarriaga, M., Spergel, D. N. & Seljak, U. 1997, ApJ, 488, 1

SURVEYS OF EXTRAGALACTIC SOURCES WITH PLANCK

G. De Zotti
INAF - Osservatorio Astronomico di Padova
Vicolo dell'Osservatorio 5, I-35122 Padova, Italy
&
SISSA/ISAS
Via Beirut 2-4, I-34014 Trieste, Italy

C. Burigana
CNR-INAF/IASF, Sezione di Bologna
Via Gobetti 101, I-40129 Bologna, Italy

M. Negrello, S. Tinti and R. Ricci
SISSA/ISAS
Via Beirut 2-4, I-34014 Trieste, Italy

L. Silva
INAF–Osservatorio Astronomico di Trieste
Via G.B. Tiepolo 11, I-34131 Trieste, Italy

J. González-Nuevo and L. Toffolatti
Departamento de Física, Universidad de Oviedo
Calvo Sotelo s/n, 33007 Oviedo, Spain

Abstract Although the primary goal of ESA's PLANCK mission is to produce high resolution maps of the temperature and polarization anisotropies of the Cosmic Microwave Background (CMB), its high-sensitivity all-sky surveys of extragalactic sources at 9 frequencies in the range 30–860 GHz will constitute a major aspect of its science products. In particular, PLANCK surveys will provide key information on several highly interesting radio source populations, such as Flat Spectrum Radio Quasars (FSRQs), BL Lac objects, and, especially, extreme GHz Peaked Spectrum (GPS) sources, thought to correspond to the very earliest phases of the evolution of radio sources. Above 100 GHz, PLANCK will provide the first all-

J. C. del Toro Iniesta (eds.), The Many Scales in the Universe, 45–54.

sky surveys, that are expected to supply rich samples of highly gravitationally amplified dusty proto-galaxies and large samples of candidate proto-clusters at $z \simeq 2$–3, thus shedding light on the evolution of large scale structure across the cosmic epoch when dark energy should start dominating the cosmic dynamics.

Keywords: Cosmology, Galaxy evolution, Radio galaxies

1. Introduction

The PLANCK satellite will carry our high sensitivity all sky surveys at 9 frequencies in the poorly explored range 30–860 GHz (see the contribution by J. Tauber, this volume). At low frequencies, PLANCK will go several times deeper (and will detect about ten times more sources) than WMAP, that has provided the first all-sky surveys at frequencies of tens of GHz, comprising about 200 objects (Bennett et al. 2003).

Above 100 GHz, PLANCK surveys will be the first and will remain the only all sky surveys available for many years to come. They will fill an order of magnitude gap in our knowledge of the spectrum of bright extragalactic sources and may discover new populations, not represented, or not recognized, in lower or higher frequencies surveys.

Rather than presenting a comprehensive review of the expected scientific results from PLANCK measurements of extragalactic sources (see, e.g., De Zotti et al. 1999) we will focus on a couple of frequencies, one of the Low Frequency Instrument, namely 30 GHz, and one of the High Frequency Instrument, namely 350 GHz. The relatively shallow but all-sky PLANCK surveys will be ideal to study populations which are both very powerful at mm/sub-mm wavelengths, and very rare, such as radio sources with inverted spectra up to ≥ 30 GHz [extreme GHz Peaked Spectrum (GPS) sources or High Frequency Peakers (HFP)], thought to be the most recently formed and among the most luminous radio sources, and ultra-luminous dusty proto-spheroidal galaxies, undergoing their main and huge episode of star formation at typical redshifts ≥ 2 (Granato et al. 2001, 2004). And PLANCK will observe such sources with an unprecedented frequency coverage.

To estimate the detection limit, and the number of detectable sources, we need to take into account, in addition to the instrument noise, the fluctuations due to Galactic emissions, to the Cosmic Microwave Background (CMB), and to extragalactic sources themselves. These fluctuations will be briefly reviewed in Sect. 2, while in Sects. 3 and 4 we will discuss the expected impact of PLANCK data on our understanding of HFPs and of ultra-luminous proto-spheroidal galaxies, respectively. Our main conclusions are summarized in Sect. 5.

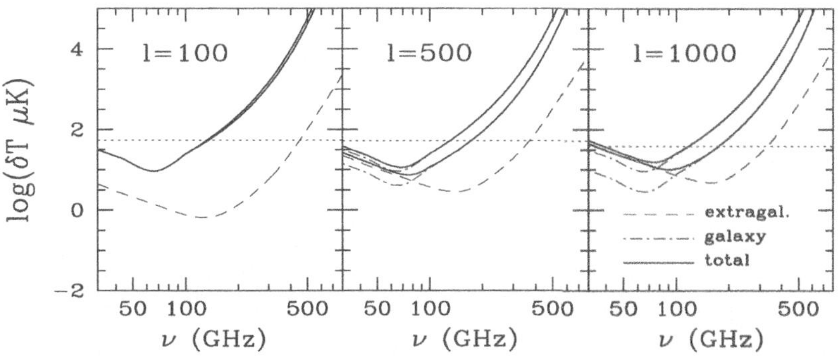

Figure 1. Galactic (dot-dashed) and extragalactic (dashed) contributions to the power spectrum of foreground fluctuations, compared with the CMB (dotted horizontal line) for three values of the multipole number ℓ. The solid lines show the sum, in quadrature, of the two contributions. At $\ell = 100$ the dot-dashed line essentially coincides with the solid line; the two lines largely overlap at high frequencies also for higher ℓ's. The Galactic contributions are averages for $|b| \geq 20°$, after having applied the Kp0 mask which include the point source removal, and comprise synchrotron, free-free and thermal dust emissions, whose power spectra are normalized to the K-band (22.8 GHz), V-band (60.8 GHz), and W-band (93.5 GHz) WMAP data, respectively (where each component is best measured). The extrapolation in frequency has been done adopting, for free-free, the antenna temperature spectral index ($T_A \propto \nu^{\beta}$) $\beta_{\rm ff} = -2.15$, and for synchrotron the expression proposed by Jackson & Wall (2002) for low-luminosity radio sources ($\log S_\nu$ = const $- 0.6424 \log(\nu) - 0.0692(\log(\nu)^2)$, with ν in GHz); this formula, which allows for the high-frequency steepening of the synchrotron spectrum due to electron energy losses, is consistent with the steepening observed in WMAP data (Fig. 9 of Bennett et al. 2003). As for thermal dust we have considered two cases: $\beta_d = 2.2$, the best fit value of Bennett et al. (2003), and the more usual value $\beta_d = 2$. With these spectra, an additional component (spinning dust?) is necessary to account for the foreground signal detected by WMAP particularly in the Q-band (40.7 GHz); the solid lines include this component. Power spectra at $\ell = 100$ were derived directly from WMAP data. At higher ℓ's we assume $C_\ell = C_{100}(\ell/100)^{-\gamma}$ with $\gamma = 2$ or $\gamma = 3$. The upper dot-dashed curve corresponds to $\gamma = 2$ and $\beta_d = 2.2$, the lower one to $\gamma = 3$ and $\beta_d = 2$. The dashed curve includes, summed in quadrature, the contributions of all classes of extragalactic sources, based on models by De Zotti et al. (2004), including canonical radio sources, starburst galaxies, proto-spheroidal galaxies and Sunyaev-Zeldovich effects. The effect of clustering of proto-spheroidal galaxies has been taken into account as in Negrello et al. (2004a).

2. Power spectra of foreground emissions

For a very high sensitivity experiment, like PLANCK, the main limitation to the capability of mapping the CMB is set by contamination by astrophysical sources ("foregrounds"), while CMB fluctuations may be the highest "noise" source for the study of astrophysical emissions at mm wavelengths. The most intense foreground source is our own Galaxy. Because of the different power spectra of the various emission components, the frequency of minimum fore-

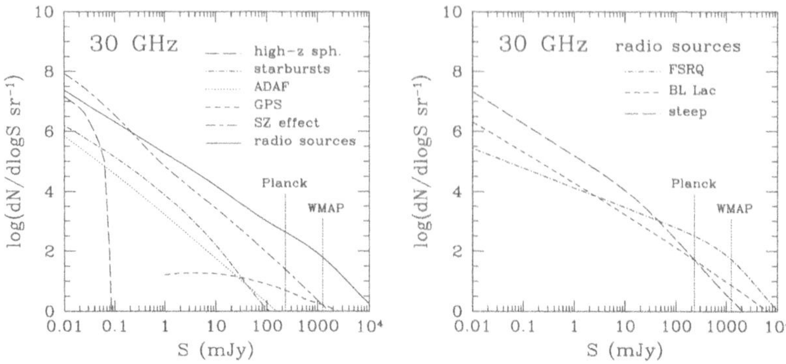

Figure 2. Predicted 30 GHz differential counts. The left-hand panel shows the counts of all the main populations (see De Zotti et al. 2004 for details). The right-hand panel details the contributions of three sub-classes of canonical radio sources: FSRQs, BL Lac objects, and steep-spectrum sources.

ground fluctuations depends to some extent on the angular scale (see Fig. 1, where δT are fluctuations of the CMB thermodynamic temperature, in μK, related to the power spectrum C_ℓ by $\delta T = [\ell(\ell+1)C_\ell/(2\pi)]^{0.5}(e^x-1)^2/(x^2e^x)$, with $x = h\nu/kT_{\rm CMB}$). So long as diffuse Galactic emissions dominate the fluctuations ($\theta \gtrsim 30'$; see De Zotti et al. 1999), they have a minimum in the 60–80 GHz range (depending also on Galactic latitude; cf. Bennett et al. 2003).

But the power spectra of diffuse Galactic emissions decline rather steeply with increasing multipole number (or decreasing angular scale). Thus, on small scales, fluctuations due to extragalactic sources, whose Poisson contribution has a white-noise power spectrum (on top of which we may have a, sometimes large, clustering contribution) take over, even though their integrated emission is below the Galactic one. At high frequencies, however, Galactic dust may dominate fluctuations up to ℓ values of several thousands. Unlike the relatively quiescent Milky Way, the relevant classes of extragalactic sources have strong nuclear radio activity or very intense star formation, or both. Thus, although in many cases their SEDs are qualitatively similar to that of the Milky Way, there are important quantitative differences. In particular, dust in active star forming galaxies is significantly hotter and the radio to far-IR intensity ratio of the extragalactic background is much higher than that of the Milky Way. Both factors, but primarily the effect of radio sources, cooperate to move the minimum of the SED to 100–150 GHz.

3. 30 GHz counts

Figure 2 provides a synoptic view of the contributions of different source classes to the global counts of extragalactic sources. Shallow surveys, such

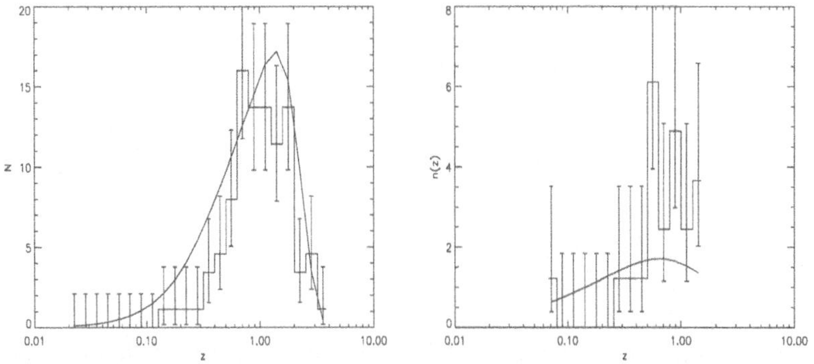

Figure 3. Redshift distributions of WMAP FSRQs (left-hand panel) and BL Lacs (right-hand panel) compared with the model by De Zotti et al. (2004, solid line).

as those by WMAP and PLANCK, mostly detect canonical radio sources. As shown by the right-hand panel of Fig. 2, detected sources will be mostly flat-spectrum radio quasars (FSRQs), while the second more numerous population are BL Lac objects. PLANCK will detect about ten times more sources than WMAP, thus allowing a substantial leap forward in the understanding of evolutionary properties of both populations at high frequencies, only weakly constrained by WMAP data (Fig. 3).

PLANCK will also provide substantial complete samples of sources not (yet) represented in the WMAP catalog, such as Sunyaev-Zeldovich (1972) signals and extreme GPS sources or HFPs (Dallacasa et al. 2000).

GPS sources are powerful ($\log P_{1.4\,\mathrm{GHz}} \gtrsim 25\,\mathrm{W\,Hz^{-1}}$), compact ($\lesssim 1$ kpc) radio sources with a convex spectrum peaking at GHz frequencies. It is now widely agreed that they correspond to the early stages of the evolution of powerful radio sources, when the radio emitting region grows and expands within the interstellar medium of the host galaxy, before plunging in the intergalactic medium and becoming an extended radio source (Fanti et al. 1995, Readhead et al. 1996, Begelman 1996, Snellen et al. 2000). Conclusive evidence that these sources are young came from measurements of propagation velocities. Velocities of up to $\simeq 0.4c$ were measured, implying dynamical ages $\sim 10^3$ years (Polatidis et al. 1999, Taylor et al. 2000, Tschager et al. 2000). The identification and investigation of these sources is therefore a key element in the study of the early evolution of radio-loud AGNs.

There is a clear anti-correlation between the peak (turnover) frequency and the projected linear size of GPS sources. Although this anti-correlation does not necessarily define the evolutionary track, a decrease of the peak frequency as the emitting blob expands is indicated. Thus high-frequency surveys may be able to detect these sources very close to the moment when they turn on.

The self-similar evolution models by Fanti et al. (1995) and Begelman (1996) imply that the radio power drops as the source expands, so that GPS's evolve into lower luminosity radio sources, while their luminosities are expected to be very high during the earliest evolutionary phases, when they peak at high frequencies. De Zotti et al. (2000) showed that, with a suitable choice of the parameters, this kind of models may account for the observed counts, redshift and peak frequency distributions of the samples then available. The models by De Zotti et al. (2000) imply, for a maximum rest-frame peak frequency $\nu_{p,i} = 200\,\mathrm{GHz}$, about 10 GPS quasars with $S_{30\mathrm{GHz}} > 2\,\mathrm{Jy}$ peaking at $\geq 30\,\mathrm{GHz}$ over the 10.4 sr at $|b| > 10°$.

Although the number of *candidate* GPS quasars (based on the spectral shape) in the WMAP survey is consistent with such expectation, when data at additional frequencies (Trushkin 2003) are taken into account the GPS candidates look more blazars caught during a flare optically thick up to high frequencies. Furthermore, Tinti et al. (2004) have shown that most, perhaps two thirds, of the quasars in the sample of HFP candidates selected by Dallacasa et al. (2000) are likely blazars.

Thus, WMAP data are already providing strong constraints on the evolution of HFPs. PLANCK will substantially tighten such constraints and may allow us to directly probe the earliest phases (ages $\sim 100\,\mathrm{yr}$) of the radio galaxy evolution, hopefully providing hints on the still mysterious mechanisms that trigger the radio activity.

We note, in passing, that contrary to some claims, we do not expect that PLANCK can detect the late phases of the AGN evolution, characterized by low accretion/radiative efficiency (ADAF/ADIOS sources).

At faint flux densities, other populations come out and are expected to dominate the counts. In addition to SZ effects, we have active star-forming galaxies, seen either through their radio emission, or through their dust emission, if they are at substantial redshift. The latter is the case for the sub-mm sources detected by the SCUBA surveys if they are indeed at high redshifts (see below).

Such sources may be relevant in connection with the interpretation of the excess signal on arc-minute scales detected by CBI (Mason et al. 2003, Readhead et al. 2004) and BIMA (Dawson et al. 2002) experiments at 30 GHz, particularly if, as discussed below, they are highly clustered, so that their contribution to fluctuations is strongly super-Poissonian (Toffolatti et al. 2004). In fact, to abate the point source contamination of the measured signals, the CBI and BIMA groups could only resort to existing or new low frequency surveys. But the dust emission is undetectable at low frequencies. Although our reference model (Granato et al. 2004), with its relatively warm dust temperatures yielded by the code GRASIL (Silva et al. 1998), imply dusty galaxy contributions to small scale fluctuations well below the reported signals, the (rest-frame) mm emission of such galaxies is essentially unknown and may be

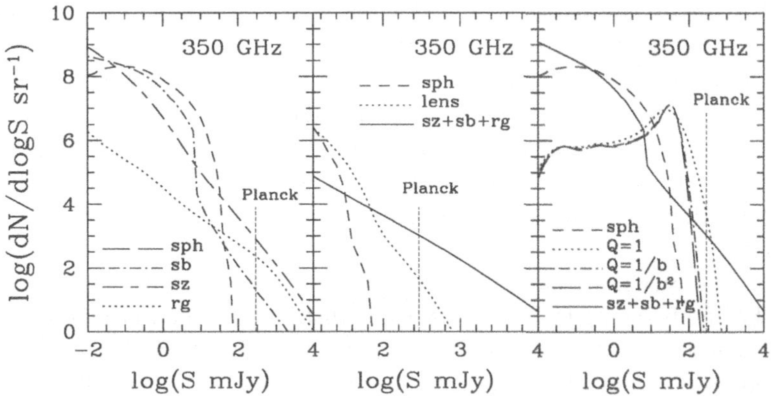

Figure 4. Left-hand panel: contributions of different populations to the 350 GHz counts. Central panel: effect of lensing on counts of proto-spheroidal galaxies. Right-hand panel: estimated counts of "clumps" of proto-spheroids observed with PLANCK resolution.

higher than predicted, e.g. in the presence of the extended distribution of cold dust advocated by Kaviani et al. (2003) or of a widespread mm excess such as that detected in several Galactic clouds (Dupac et al. 2003) and in NGC1569 (Galliano et al. 2003). This is another instance of the importance of a multi-frequency approach, like PLANCK's, capable of keeping under control all the relevant emission components, with their different emission spectra.

4. 350 GHz counts

The 350 GHz counts of extragalactic sources have been determined in the range from $\simeq$ 10 mJy to $\simeq$ 0.25 mJy by surveys with the SCUBA camera, covering small areas of the sky (overall, a few tenths of a square degree). These surveys have led to the discovery of very luminous high-z galaxies, with star-formation rates $\sim 10^3\, M_\odot$/yr, a result confirmed by 1.2mm surveys with MAMBO on the IRAM 30m telescope. These data proved to be extremely challenging for semi-analytic galaxy formation models, and have indeed forced to reconsider the evolution of baryons in dark matter halos.

The bright portion of observed counts appears to be declining steeply with increasing flux density, probably reflecting the exponential decline of the dark-halo mass function at large masses implied by the Press & Schechter formula, so that one would conclude that PLANCK cannot do much about these objects, but rather detect brighter sources such as blazars and relatively local star-forming galaxies, or SZ signals. There are, however, two important effects to be taken into account, that may change this conclusion: gravitational lensing and clustering.

We refer here to the model by Granato et al. (2004) according to which SCUBA sources are large spheroidal galaxies in the process of forming most of their stars. Forming spheroidal galaxies, being located at relatively high redshift, have a substantial optical depth for gravitational lensing, and the effect of lensing on their counts is strongly amplified by the steepness of the counts. This is illustrated by the left-hand panel of Fig. 4, based on calculations by Perrotta et al. (2003). Strong lensing is thus expected to bring a significant number of high-z forming spheroids above the estimated PLANCK 5σ detection limit.

If indeed SCUBA galaxies are massive spheroidal galaxies at high z, they must be highly biased tracers of the matter distribution, and must therefore be highly clustered. There are in fact several, although tentative, observational indications of strong clustering with comoving radius $r_0 \simeq 8\mathrm{h}^{-1}$ Mpc (Smail et al. 2004, Blain et al. 2004, Peacock et al. 2000), consistent with theoretical expectations.

But if massive spheroidal proto-galaxies live in strongly over-dense regions, low resolution experiments like PLANCK unavoidably measure not the flux of individual objects but the sum of fluxes of all physically related sources in a resolution element.

This is an aspect of the "source confusion" problem, whereby the observed fluxes are affected by unresolved sources in each beam. The problem was extensively investigated in the case of a Poisson distribution, particularly by radio astronomers (Scheuer 1957, Murdoch et al. 1973, Condon 1974, Hogg & Turner 1998). The general conclusion is that unbiased flux measurements require a $S/N \geq 5$.

Not much has been done yet on confusion in the presence of clustering (see however Hughes & Gaztanaga 2000). The key difference is that, for a Poisson distribution, a bright source is observed on top of a background of unresolved sources that may be either above or below the all-sky average, while in the case of clustering, sources are preferentially found in over-dense regions.

Clearly, the excess signal (over the flux of the brightest source in the beam) depends on the angular resolution. For a standard $\xi(r) = (r/r_0)^{-1.8}$ the mean clustering contribution is $\propto r_0^{1.8} r_{\mathrm{beam}}^{1.2}$. The PLANCK beam at this frequency corresponds to a substantial portion of the typical clustering radius at $z \simeq 2$–3, so that PLANCK will actually measure a significant fraction of the flux of the clump, which may be substantially larger than the flux of any member source. The effect on counts depends on the joint distribution of over-densities and of M/L ratios. The former depends on both the two- and the three-point correlation function, while the latter depends on the luminosity function.

Preliminary estimates of the distribution of excess luminosities due to clustering around bright sources have been obtained by Negrello et al. (2004b). The right-hand panel of Fig. 4 shows the estimated counts of clumps observed with PLANCK resolution for three models for the evolution of the coefficient Q of the three-point correlation function. Obviously PLANCK can provide information only on the brightest clumps, and, except in the extreme case of $Q = 1$ at all cosmic times, the clumps will only show up as $< 5\sigma$ fluctuations. On the other hand, such fluctuations will provide a rich catalogue of candidate proto-clusters at substantial redshifts (typically at $z \simeq$ 2–3), very important to investigate the formation of large scale structure and, particularly, to constrain the evolution of the dark energy thought to control the dynamics of the present day universe.

5. Conclusions

Although extragalactic surveys are not the primary goal of the mission, PLANCK will provide unique data for several particularly interesting classes of sources. Examples are the FSRQs, BL Lac objects, but especially extreme GPS sources that may correspond to the earliest phases of the life of radio sources, and proto-spheroidal galaxies. Thus PLANCK will investigate not only the origin of the universe but also the origin of radio activity and of galaxies. Sub-mm surveys will provide large samples of candidate proto-clusters, at $z \simeq$ 2–3, shedding light on the evolution of the large scale structure (and in particular providing information on the elusive three-point correlation function) and of the dark energy, across the cosmic epoch when it is expected to start dominating the cosmic dynamics.

Acknowledgments

Work supported in part by MIUR through a PRIN grant and by ASI.

References

Begelman, M.C. 1996, in *Cygnus A – Study of a Radio Galaxy*, Eds. C.L. Carilli & D.E. Harris, Cambridge University Press, 209

Bennett, C.L., et al. 2003, ApJS, 148, 97.

Blain, A. W., S.C. Chapman, I. Smail, & R. Ivison 2004, ApJ, 611, 725

Condon, J.J. 1974, ApJ, 188, 279.

Dallacasa, D., C. Stanghellini, M. Centonza, & R. Fanti 2000, A&A, 363, 887.

Dawson, K.S., et al. 2002, ApJ, 581, 86.

De Zotti, G., et al. 1999, in *Proceedings of the EC-TMR Conference "3K Cosmology", AIP CP*, 476, Eds. L. Maiani, F. Melchiorri & N. Vittorio, 204.

De Zotti, G., G.L. Granato, L. Silva, D. Maino, & L. Danese 2000, A&A, 354, 467.

De Zotti, G., R. Ricci, D. Mesa, L. Silva, P. Mazzotta, L. Toffolatti, & J. González-Nuevo 2004, A&A, in press

Dupac, X., et al. 2003, A&A, 404, L11.

Fanti, C., et al. 1995, A&A, 302, 317.

Galliano, F., et al. 2003, A&A, 407, 159.

Granato, G.L., L. Silva, P. Monaco, P. Panuzzo, P. Salucci, G. De Zotti, & L. Danese 2001, MNRAS, 324, 757.

Granato, G.L., G. De Zotti, L. Silva, A. Bressan, & L. Danese 2004, ApJ, 600, 580.

Hogg, D.W., & E.L. Turner 1998, PASP, 110, 727.

Hughes, D.H., & E. Gaztanaga 2000, In *Star formation from the small to the large scale*, proc. 33rd ESLAB symp.,Eds. F. Favata, A. Kaas, & A. Wilson, ESA, SP 445, 29.

Jackson, C.A., & J.V. Wall 2001, In *Particles & Fields in Radio Galaxies*, Eds. R.A. Laing & K.M. BlundellASP Conf. Proc. 250, 400.

Kaviani, A., M.G. Haehnelt, & G. Kauffmann 2003, MNRAS, 340, 739.

Mason, B.S., et al. 2003, ApJ, 591, 540.

Murdoch, H.S., D.F. Crawford, & D.L. Jauncey 1973, ApJ, 183, 1.

Negrello, M., M. Magliocchetti, L. Moscardini, G. De Zotti, G.L. Granato, & L. Silva 2004a, MNRAS, 352, 493.

Negrello, M., M. Magliocchetti, L. Moscardini, G. De Zotti, & L. Danese 2004b, MNRAS, submitted

Peacock, J.A., et al. 2000, MNRAS, 318, 535.

Perrotta, F., M. Magliocchetti, C. Baccigalupi, M. Bartelmann, G. De Zotti, G.L. Granato, L. Silva, & L. Danese 2003, MNRAS, 338, 623.

Polatidis, A., et al. 1999, NewAR, 43, 657.

Readhead, A.C.S., G.B. Taylor, T.J. Pearson, & P.N. Wilkinson 1996, ApJ, 460, 634.

Readhead, A.C.S., et al. 2004, ApJ, 609, 498.

Scheuer, P.A.G. 1957, Proc. Cambridge Phil. Soc., 53, 764.

Silva, L., G.L. Granato, A. Bressan, & L. Danese 1998, ApJ, 509, 103.

Smail, I., S.C. Chapman, A.W. Blain, & R.J. Ivison 2004, in *The Dusty and Molecular Universe: A Prelude to Herschel and ALMA*, Ed. A. Wilson , ESA Conf. Ser., p. 16.

Snellen, I.A.G., et al. 2000, MNRAS, 319, 445.

Sunyaev, R. A., & Ya.B. Zeldovich 1972, Comm. Ap. Space Phys., 4, 173.

Taylor, G.B., J.M. Marr, T.J. Pearson, & A.C.S. Readhead 2000, ApJ, 541, 112.

Tinti, S., D. Dallacasa, G. De Zotti, A. Celotti, & C. Stanghellini 2004, A&A, in press.

Toffolatti, L., J. González-Nuevo, M. Negrello, G. De Zotti, L. Silva, G.L. Granato, & F. Argüeso 2004, A&A, submitted

Trushkin, S.A. 2003, Bull. Spec. Astrophys. Obs. N. Caucasus, 55, 90, astro-ph/0307205.

Tschager, W., et al. 2000, A&A, 360, 887.

NEW PHYSICS IN CLUSTERS OF GALAXIES

A New Dark Matter Component?

M. Douspis

Laboratoire d'sAtrophysique de Toulouse Tarbes
France
douspis@ast.obs-mip.fr

Abstract The latest cosmological observables analyses seem to converge to a concordant view of the cosmological model: namely the power law Λ-CDM. The recent WMAP results comfort this new standard model. Nevertheless, some degeneracy in the CMB physics do not allow one to exclude alternative models. A combined analysis with other cosmological observations is thus needed. An example of such work is shown here, focusing on the abundance of local clusters. The latter is a traditional way to derive the amplitude of matter fluctuations, but which suffers from a lack of accurate knowledge of their masses. Here we show that the mass temperature relation can be specified for any cosmological model from consistency arguments, removing most of this uncertainty. This allows to obtain an estimation of the amplitude of matter fluctuations with an accuracy of 5%. Quite remarkably, this amplitude can be also tightly constrained from existing CMB measurements. However, the amplitude inferred in this way in a concordance model (Λ-CDM) is significantly larger than the value derived from X-ray clusters. Such a discrepancy may reveal the existence of a new dark component in the Universe. It may alternatively reveal a significant depletion of the gas during the formation of clusters. In all cases, an essential element of clusters formation history seems to be missing.

Keywords: Cosmology, CMB, Clusters of galaxies, Parameter estimation

1. Introduction

The cosmological scenario seems more and more well defined. Recent cosmological analyses have lead to the so-called concordance model. The latter passes a lot of cosmological tests. Nevertheless, some observations seem not to be concordant. Furthermore, constraints on the concordance model are obtained in some (well motivated) particular framework. Some alternatives are thus still possible and the conclusions may be drawn with care. In a first section

J. C. del Toro Iniesta (eds.), The Many Scales in the Universe, 55–70.

we will examine the concordance model through the constraints and degeneracies given by the study of CMB anisotropies on cosmological parameters. We then show how some combined analyses of different cosmological observations could lead to a more precise view of the concordance model. After few remarks on the dark matter candidates, we will focus on the amplitude of fluctuations and the need for new physics (or dark matter component) to reconcile clusters and CMB constraints.

2. The concordance model

From CMB

Constraints. The CMB anisotropies observations are used since their discovery in the early nineties by COBE-DMR to put constraints of cosmological parameters. The detection of the Sachs-Wolfe plateau in the angular power spectrum (APS hereafter) in 1992 allows one to put constraints on the shape of the initial power spectrum (IPS hereafter) of fluctuations by constraining its amplitude and slope (Gorski et al. 2004). Then the first hint for the presence of a first acoustic peak in the APS around the degree scale (multipole moment $\ell \sim 200$) was enough for excluding low density models (open models) and comforting inflationary predictions for a flat Universe (no curvature) (Lineweaver et al. 1997). Then new observations were made from ground, balloon and recently satellite again, with a large range of instruments, strategies, frequencies, resolutions. All these data were used progressively to make better and better constraints on cosmological parameters. Today around thirty experiments have observed the microwave sky and the last one opens a new generation of observation. WMAP 2003 release (first release) marked a step in CMB anisotropies studies (WMAP). The APS derived from full sky maps both in temperature and in cross-polarization helped in constructing a better convergence to the new standard Λ-CDM model. WMAP becomes a reference by constraining most of the cosmological parameter by itself. Such result, shown in Table 1 and Fig. 1, was obtained in the following framework: a cosmic scenario in which the anisotropies are due to some Gaussian random and adiabatic fluctuations evolving in a cold dark matter and dark energy dominated (topologically trivial) Universe, in which the initial power spectrum is a power law caracterised by its amplitude (A) and slope (n). Further hypothesis (or priors on the parameters) assumed are summarized in Table 1. In such (well motivated) scenario, most of the errors on the cosmological parameters are smaller than 10%.

Degeneracies. Even in such well defined scenario some degeneracies remain. Some parameters have the property of changing the APS, when varying, in an opposit manner from other (combinations of) parameters at first order.

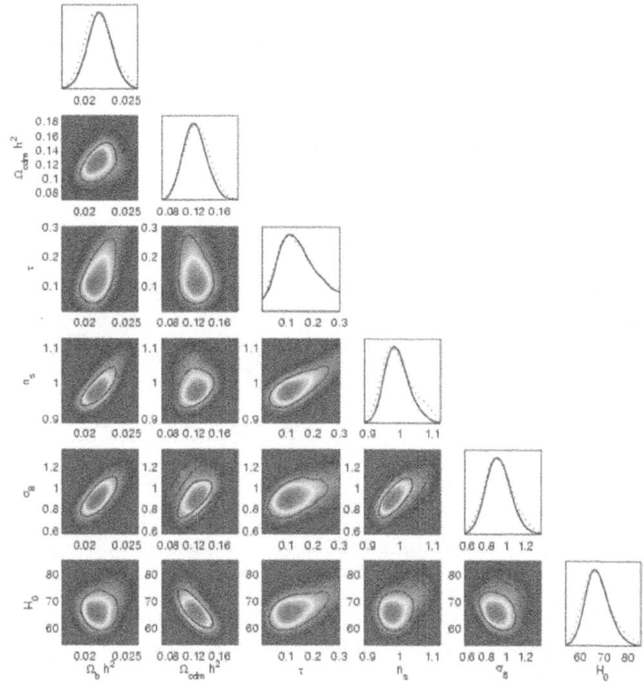

Figure 1. Constraints from WMAP as 2 dimensional likelihood contours (from Fosalba & Szapudi 2004).

Table 1. Parameters from WMAP

Parameter	*symbol*	*Value with errors*	*Hypothesis*
Baryon density	$\Omega_b h^2$	0.024 ± 0.001	$\tau < 0.3$
Dark matter density	$\Omega_b h^2$	0.14 ± 0.02	$h > 0.5$
Hubble constant	h	0.72 ± 0.05	$\Omega_{tot} = 1$
Optical depth	τ	$0.166^{0.076}_{0.071}$	$P_0(k) \propto k^n$
Spectral index	n	0.99 ± 0.04	

For example, an early reionisation will suppress power at smale scales in the temperature APS, whereas a blue power spectrum (index n bigger than 1) will enhance the power at small scale relative to the large scales. Even at WMAP sensitivity, such effects are still visible (elongated ellipses in Fig. 1).

If some of the hypothesis made in the above scenario are relaxed more degeneracies appear. That is the case for the one between the Hubble constant H_0 and the total density if the Universe (or curvature) Ω_{tot}. The Fig. 2 shows in blue the confidence intervalls obtained with WMAP in the H_0, Ω_{tot} plane without assumptions on these parameters. In red are the contours when a prior $H_0 > 50$ km/s/Mpc is assumed. Such example shows that the CMB by itself do not prove the flatness of the Universe. Nevertheless the CMB data excluded drastically open low density models.

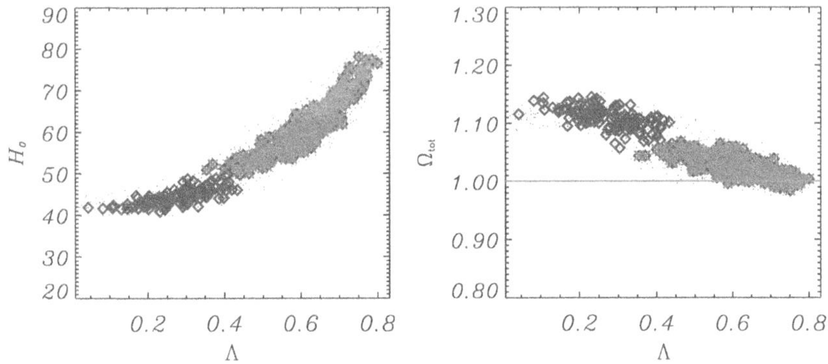

Figure 2. Curvature degeneracy: constraints in the (H_0, Λ) and (Ω_{tot}, Λ) planes. In blue, analyses without priors on H_0 nor Ω_{tot}. In red, analyses with $H_0 > 50\mathrm{km/s/Mpc}$ prior. In green, analyses with $\Omega_{tot} = 1$ prior.

Another kind of degeneracies appear when more freedom is let to the initial conditions. The APS is the result of the evolution in a particular cosmology of some initial conditions such as the IPS. The APS could be written as follows:

$$C_\ell = \int P_0(k)\Delta^2_{\ell,k}\frac{dk}{k} \tag{1}$$

where $P_0(k)$ is the IPS and Δ is the transfer function depending on the cosmological parameters such as the densities, the Hubble constant, etc. As we have seen the "standard" scenario considers power law IPS. If such hypothesis is relaxed, then the constraints on cosmological parameters weaken. Figure 3 shows an example from Blanchard et al. (2003) where two APS are drawn over WMAP data. The dotted model is the best power law Λ-CDM model obtained in (Spergel et al. 2003), whereas the blue solid line model in a pure matter dominated flat model ($\Omega_\Lambda = 0$) with a IPS caracterised with two different slopes at large and small scales. These two models are thus equiprobable as seed of the WMAP observed APS.

As noted by the WMAP team, few outliers in the observed APS make some strong contributions in the global χ^2 which is consequently subsequently high. These outliers, expected to be due to some experimental effects not well substracted could also have cosmic origin. Some features in the IPS could for example produce such deviations from the theoretical APS. Under the assumption

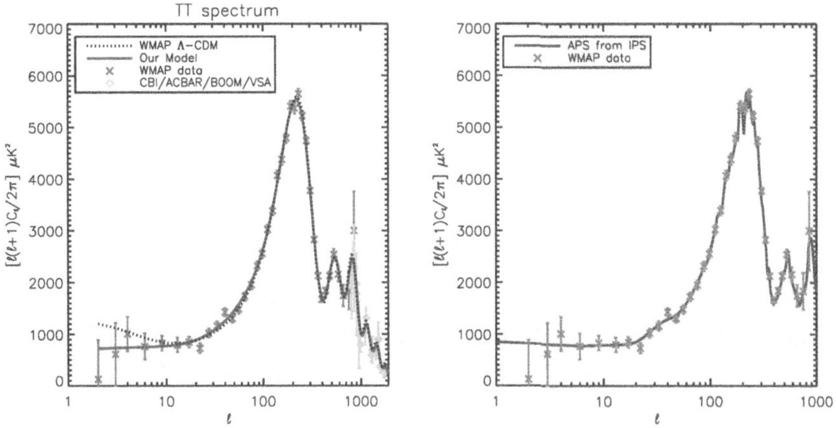

Figure 3. left: Angular power spectra from power law Λ-CDM and broken power law de Sitter Universe. right: Angular power spectrum from the reconstructed initial power spectrum shown in Fig. 4 (left).

that these outliers are cosmological signal, Tocchini–Valentini et al. (2004) have inverted the relation in Equ. 1 to retrieve the IPS from the APS (by assuming concordance cosmological parameters). The results is plotted in Fig. 4 (left). The IPS derived from WMAP $C_\ell's$ (in blue with error bars in green) exhibit deviations from a simple power law (black $\sim$ horizontal line). If such an IPS is now evolved to present time one have the corresponding matter power spectrum that can be confronted to observations. As shown in Fig. 4 (center), the deviations in the IPS produce deviations in the matter power spectrum that are in agreement with the features detected by the Sloan Digital Sly Survey (SDSS, SDSS) comforting the idea of cosmological origin of the deviations in the WMAP APS. The method used by Tocchini-Valentini et al. could be applied whatever the cosmology is. Given the degeneracy illustrated by Equ. 1, one can retrieve an IPS able to fit WMAP $C_\ell's$ for a purely matter dominated Universe. Figure 4 (right) shows the IPS retrieved in such a cosmology ($\Omega_\Lambda = 0$).

The CMB anisotropies studies can bring a lot of information on the cosmological scenario. Strong constraints could be put if a reasonable framework is assumed. Nevertheless, degeneracies remain in more general approaches. The physics of the CMB anisotropy imply degeneracies in the shape of the APS. Moreover, the initial conditions remain degenerated with the evolution of the Universe (see also Bucher et al. 2004).

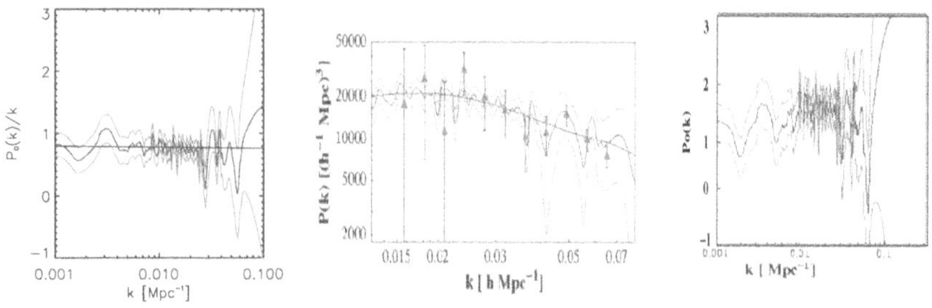

Figure 4. Reconstructed IPS (left) from WMAP APS under concordance prior and corresponding matter power spectrum compared to SDSS data (middle). Example of reconstructed IPS from WMAP with $\Omega_M = 1$ prior (right).

From combined analysis

In order to break these degeneracies, one can combine constraints from different observations. Angular distances, shape of the matter power spectrum, Hubble law are not sensitive to the cosmological parameters in the same way as the APS of CMB anisotropies. This allows one to break degeneracies intrinsic to one of the observable. Figure 5 summarizes such a combination. Most of studies implying this kind of combinations converge to the now standard "concordant model" (Douspis et al. 2003; Tegmark et al. 2004; Spergel et al. 2003). Some observations remain nevertheless inconsistent and combining them may lead to other conclusions (Kochanek & Scechter 2004; Blanchard & Douspis 2003;). Such an example is given in the next sections.

3. Dark matter

Dark matter is now a major component of the cosmological model. We have direct proofs of existence (rotation curves of galaxy, mass luminosity ratios, or lensing of background galaxy by clusters). This dark matter is probably non baryonic (massive compact halo objects search seems to indicate that such dark matter could only contribute to 20% of the total mass of clusters Afonso et al. 2003; Yoo, Chanamé & Gould 2004) and may contribute to 25% of the total density of the Universe (with 5% of baryons and 70% of dark energy in the concordance model).

This dark matter could be cold (few GeV, and non relativistic at galaxy formation), warm (few keV) or hot (few eV and relativistic at galaxy formation). Few candidates can be classified as follows (Gondolo 2004): existing (light neutrinos),well motivated (heavy neutrinos, neutralinos, axions) and other (self-interactive particles, wimpzillas, ...). Recent experiments have put a lower limit on the mass of the neutrinos (Klapdor-Kleingrothaus et al.

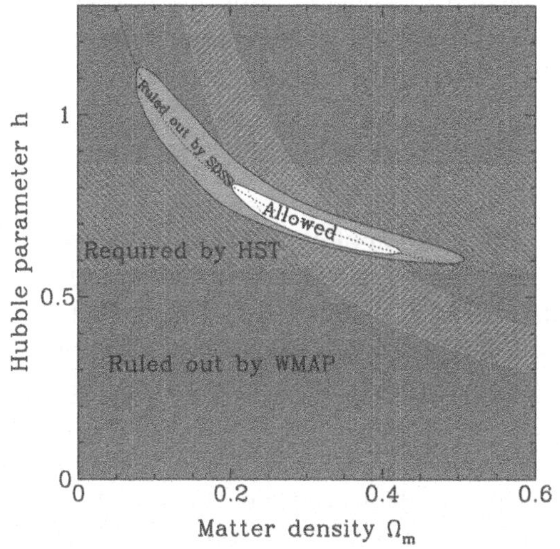

Figure 5. Constraints ion the Hubble constant and matter density in combined analyses (from Tegmark et al. 2003).

2004). On the other side, heavy neutrinos have not been detected in laboratory, and the shape of the matter power spectrum from CMB and galaxy surveys (Tegmark et al. 2004) sets an upper bound to the mass of such particles. From these limits one can deduce the cosmological contribution of neutrinos in the energy density budget:

$$0.0005 < \Omega_\nu h^2 < 0.03$$

whereas the total dark matter contribution is $\Omega_{DM} h^2 \sim 0.12$. Neutrinos (light or heavy) can not therefore explain the full dark matter content. The next well motivated candidate is the neutralino (cold dark matter). It is the lightest supersymetric particle. It has not been detected yet but recent experiments (from cosmology or particle physics) could not exclude it. See (Gondolo 2004) for a better an more complete review of non baryonic dark matter and (Boehm, Fayet & Silk 2004) for alternatives to the supersymetric candidates.

4. The amplitude of fluctuations

The amplitude of matter fluctuations in the present-day universe is an important quantity of cosmological relevance. The abundance of clusters is an efficient way to evaluate this quantity, commonly expressed by σ_8, the *r.m.s.* amplitude of the matter fluctuations on the $8h^{-1}$Mpc scale. This quantity can also be determined by the study of weak lensing shear or as a derived parameter from CMB analysis. The determination of σ_8 has gained some tension because the amplitude of matter obtained from clusters with hydrostatic equation leads to low values $\sigma_8 \sim 0.7 \pm 0.06$ (Markevitch 1998; Reiprich & Böhringer 2002; Seljak 2002)while WMAP obtained recently $\sigma_8 \sim 0.9 \pm 0.1$ (Spergel et al. 2003).

From CMB

WMAP team determination of cosmological parameter lead to the concordance model. Such a model, normalized to CMB scales imply a value of the amplitude of fluctuations $\sigma_8 \sim 0.9 \pm 0.1$. This quantity is related to the amplitude of the APS in temperature around $\ell \sim 1000$, whatever the shape of the IPS (Blanchard & Douspis 2003). However, some degeneracies may alter this "first order" assumption. Some physical processes which lower the power of fluctuations at small scale may lower the value of σ_8. This is the case of the reionisation of the Universe (or the presence of neutrinos). Figure 7 shows the degeneracy between the optical depth of reionisation (τ) and the amplitude of fluctuations (σ_8) when only the temperature APS is considered (which means no constraints on τ). For a Universe in a concordance model but without reionisation, the amplitude of fluctuation can be as low as $\sigma_8 \sim 0.65$. The cross power spectrum (temperature–polarisation E) of WMAP contains a signature at large scale of an early reionisation. This suggests a value of quite high leading to the value of σ_8 quoted by Spergel et al. (2003). This signature is nevertheless weak because some galactic foregrounds contamination may play a role at these scales. The full polarized spectrum (E–E), released soon, may clarify this.

From weak lensing shear

Weak lensing shear is another way to derive the amplitude of fluctuations. Recent observations have lead to an ensemble of consistent relations between the matter content and the amplitude of fluctuations (Bacon et al. 2003; Brown et al. 2003; Bucher et al. 2004; Hamana et al. 2003; Heymans et al. 2004; Hoekstra, Yee & Gladders 2002; Jarvis et al. 2003; Massey et al. 2004; Refregier 2003; Refregier, Rhodes & Groth 2002; Rhodes et al. 2004; Van Waerbeke, Mellier & Hoekstra 2004) which can be averaged by: $\sigma_8 \left(\frac{\Omega_M}{0.3}\right)^{0.47} = 0.836 \pm 0.035$.

From clusters of galaxies

A statistical precision of few % on σ_8 is possible from existing samples of x-ray clusters, but in practice the relation between mass and temperature is needed for such evaluation:

$$T = A_{TM} M_{15}^{2/3} (\Omega_M (1 + \Delta_v)/179)^{1/3} h^{2/3} (1 + z) \text{ keV} \tag{2}$$

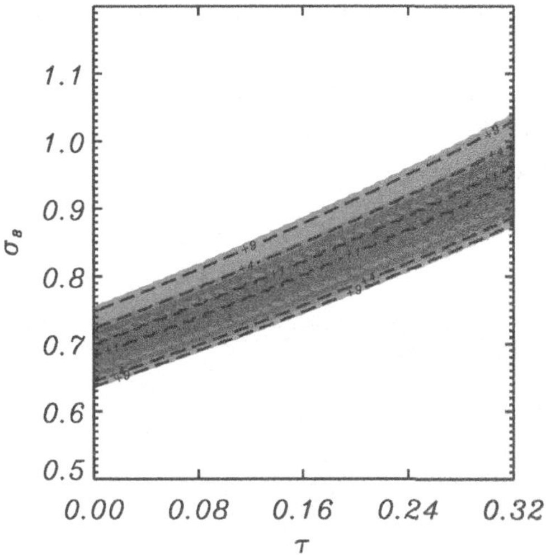

Figure 6. Degeneracy between the amplitude of fluctuations and the optical depth from WMAP temperature APS.

This value of A_{TM} has been estimated from x-ray properties of clusters by different methods, essentially hydrostatic equation on one side and numerical simulations on the other side, which lead to sensitively different normalizations (from $\sigma_8 \sim 0.6$ to $\sigma_8 \sim 1.$).

We have shown in (Blanchard & Douspis 2004) a third way to derive the value of A_{TM} that we called self-consistent.

Clusters are useful cosmological probes in several important ways. Their baryonic fraction f_b can be inferred from observations: $f_b = \frac{M_b}{M_{tot}}$. Under the assumption that the baryonic and dark matter amounts are representative of the universe, the baryon fraction can be related to the cosmological parameters density Ω_b and Ω_m: $f_b = \Upsilon \frac{\Omega_b}{\Omega_m}$ where Υ is a numerical factor that has to be introduced in order to correct for the depletion of gas during cluster formation and which can be determined only from numerical simulations (White et al. 1993). In practice a good working value, at least in the outer part of clusters, is $\Upsilon = 0.925$ (Frenk & White 1999). The baryonic content of the Universe is now known quite accurately through WMAP and other CMB measurements ($\omega_b = \Omega_b h^2 = 0.023 \pm 0.002$, Spergel et al. 2003) the statistical uncertainty being doubled in order to account for differences in various priors) essentially consistent with the abundance of Deuterium (Kirkman et al. 2003) and with the baryonic content of the IGM (Tytler et al. 2004). We can furthermore assume the following relation given by the position of the first Doppler peak in the CMB APS (Page et al. 2003): $\Omega_m h^{3.4} = 0.086 \pm 0.006$. Then, if we consider a sample of clusters around 4 keV from which we know the baryon mass (Vikhlinin, Forman & Jones 1999) we can rewrite Eq. 2 as: $A_{TM} = f(\Omega_M)$, plotted in Fig. 7 (red solid line with error bars envelops). From there we conclude that A_{TM} varies with Ω_M and is consistent with the baryon fraction in clusters whatever the cosmology is, contrarily to the previous determinations (horizontal shaded regions).

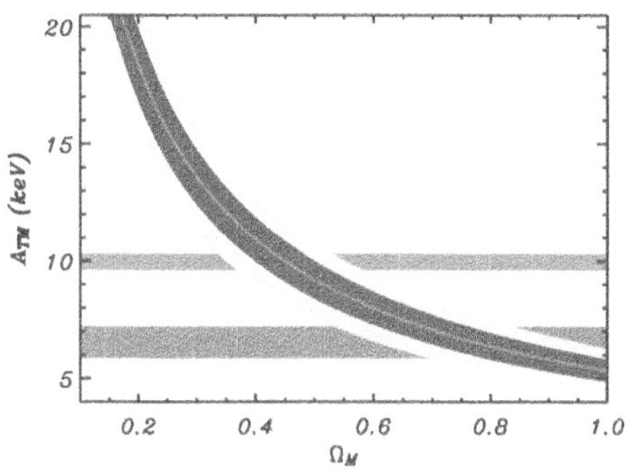

Figure 7. The red line is the central value of $A_{\rm TM}$ which is the normalization of the mass temperature relation Eq. 1. The WMAP relation between H_0 and Ω_0 has been used, as well as the constraint on the baryon content of the Universe. One and two σ uncertainties arising from uncertainty on Ω_b ($= 0.023 \pm 0.002$) are shown as blue and yellow areas. Horizontal areas correspond to estimations of $A_{\rm TM}$ from hydrostatic methods (light green) obtained by Roussel, Sadat & Blanchard (2000) and Markevitch (1998) and from numerical simulations (light blue) obtained from (Bryan & Norman 1998; Evrard, Metzler & Navarro 1996).

Knowing the mass-temperature relation and its uncertainty we can determine the amplitude of matter fluctuations by fitting the local temperature distribution function and assuming a Γ-like spectrum with $\Gamma = 0.2$. Here we use the Sheth and Tormen (1999) mass function and a sample of x-ray selected local clusters ($f_x \leq 2.210^{-11}$ erg/s/cm^2 and $|b| \leq 20\,$deg, Blanchard et al. 2000, updated from BAX, Sadat et al. 2004). The result is shown as lines in Fig. 8. As one can see, at a given value of Ω_M the amplitude of σ_8 is well constrained. Furthermore to the first order the best σ_8 is independent of Ω_M ($\sigma_8 \sim 0.63 \pm 4.5\%$ for $\Omega_\Lambda = 0.7$. interestingly close to the value obtained by Viana, Nichol & Loddle 2002: $\sigma_8 \sim 0.61$). Our conclusion appears somewhat surprising as it differs from standard analyzes based on a fixed normalization A_{TM}, which cannot account simultaneously for the baryon fraction in a consistent way for arbitrary Ω_M.

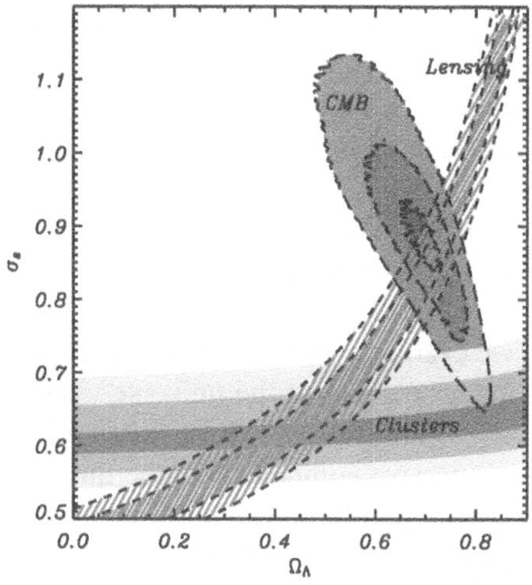

Figure 8. The amplitude of matter fluctuations from clusters abundance using the mass–temperature relation found in the present analysis compared to the amplitude of matter fluctuations derived from CMB data (Hinshaw et al. 2003; Grainge et al. 2004; Pearson et al. 2003; Ruhl et al. 2003; Kuo et al. 2004) grey area correspond to 1,2,3 σ contours on two parameters, dashed lines are contours on one parameter. The one and two sigma amplitudes obtained from an average of recent weak shear measurements are also shown as dashed regions (see text for references).

Evidence of new physics

In the following we use the constraint on σ_8 in a concordance model obtained from the CMB fluctuations analysis including the temperature polarization cross power spectrum (TE) by the WMAP team. The comparison of the value of σ_8 from CMB data with the one from clusters is revealing a critical discrepancy among the two measurements (Fig. 8). It is clear that within any model with $\Omega_\Lambda \sim 0.7$ the amplitude of σ_8 we derived from clusters $\sigma_8 = 0.63 \pm 0.03$ is significantly smaller than what is expected from the CMB alone ($\sigma_8 = 0.88 \pm 0.035$).

As mentioned before, an accurate knowledge of τ is critical to properly evaluate the amplitude of matter fluctuations in the concordance model.

Here above, we have considered models in which the dark matter is only made of cold dark matter, the dark energy being a pure cosmological constant (in term of the equation of state of vacuum $p = w\rho$, this means $w = -1$), and that X-ray gas and known stars are the only existing baryons in clusters.

A first possibility to investigate is to examine whether a different equation of state for the vacuum, so-called quintessence, might solve this discrepancy. We have therefore investigated flat models with arbitrary w and quintessence content Ω_Q. Indeed combinations of CMB and clusters data are known to provide tight constraints on such models (Douspis et al. 2003). With the approach developed here, models which were found to match CMB and clusters were found to satisfy the following constraints: $0.46 < \Omega_Q < 0.54$ and $-0.5 < w < -0.4$. Such models are currently at odds with constraints on quintessential models (Douspis et al. 2003; Tegmark et al. 2004; Riess et al. 2004) resulting from combination of various data including the supernovae type Ia data. We have therefore to turn toward more drastic paths to solve the above issue.

In the following, we examine whether the introduction of an additional component of the dark matter content of the universe would allow to remove the above discrepancy. Neutrinos are known to exist and to be massive, so perhaps the most natural massive component of the universe to be introduced is in the form of a neutrino contribution. This solution has already been advocated in order to solve this discrepancy in an Einstein de Sitter Universe (Elgaroy & Lahav 2003; Blanchard & Douspis 2003). Indeed the presence of a light, but non-zero, component of the dark matter modifies significantly the transfer function of primordial fluctuations which results in a lower amplitude on small scales. Given existing measurements of mass differences we consider only the case where the masses are equal. Within a concordance model ($\Omega_\Lambda = 0.7$, $\Omega_m = 0.3$) by combining the constraints from CMB and clusters data, and marginalizing on (ω_b, H_0, n, τ) we found that a contribution of $\Omega_\nu = 0.015 \pm 0.01$ is preferred with a significance level, well above 3σ (see Fig. 9a), improving the significance of such possible evidence compared to Allen et al. (2003). This confirms that the presence of a small contribution of neutrinos, with a typical mass of .25 eV, to the density of the universe allows to reconcile the amplitude of matter fluctuations from clusters with the one inferred from CMB data. We notice that such value is above the upper limit inferred by the WMAP team using a combination of several astronomical data (Spergel et al. 2003). Finally, weak shear estimations have provided measurements of the amplitude of matter fluctuations which not favor such a solution.

We are therefore left with the conclusion that our initial assumption that baryons in clusters are fairly representative of baryons in the Universe is unlikely and therefore that the observed amount of baryon in clusters does not reflect the actual primordial value (a possibility that has been advocated by Ettori 2003). Several mechanisms could lead to this situation: the most direct way could be the fact that a significant fraction of the baryons are in a dark form, either in the Universe or in clusters (for instance either in the form of Macho's, or in a large gaseous unidentified component, Bonamente, Joy & Lieu 2003), or that a significant fraction of the baryons has been expelled from clusters during their formation process. In such cases, the observed M_b is

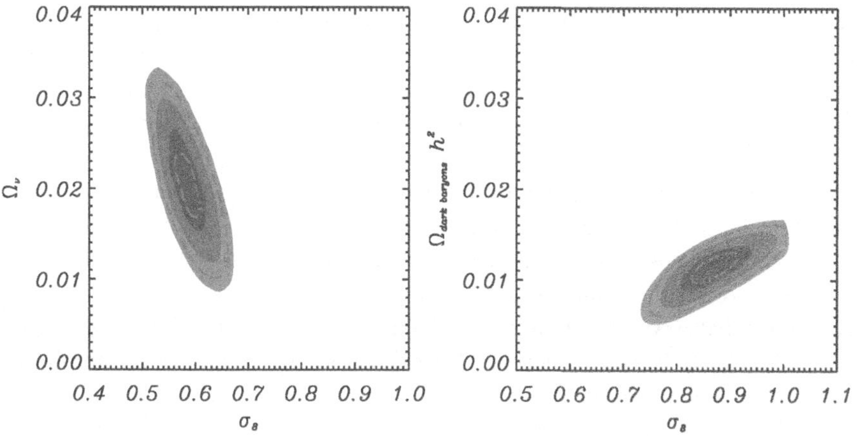

Figure 9. Constraints on Ω_ν and $\Omega_{dark\ baryon}h^2$ from a combined analyses of CMB and clusters, by fixing $\Omega_\Lambda = 0.7$.

biased low. The actual mass of clusters can then be obtained assuming a depletion factor $1 - f$ implying that $f\Omega_b$ represents the missing baryons. Again the combination of CMB and clusters constraints allow to evaluate the amplitude of $f\Omega_b$. From Fig. 9b, one can see such a component, $f\Omega_b \sim 0.023$ should represent nearly half ($\Upsilon \sim 0.5$) of the primordial baryon in order to solve the discrepancy. Although heating processes are advocated in order to account for observed properties of X-ray clusters, they currently do not lead to such a high level of depletion (Bialek, Evrard, & Mohr 2001).

5. Summary

We have seen how the concordance model seems to fit pretty well the CMB power spectrum data, as well as a combination of other cosmological observations. But it is also noticeable how much some degeneracies may be unable one to give strong constraints on cosmological parameters. Furthermore, all the observations are not concordant. The determination of the amplitude of matter fluctuations within pure cold dark matter models, using two methods, namely the CMB and the local clusters abundance, leads to two surprising significantly different values. There are several ways to solve out this discrepancy, although each of them represents a noticeable difference with the standard concordance model. The existence of a non-baryonic dark component, like a neutrino contribution, would allow to solve this discrepancy, although such a solution leads to a low value of σ_8 which is not favored by some other evidences. If the actual value is actually larger, $\sigma_8 \sim 0.8 - 0.9$, the unavoidable conclusion is that the baryonic content of clusters is not representative of the Universe. In this case, an astrophysical solution could be that baryons in clusters could be in a dark form, or at least undetected until now. Alternatively, baryon in clusters could have been severely depleted implying that the actual value Υ is much smaller than the value we used above, the apparent baryon fraction being biased low compared to the actual primordial value. Finally, several observations might help to clarify this issue: the above conclusion relies on the actual value of the optical depth τ found by WMAP. Other sources of informations on σ_8 will also obviously put light on this issue: weak lensing can potentially allow to measure directly the actual amplitude of matter fluctuations with a similar precision to what has been obtained here with clusters, provided that systematic uncertainties are fully understood. Other direct measurements of the amplitude of matter fluctuations like those derived from the Lyman–α forest power spectrum (Croft et al. 2004) could also bring light on this issue. It is remarkable that some of these observations that are expected in the near future will potentially bring fundamental informations on clusters physics or alternatively may reveal the existence of a previously unidentified type of dark matter with Ω_{DM} as low as 0.01.

Acknowledgments

This research has made use of the X-Ray Cluster Database (BAX) which is operated by the Laboratoire d'Astrophysique de Toulouse-Tarbes (LATT), under contract with the Centre National d'Etudes Spatiales (CNES).

References

Afonso, C., et al. 2003, A&A, 400, 951

Allen, S. W., Schmidt, R. W., & Bridle, S. L. 2003, MNRAS, 346, 593

Bacon D., Massey R., Refregier A., Ellis R., 2003, MNRAS, 344, 673

Bialek, J. J., Evrard, A. E., & Mohr, J. J. 2001, ApJ, 555, 597

Blanchard, A., & Douspis, M. 2004, A&A in press, astro-ph/0405489

Blanchard, A., Douspis, M., Rowan-Robinson, M., & Sarkar, S. 2003, A&A, 412, 35

Blanchard, A., Sadat, R., Bartlett, J. G., & Le Dour, M. 2000, A&A, 362, 809

Boehm, C., Fayet, P., & Silk, J. 2004, Phys. Rev. D, 69, 101302

Bonamente, M., Joy, M. K., & Lieu, R. 2003, ApJ, 585, 722

Brown M., Taylor A., Bacon D., Gray M., Dye S., Meisenheimer K., Wolf C., 2003, MNRAS, 341, 100

Bryan, G.L. & Norman, M.L. 1998, ApJ, 495, 80

Bucher, M., Dunkley, J., Ferreira, P. G., Moodley, K., & Skordis, C. 2004, Phys. Rev. Letters, 93, 081301

Chang, T.-C., Refregier, A., Helfand, D. J., Becker, R. H., White, R. L., 2004, ApJ submitted, astroph/0408548

Croft, R.A.C. Weinberg, D.H., Katz, N., Hernquist, L. 1998, ApJ, 495, 44

Douspis, M., Riazuelo, A., Zolnierowski, Y., & Blanchard, A. 2003, A&A, 405, 409

Elgaroy, Ø. & Lahav, O. 2003, Journal of Cosmology and Astro-Particle Physics, 4, 4

Ettori, S. 2003, MNRAS, 344, L13

Evrard A. E., Metzler C. A. & Navarro J. F., 1996, ApJ, 469, 494

Fosalba, P. & Szapudi, I., 2004, ApJ in press, astro-ph/0405589

Frenk, C.S & White, S.D.M et al. 1999, ApJ, 525, 554

Gondolo, P., 2004, astro-ph/0403064

Gorski, K. M., Hinshaw, G., Banday, A. J., Bennett, C. L., Wright, E. L., Kogut, A., Smoot, G. F., & Lubin, P. 1994, ApJL, 430, L89

Grainge, K. et al. 2003, MNRAS, 341, L23

Hamana T., Miyazaki S., Shimasaku K., Furusawa H., Doi M., Hamabe M., Imi K., Kimura M., Komiyama Y., Nakata F., Okada N., Okamura S., Ouchi M., Sekiguchi M., Yagi M. & Yasuda N., 2003, ApJ, 597, 98

Heymans C. et al., 2004, astroph/0406468

Hinshaw, G., et al. 2003, ApJS, 148, 135

Hoekstra H., Yee H. & Gladders M., 2002, ApJ, 577, 595

Jarvis M., Bernstein G., Jain B., Fischer P., Smith D., Tyson J., Wittman D., 2003, ApJ, 125, 1014

Kirkman, D., Tytler, D., Suzuki, N., O'Meara, J. M., & Lubin, D. 2003, ApJS, 149, 1

Klapdor-Kleingrothaus, H. V., Krivosheina, I. V., Dietz, A., & Chkvorets, O. 2004, Phys. Letters B, 586, 198

Kochanek, C. S. & Schechter, P. L. 2004, *Measuring and Modeling the Universe*, 117

Kuo, C. L., et al. 2004, ApJ, 600, 32

Lineweaver, C. H., Barbosa, D., Blanchard, A., & Bartlett, J. G. 1997, A&A, 322, 365

Massey, R., Refregier, A. Bacon, D. & Ellis, R. 2004, astroph/0405457

Markevitch, A. 1998, ApJ, 504, 27

Navarro, J. F., Frenk, C. S. & White, S. D. M. 1995, MNRAS, 275, 720 (NFW)

Page, L., et al. 2003, ApJS, 148, 233

Pearson, T. J., et al. 2003, ApJ, 591, 556

Refregier, A. 2003, ARA&A, 41, 645

Refregier A., Rhodes J., Groth E. J., 2002, ApJL, 572, L131

Reiprich, T. H., & Böhringer, H. 2002, ApJ, 567, 716

Riess, A.G. et al. 2004, ApJ, 607, 665

Rhodes J., Refregier A., Collins N. R., Gardner J. P., Groth E. J., Hill R. S., 2004, ApJ, 605, 29

Roussel, H., Sadat, R. & Blanchard, A. 2000, A&A, 361, 429

Ruhl, J. E., et al. 2003, ApJ, 599, 786

Sadat, R., & Blanchard, A. 2001, A&A, 371, 19

Sadat, R., Blanchard, A., Kneib, J.-P., Mathez, G., Madore, B., & Mazzarella, J. M. 2004, A&A, 424, 1097

SDSS Web page, *www.sdss.org*

Seljak, U. 2002, MNRAS, 337, 769

Sheth, R. K. & Tormen, G. 1999, MNRAS, 308, 119

Spergel, D. N. et al. 2003, ApJS, 148, 175

Tegmark, M. et al., 2004, Phys. Rev. D, 69, 103501

Tocchini–Valentini, D., Douspis, M., & Silk, J., 2004, MNRAS in press, astro-ph/0402583

Tytler, D. et al., 2004, astro-ph/0403688

Vikhlinin, A., Forman, W. and Jones, C. 1999, ApJ, 525, 47

Van Waerbeke, L., Mellier, Y., Hoekstra, H. 2004, A&A, 393, 369
Vauclair, S. C., et al. 2003, A&A, 412, L37
Viana, P. T. P., Nichol, R. C., & Liddle, A. R. 2002, ApJL, 569, L75
White, S.D.M., Navarro, J.F., Evrard, A.E. & Frenk, C. 1993, Nature, 366, 429
WMAP at lambda.gsfc.nasa.gov/product/map
Yoo, J., Chanamé, J., & Gould, A. 2004, ApJ, 601, 311

EMISSION LINE GALAXIES IN CLUSTERS

Bianca M. Poggianti
INAF-Osservatorio Astronomico di Padova
Italy
poggianti@pd.astro.it

Abstract At the present epoch, clusters of galaxies are known to be a hostile environment for emission-line galaxies, which are more commonly found in low density regions outside of clusters. In contrast, going to higher redshifts the population of emission-line galaxies in clusters becomes progressively more conspicuous, and large numbers of star-forming late-type galaxies are observed. I present an overview of the observational findings and the theoretical expectations regarding the evolution of emission-line galaxies in dense environments, discussing the properties of these galaxies and the current evidence for environmental influences on their evolution.

Keywords: Galaxy evolution, Star-forming galaxies, Clusters of galaxies

1. Introduction

This review focuses on galaxies with emission lines in their optical spectra. In the great majority of cases, at least in clusters, an emission-line galaxy can be assumed to be currently forming stars[1], thus the subject of this contribution is essentially the whole population of cluster galaxies with ongoing star formation at any given epoch. One of the major goals of current studies of galaxy formation and evolution is to understand when each galaxy formed or will form its stars, and why. Naturally, star-forming galaxies are at the heart of this investigative effort. Knowing their frequency and their evolution as a function of the environment is fundamental to trace the evolution of the star formation activity with redshift and to start gaining insight on the processes that regulate such activity.

This is an area of research that really encompasses "many scales" of the Universe, from the star formation on the scale of a single star-forming region within a galaxy, to the integrated properties on a galaxy scale, to a galaxy-cluster scale and to the evolution of the global star formation history on a

[1] In the following, any contribution to the emission originating from an eventual AGN will be disregarded.

J. C. del Toro Iniesta (eds.), The Many Scales in the Universe, 71–85.

cosmic scale. I will briefly touch upon each of these scales and I will focus on the following main questions: how many emission line galaxies are there in clusters, at different redshifts? What are their properties (star formation rates, Hubble types, gas content)? What is their fate? How do they evolve and why? And, finally, what can they teach us about galaxy formation and evolution in general?

2. High redshift

Star formation activity

At any given redshift, the properties of cluster galaxies display a large cluster to cluster variance. The "average galaxy" in the Coma cluster looks quite different from the "average galaxy" in the Virgo cluster, the first cluster being dominated by passively evolving, early-type galaxies, and the second one having a larger population of star-forming spirals. Moreover, clusters of galaxies, far from being closed boxes, continuously "form" and evolve accreting single galaxies, pairs, groups, or merging with other clusters. The galaxy content of a cluster thus changes with redshift. Keeping these two things in mind, it is possible to look for evolutionary trends with redshift.

Historically, the first evidence for a higher incidence of star-forming galaxies in distant clusters compared to nearby clusters came from photometric studies (Butcher & Oemler 1978, 1984, Ellingson et al. 2001, Kodama & Bower 2001). Spectroscopy is of course the most direct way to identify emission–line galaxies. For distant galaxies, the Hα line is redshifted at optical wavelengths that are severely affected by sky or in the near-IR, thus the feature most commonly used is the [OII]λ3727 line.

In the MORPHS sample of 10 clusters at $z \sim 0.4 - 0.5$, the fraction of emission–line galaxies is $\sim$ 30% for galaxies brighter than $M_V = -19 + 5\log h^{-1}$ (Dressler et al. 1999, Poggianti et al. 1999). In the CNOC cluster sample, at an average redshift $z \sim 0.3$, this fraction is about 25% (Balogh et al. 1999). This incidence is much higher than it is observed in similarly rich clusters at $z = 0$ (Dressler, Thompson & Shectman 1988). Significant numbers of emission-line galaxies have been reported in virtually all spectroscopic surveys of distant clusters (Couch & Sharples 1987, Fisher et al. 1998, Postman et al. 1998, 2001).

The importance of the [OII] emission can also be assessed from cluster composite spectra, that are obtained summing up the light from all galaxies in a given cluster to produce a sort of "cluster integrated spectrum" (Fig. 1, Dressler et al. 2004). As expected, the strength of [OII] in these composite spectra displays a large cluster–to–cluster variation at any redshift, but there is a tendency for the $z = 0.5$ clusters to have on average a stronger composite EW([OII]) than the clusters at $z = 0$.

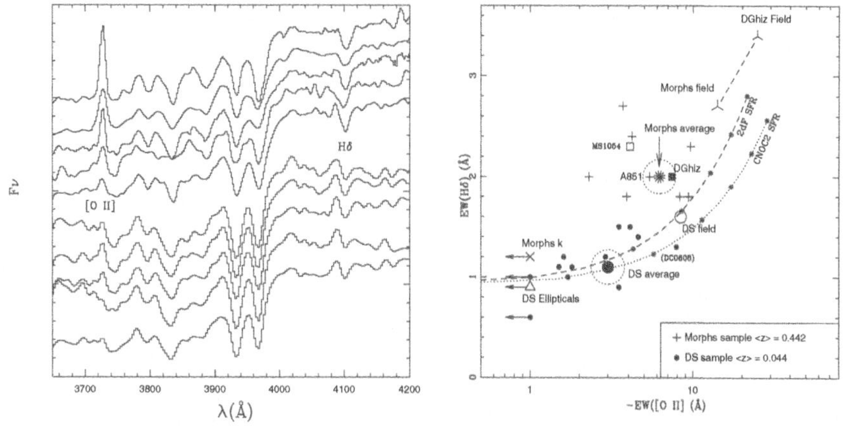

Figure 1. From Dressler et al. (2004). *Left.* Composite spectra of five clusters at $z \sim 0.5$ (top five) and five clusters at $z \sim 0$ (bottom five). The [OII] line is generally more prominent in the high-z spectra. *Right.* Equivalent widths of [OII] versus Hδ as measured from composite spectra of clusters at $z \sim 0.4 - 0.5$ (crosses) and clusters at $z \sim 0$ (filled dots).

If numerous observations indicate that emission–line galaxies were more prominent in clusters in the past than today, and if these results are unsurprising given the evolution with z of the star formation activity in the general "field", *quantifying* this evolution in clusters has proved to be very hard. The fact that the emission–line incidence varies strongly from a cluster to another at all redshifts, and the relatively small samples of clusters studied in detail at different redshifts, have so far hindered our progress in measuring how the fraction of emission–line galaxies evolves with redshift as a function of the cluster properties.

Discriminating between cosmic evolution and cluster–to–cluster variance is a problem also for Hα cluster–wide studies. Using narrow–band imaging or multiplex multislit capabilities, a handful of clusters have been studied to date at $z \geq 0.2$ (Couch et al. 2001, Balogh et al. 2002a, Finn et al. 2004, and submitted, Kodama et al. 2004, Umeda et al. 2004). These studies have confirmed that the fraction of emission–line (Hα–detected, in this case) galaxies is lower in clusters than in the field at similar redshifts, and have shown that the bright end of the Hα luminosity function does not seem to depend strongly on environment. As shown in Fig. 2, the number of clusters studied is still insufficient to pin down the star formation rate (SFR) per unit of cluster mass as a function of redshift AND of global cluster properties such as the cluster velocity dispersion.

A word of caution is compulsory when using emission–lines and assuming they provide an unbiased view of the evolution of the star formation activity

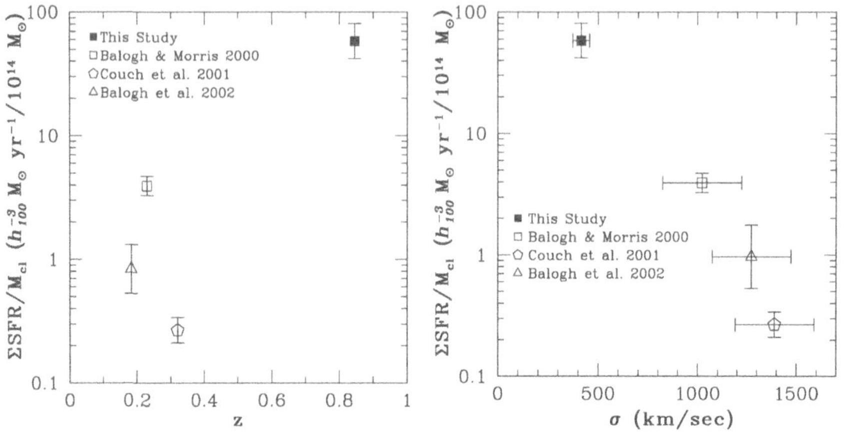

Figure 2. From Finn et al. (2004). Star formation rate per unit of cluster mass, as measured from Hα narrow-band imaging, as a function of redshift (left) and cluster velocity dispersion (right).

in cluster galaxies. There are several indications that dust extinction is in fact important and strongly distorts our view of the star formation activity in at least some cluster galaxies. Evidence for dust arises from optical spectroscopy itself, which finds many dusty starbursting or star–forming galaxies with relatively weak emission–lines both in distant clusters and in the field at similar redshifts (Poggianti et al. 1999, Shioya et al. 2000, Poggianti et al. 2001, Bekki et al. 2001). The radio–continuum detection of galaxies with no optical emission lines (Smail et al. 1999, Miller & Owen 2002) and mid-IR estimates of the star formation rate (Duc et al. 2002, Coia et al. 2004 submitted, Biviano et al. 2004) indicate that even the majority or all of the star formation activity of some cluster galaxies can be obscured at optical wavelengths. Whether taking into account dust obscuration changes significantly the evolutionary picture inferred from emission lines is still a critical open question.

Finally, precious informations about emission–line galaxies can be obtained from absorption–line spectra. In distant clusters, the presence of galaxies with strong Balmer lines in absorption in their spectra, and no emission lines, testifies that these are post-starburst/post-starforming galaxies observed soon after their star formation activity was interrupted and observed within 1-1.5 Gyr from the halting (Dressler & Gunn 1983, Couch & Sharples 1987, see Poggianti 2004 for a review). These galaxies have been found to be proportionally more numerous in distant clusters than in the field at comparable redshifts (Dressler et al. 1999, Poggianti et al. 1999, Tran et al. 2003, 2004). Their spectral characteristics and their different frequency as a function of the environment are a strong indication for a truncation of the star formation activity related to the dense environment.

Galaxy morphologies

In this section I summarize the results concerning the morphologies of emission line galaxies in distant clusters obtained with the Wide Field and Planetary Camera 2. At the time of writing, high–quality data of distant clusters have been obtained with the Advanced Camera for Surveys by several groups, and results should appear soon (Desai et al. in prep., Postman et al. in prep.).

The HST images have revealed the presence of large numbers of spiral galaxies in all distant clusters observed. Comparing HST morphologies and spectroscopy, it has been shown that emission–line galaxies in distant clusters are for the great majority *spirals* (Dressler et al. 1999). The viceversa is not always true: several of the cluster spirals, in fact, do not display any emission line in their spectra, and both their spectra and their colors indicate a lack of current star formation activity (Poggianti et al. 1999, Couch et al. 2001, Goto et al. 2003).

These "passive spirals" might be an intermediate stage when star–forming spirals are being transformed into passive S0 galaxies. A strong reason to believe that a significant fraction of the spirals in distant clusters evolve into S0s comes from the evolution of the Morphology-Density (MD) relation. The MD relation is the observed correlation between the frequency of the various Hubble types and the local galaxy density, normally defined as the projected number density of galaxies within an area including its closest neighbours. In clusters in the local Universe, the existence of this relation has been known for a long time: ellipticals are frequent in high density regions, while the fraction of spirals is high in low density regions (Oemler 1974, Dressler et al. 1980). At $z = 0.4 - 0.5$, an MD relation is already present, at least in concentrated clusters, but it is *quantitatively* different from the relation at $z = 0$: the fraction of S0 galaxies at $z = 0.5$ is much lower, at all densities, than in clusters at $z = 0$ (Dressler et al. 1997). The fraction of S0s in clusters appears to increase towards lower redshifts, while the proportion of spirals correspondingly decreases (Dressler et al. 1997, Fasano et al. 2000). Interestingly, ellipticals are already as abundant at $z = 0.5$ as at $z = 0$. Adopting a more conservative distinction between "early-type" (Es+S0s) and late-type (spirals) galaxies, a similar evolution is found, with the early-type fraction decreasing at higher redshifts (van Dokkum et al. 2000, Lubin et al. 2002). First results at $z \sim 0.7 - 1.3$ seem to indicate that between $z = 0.5$ and $z = 1$ what changes in the MD relation is only the occurrence of early-type galaxies in the very highest density regions (Smith et al. 2004).

Alltogether, the findings described in this and the previous section suggest that many galaxies have stopped forming stars in clusters quite recently, as a consequence of environmental conditions switching off their star formation activity, and that many galaxies have morphologically evolved from late to early type galaxies. What can be the cause/causes for these changes?

3. Physical processes

The physical mechanisms that are usually considered when trying to assess the influence of the environment on galaxy evolution can be grouped in four main families:

1 Mergers and strong galaxy-galaxy interactions (Toomre & Toomre 1972, Hernquist & Barnes 1991, see Mihos 2004 for a review). These are most efficient when the relative velocities between the galaxies are low, thus are expected to be especially efficient in galaxy groups.

2 Tidal forces due to the cumulative effect of many weaker encounters (also known as "harassment") (Richstone 1976, Moore et al. 1998). These are expected to be especially important in clusters, and particularly on smaller / lower mass galaxies.

3 Gas stripping - Interactions between the galaxy and the inter-galactic medium (IGM) (Gunn & Gott 1972, Quilis et al. 2000). The interstellar medium of a galaxy can be stripped via various mechanisms, including viscous stripping, thermal evaporation and – the most famous member of this family – ram pressure stripping. Ram pressure can be efficient when the IGM gas density is high and the relative velocity between the galaxy and the IGM is high, and such conditions are expected to be met especially in the very central regions of cluster cores.

4 Strangulation (also known as starvation, or suffocation) (Larson, Tinsley & Caldwell 1980, Bower & Balogh 2004). Assuming galaxies possess an envelope of hot gas that can cool and feed the disk with fuel for star formation, the removal of such reservoir of gas is destined to inhibit further activity once the disk gas is exhausted. In semi-analytic models, for example, the gas halo is assumed to be removed when a galaxy enters as satellite in a more massive dark matter halo.

Note that while stripping gas from the disk induces a truncation of the star formation activity on a short timescale ($\sim 10^7$ yrs), strangulation is expected to affect a galaxy star formation history on a long timescale (> 1 Gyr) provoking a slowly declining activity which consumes the disk gas after the supply of cooling gas has been removed.

The former two of these families of processes affect the galaxy structure, thus morphology, in a direct way: the merger of two spirals can produce an elliptical galaxy, and repeated tidal encounters can change a late–type into an early-type galaxy. The latter two families, instead, act on the gas content of galaxies, hence their star formation activity, and can modify their morphologies in an indirect way: once star formation is halted in a disk, this can fade significantly, the bulge-to-disk relative importance can change and the galaxy appearance and morphology can appear significantly modified.

4. Low redshift

Numerous excellent works have been carried out on a single cluster or samples of clusters in the local Universe. Summarizing them is beyond the scope of this paper, and the reader can find a comprehensive review in Gavazzi & Boselli (in prep.). I have chosen to mention two low-redshift results here, to compare and contrast them with the results at higher redshifts: the observed trends of star formation with local environment, and the gas/SF distribution within galaxies.

Trends of star formation with local environment

It has been known for a long time that in the nearby Universe also the average star formation activity correlates with the local density: in higher density regions, the mean star formation rate per galaxy is lower. This is not surprising, given the existence of the MD relation: the highest density regions have proportionally more early-type galaxies devoid of current star formation.

Interestingly, the correlation between mean SF and local density extends to very low local densities, comparable to those found at the virial radius of clusters, and such a correlation exists also outside of clusters (Lewis et al. 2002, Gomez et al. 2003). Again, this seems to parallel the fact that an MD relation is probably existing in all environments, and it has been observed in clusters of all types (Dressler et al. 1980) and groups (Postman & Geller 1984) – though the MD relation is *not the same* in all environments, e.g. in concentrated vs. irregular, high- vs. low- L_X clusters (Dressler et al. 1980, Balogh et al. 2002b).

A variation in the mean SF/galaxy with density can be due either to a difference in the *fraction of star-forming galaxies*, or in the *star formation rates* of the star-forming galaxies, or a combination of both. In a recent paper, Balogh et al. (2004a) have shown that the distribution of Hα equivalent widths (EW) in star-forming galaxies does not depend strongly on the local density, while the fraction of star-forming galaxies is a steep function of the local density, in all environments. Again, a dependence on the *global environment* is observed, in the sense that, at a given local density, the fraction of emission-line galaxies is slightly lower in environments with high density on large scales (~ 5 Mpc) (but see Kauffmann et al. 2004 for an opposite result).

The fact that a relation between star formation and density is observed also outside of clusters has often been interpreted as a sign that the environment starts affecting the star formation activity of galaxies (provoking a decline in star formation in galaxies that if isolated would continue forming stars) at relatively low densities, when a galaxy becomes part of a group. Personally, I believe the existence of such a correlation is more probably the result of a correlation between initial conditions (galaxy mass and/or local environment very early on, at the time the first stars formed in galaxies) and type of galaxy formed. The exact shape of the correlation, instead, is probably influenced by transformations happening in galaxies when they enter a different environment.

Gas content and gas/SF distribution within galaxies

In order to understand what happens to galaxies in clusters, two crucial pieces of information are 1) the gas content of cluster galaxies and 2) the spatial distribution of the gas and of the star formation activity within each galaxy.

It has been several years since it became evident that many spirals in clusters are deficient in HI gas compared to similar galaxies in the field (Giovanelli & Haynes 1985, Cayatte et al. 1990, see van Gorkom 2004 for a review). Most (but not all) of the HI deficient spirals are found at small distances from the cluster centre. In the central regions of clusters, the sizes of the HI disks are smaller than the optical disks, and a spatial displacement between the HI and the optical occurs in several cases (Bravo-Alfaro et al. 2000). The fraction of HI–deficient spirals increases going towards the cluster centre, and a correlation is observed between deficiency and orbital parameters: more deficient galaxies tend to be on radial orbits (Solanes et al. 2001).

All of these findings strongly suggest that ram pressure stripping, or at least gas stripping in general, plays an important role (see also Bravo-Alfaro and Solanes contributions in these proceedings). On the other hand, the work from Solanes et al. (2001) has unexpectedly shown that the HI deficiency is observed out to 2 Abell radii. This result has raised the question whether the origin of the HI deficiency in the cluster outskirts can be consistent with the ram pressure scenario and whether can be simply due to effects such as large distance errors or rebounding at large clustercentric distances of galaxies that have gone through the cluster center (Balogh, Navarro & Morris 2000, Mamon et al. 2004, Moore et al. 2004, Sanchis et al. 2004).

Recent works have yielded a census of the spatial distribution of the star formation (as observed in Hα emission) within cluster galaxies. These works have shown that the majority of the cluster galaxies have peculiar Hα morphologies, compared to field galaxies (Moss & Whittle 1993, 2000, Koopmann & Ken-

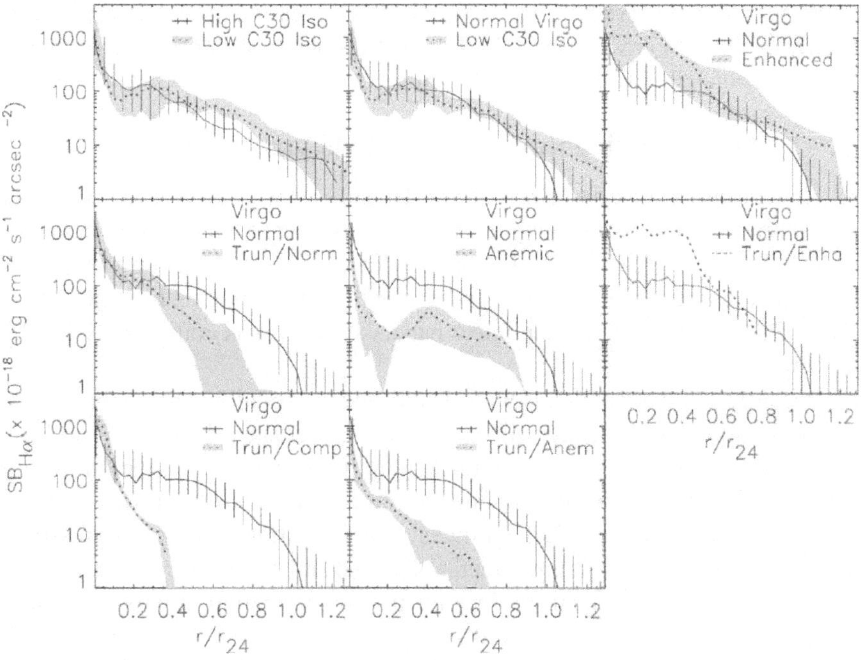

Figure 3. From Koopmann & Kenney 2004. Median Hα radial profiles for galaxies grouped in different classes according to their Hα distribution. The first two panels show profiles that can be considered "normal" (reproduced also in other panels). Truncated, anemic, enhanced classes are shown in all the other panels.

ney 2004, Vogt et al. 2004). More than half of the spirals in the Virgo cluster, for example, have Hα radial profiles truncated from a certain radius on, while others have Hα suppressed throughout the disk, and in some cases, enhanced (Fig. 3) (Koopmann & Kenney 2004). On a sample of 18 nearby clusters, similar classes of objects are observed: spirals with truncated Hα emission and HI gas on the leading edge of the disk; spirals stripped of their HI with their star formation confined to the inner regions; and quenched spirals, in which the star formation is suppressed throughout the disk (Vogt et al. 2004). From these works, gas stripping appears a very important factor in determining both the gas content and the star formation activity of cluster spirals, though tidal effects are also found to be significant (Moss & Whittle 2000, Koopmann & Kenney 2004). Spirals in clusters thus appear to be heavily affected by the environment, and to be observed in different stages of their likely transformation from infalling star-forming spirals to cluster S0s.

It is apparently hard to reconcile the peculiar Hα morphologies of spirals in nearby clusters with the fact that the distribution of Hα equivalent widths

(EWs) for blue galaxies in the 2dF and Sloan does not seem to vary significantly in clusters, groups and field (Fig. 9 in Balogh et al. 2004). This apparent contradiction still awaits an explanation.

5. Trends with galaxy mass and downsizing effect

How do the results described above depend on the galaxy mass? Does the environmental dependence of galaxy properties change with galaxy mass?

It has always been known that fainter, lower mass galaxies on average are bluer than higher mass galaxies, and on average have more active star formation. The references to old and new papers showing this could easily fill this review.

Recently, the interest in this well established observational result has grown, and its implications have been more and more appreciated. In *all* environments, lower mass galaxies have on average a more protracted star formation history. This implies that, on average, going to lower redshifts, the maximum luminosity/mass of galaxies with significant star formation activity progressively decreases. This "downsizing effect", observed in and outside of clusters, indicates an "anti-hierachical" history for the star formation in galaxies, which parallels a similar effect observed for AGNs (Cristiani et al. 2004, Shankar et al. 2004). The downsizing effect is thus another effect *besides* any environmental effect (see e.g. Fig. 4, and Kauffmann et al. 2003, 2004), and the dependence on the galaxy mass/luminosity cannot be ignored when trying to trace evolutionary effects.

In clusters, innumerable results have shown the existence of a downsizing effect (Smail et al. 1998, Gavazzi et al. 2002, De Propris et al. 2003, Tran et al. 2003, De Lucia et al. 2004, Kodama et al. 2004, Poggianti et al. 2004, to name a few). A direct observation of this effect at high redshift is shown in Fig. 5 and illustrates the consequence of downsizing on the characteristics of the color-magnitude red sequence in clusters. A deficiency of faint red galaxies is observed compared to Coma in all four clusters studied, despite of the variety of cluster properties. The red luminous galaxies are already in place on the red sequence at $z \sim 0.8$, while a significant fraction of the faint galaxies must have stopped forming stars and, consequently, moved on to the red sequence at lower redshifts.

6. Conclusions and speculations

A number of important issues have not been considered here due to page limits, but should be included in any complete review of emission-line galaxies in clusters, in particular: a description of the IR and radio-continuum studies, the Tully-Fisher relation of cluster versus field spirals, the spatial distribution and kinematics of various types of galaxies, the link between the star-formation

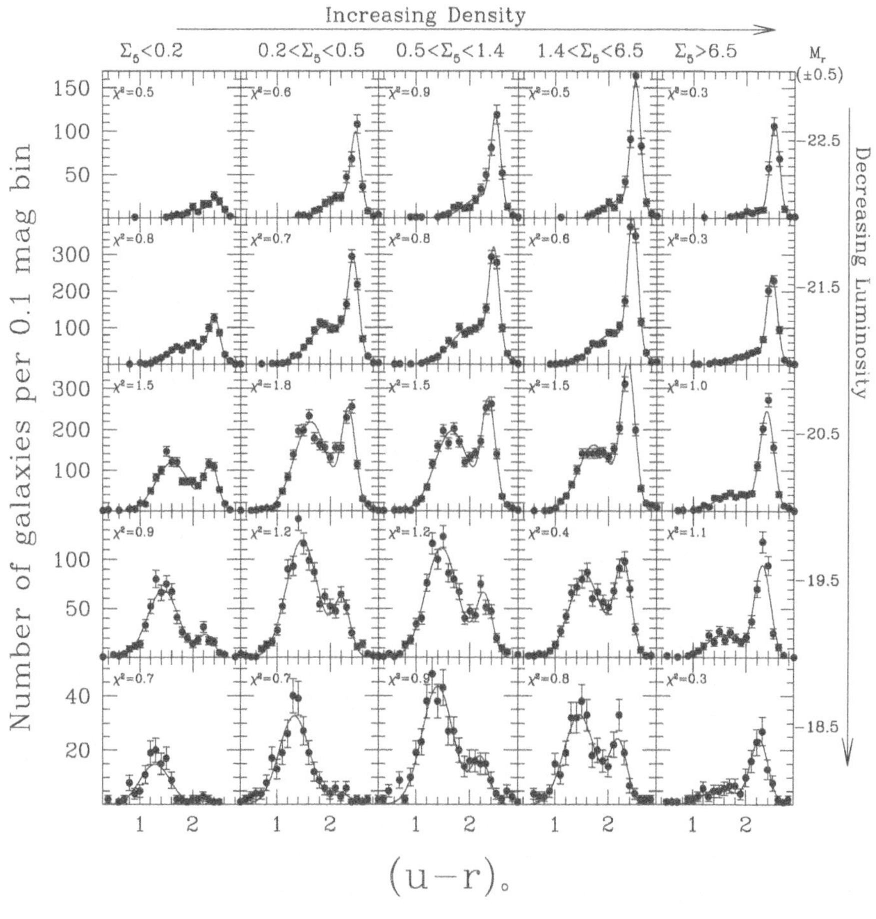

Figure 4. From Balogh et al. 2004b. Galaxy color distribution from Sloan as a function of luminosity and local density.

activity and cluster substructure, the luminosity function of emission– and non–emission line galaxies as a function of the environment, and the line emission from AGNs.

By now, mostly thanks to the scientific debate of the latest years, it is widespread wisdom that we are investigating a three-dimensional space, whose axis are redshift, environment and galaxy mass. In fact, the evolutionary histories and, in particular, the star formation activity of galaxies have similar (increasing) trends as a function of (higher) redshift, (lower) galaxy mass and (lower) density/mass of the environment. While observationally we are beginning to

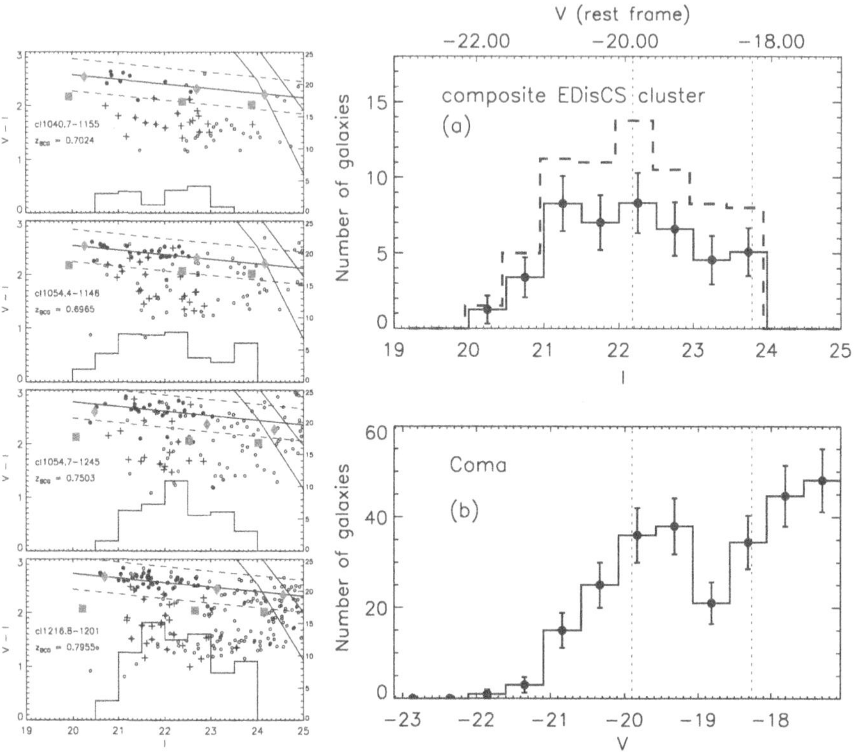

Figure 5. From De Lucia et al. 2004. *Left.* Color-magnitude diagrams of four clusters at $z = 0.7 - 0.8$ from the ESO Distant Cluster Survey. Histograms represent the magnitude distribution of galaxies within 3σ from the red sequence. *Right.* The magnitude distribution of red sequence galaxies at $z \sim 0.75$ (top) is compared with the one of red galaxies in the Coma cluster (bottom). High-z clusters exhibit a clear deficit of low-luminosity passive red galaxies compared to Coma and other nearby clusters.

fill this 3-parameter space, the great theoretical challenge is to help comprehend from a physical point of view why this is the history of our Universe.

The results mentioned above do not yet match together into one, coherent picture, and in some cases they might be apparently clashing with each other. The observational results in distant clusters, both those regarding the star formation activity and the morphologies, point to a strong evolution. Many galaxies have stopped forming stars in clusters quite recently, and the morphology of many disk galaxies must have changed. Dense environments seem to accelerate the transformation of star-forming late-type galaxies into passive early-type galaxies. The local large redshift surveys have highlighted the existence of trends of galaxy properties with local density that exist in all environments.

This, as well as several other lines of evidence, suggests that whatever induces the existence of a morphology-density and of a star-formation-density relation is not effective *solely* in clusters. On the other hand, there is now solid evidence (and the gas/star formation results in nearby clusters are most convincing in this respect) that gas stripping is at work and affects *most* of the emission-line galaxies in clusters, (or, at least, most of those galaxies in clusters that today still have emission lines).

Twenty years ago, an expression often used in this field was "nature or nurture?". Nowadays, especially within the context of a hierarchically growing Universe, the distinction between nature and nurture has become very subtle, also from a philosophical point of view.

The example of the most massive ellipticals is instructive in this respect. They are found in the cluster cores, and the current cosmological paradigm tells us that they were in the highest density peaks of the primordial fluctuations, thus have always been in the highest density regions, since very high redshifts. Therefore, at least "part" of the morphology-density relation (and of the star-formation density (SFD) relation, since their stars all formed very early on) must have been in place at $z > 3$. Is it more correct to call this a "primordial effect" – because established at very high redshift – or "environmental" – because related to the local density of the environment? but the environment of these ellipticals at $z = 10$ is strongly related to their environment today: can we speak of "galaxy destiny"?

On the other hand, we know that the morphology-density relation in clusters has evolved with redshift, at least as far as the disk galaxies are concerned, and we can directly observe its evolution. Therefore, we know that while the existence of *a* morphology-density relation can be traced back to the first epoch of galaxy formation, its evolution, and thus its exact shape at any time, depends on the subsequent evolution, likely strongly influenced by the environment. Is the *existence* of the MD (and SFD) relation primordial, and its *evolution* environmental?

Perhaps also we, as behavioural psychologists (Ridley, 2003), are coming to realize that nature and nurture are so intertwined that they become, to a certain extent, indistiguishable and unseparable.

Acknowledgments

I wish to thank Mariano Moles for the kindly provided financial support and the organizers of this JENAM meeting and, in particular, of the Session on "The life of galaxies" for their invitation and support, which allowed me to participate to a very pleasant and interesting meeting.

References

Balogh, M.L., Morris, S.L., Yee, H. K. C., Carlberg, R. G., Ellingson, E., 1999, ApJ, 527, 54
Balogh, M.L., Navarro, J.F., Morris, S.L., 2000, ApJ, 540, 113
Balogh, M.L., Couch, W.J., Smail, I., Bower, R.G., Glazebrook, K., 2002a, MNRAS, 335, 10
Balogh, M.L., Bower, R.G., Smail, I., et al., 2002b, MNRAS, 337, 256
Balogh, M. et al., 2004a, MNRAS 348, 1355
Balogh, M., Baldry, I.K., Nichol, R., Miller, C., Bower, R., et al. 2004b, ApJL, 615, L101
Bekki, K., Shioya, Y., Couch, W.J. 2001, ApJL, 547, L17
Biviano. A., et al., 2004, A&A, 425, 33
Bower, R.G., Balogh, M.L., 2004, in *Clusters of galaxies: probes of cosmological structure and galaxy evolution*, p.326, Eds. J.S. Mulchaey, A. Dressler, A. Oemler, Cambridge University press
Bravo-Alfaro, H., Cayatte, V., van Gorkom, J.H., Balkowski, C., 2000, AnJ, 119, 580
Butcher, H., Oemler, A., Jr. 1978, ApJ, 226, 559
Butcher, H., Oemler, A., Jr. 1984, ApJ, 285, 426
Cayatte, V., van Gorkom, J.H., Balkowski, C., Kotanyi, C., 1990, AnJ, 100, 604
Couch, W. J., Balogh, M. L., Bower, R. G., Smail, I., et al. 2001, ApJ, 549, 820
Couch, W. J., & Sharples, R. M. 1987, MNRAS, 229, 423
Cristiani, S. et al., 2004, ApJ, 600, L119
De Lucia, G., Poggianti, B.M., Aragon-Salamanca, A., et al., 2004, ApJ, 610, L77
De Propris, R., Stanford, S.A., Eisenhardt, P.R., Dickinson, M., 2003, ApJ, 598, 20
Dressler, A. 1980, ApJ, 236, 351
Dressler, A., & Gunn, J. E. 1983, ApJ, 270, 7
Dressler, A., et al. 1997, ApJ, 490, 577
Dressler, A., Thompson, I.B., Shectman, S.A., 1985, ApJ, 288, 481
Dressler, A., Smail, I., Poggianti, B. M., Butcher, H., Couch, W. J., et al. 1999, ApJS, 122, 51
Dressler, A., Oemler, A., Poggianti, B.M., Smail, I., et al., 2004, ApJ, 617, 867
Duc, P.-A., Poggianti, B. M., Fadda, D., Elbaz, et al. 2002, A&A, 382, 60
Ellingson, E., Lin, H., Yee, H. K. C., & Carlberg, R. G. 2001, ApJ, 547, 609
Fasano, G., Poggianti, B. M., Couch, W. J., Bettoni, D., et al. 2000, ApJ, 542, 673
Finn, R.A., Zaritsky, D., McCarthy, D.W., 2004, ApJ, 604, 141
Fisher, D., Fabricant, D., Franx, M., & van Dokkum, P. 1998, ApJ, 498, 195
Gavazzi, G., Boselli, A., Pedotti, P., Gallazzi, A., Carrasco, L., 2002, A&A, 396, 449
Giovanelli, R., Haynes, M.P., 1985, ApJ, 292, 404
Gomez, P.L., Nichol, R.C., Miller, C.J., et al., 2003, ApJ, 584, 210
Goto, T., Okamura, S., et al., 2003, PASJ, 55, 757
Gunn, J.E., Gott, J.R. 1972, ApJ, 176, 1
Hernquist, L., Barnes, J.E., 1991, Nature, 354, 210
Kauffmann, G., et al., 2003, MNRAS, 341, 54
Kauffmann, G., et al., 2004, MNRAS, 353, 713
Kodama, T., Balogh, M. L., Smail, I., Bower, R. G., Nakata, F., 2004, MNRAS, 354, 1103
Kodama, T., & Bower, R. G., 2001, MNRAS, 321, 18
Koopmann, R.A., Kenney, J.D.P., 2004, ApJ, 613, 866
Larson, R.B., Tinsley, B.M., Caldwell, C.N., 1980, ApJ, 237, 692
Lewis, I., Balogh, M., De Propris, R. et al., 2002, MNRAS, 334, 673
Lubin, L.M., Oke, J.B., Postman, M., 2002, AJ, 124, 1905
Mamon, G.A., Sanchis, T., Salvador-Sole, E., Solanes, J.M., 2004, A&A, 414, 445
Mihos, J.C., 2004, in *Clusters of galaxies: probes of cosmological structure and galaxy evolution*, p.278, Eds. J.S. Mulchaey, A. Dressler, A. Oemler, Cambridge University press

Miller, N.A, Owen, F.N., 2002, AJ, 124, 2453
Moore, B., Lake, G., Katz, N., 1998, ApJ, 495, 139
Moore, B., Diemand, J., Stadel, J., 2004, in Outskirts of galaxy clusters: intense life in the subrurbs, IAU Coll. N. 195, ed. A. Diaferio
Moss, C., Whittle, M. 1993, ApJ, 407, L17
Moss, C., Whittle, M. 2000, MNRAS, 317, 667
Oemler, A., 1974, ApJ, 194, 1
Poggianti, B. M., Smail, I., Dressler, A., Couch, W. J., Barger, A. J., Butcher, H., et al. 1999, ApJ, 518, 576
Poggianti, B.M., Bressan, A., Franceschini, A., 2001, ApJ, 550, 195
Poggianti, B.M., Bridges, T.J., Komiyama, Y. et al., 2004, ApJ, 601, 197
Poggianti, B.M., 2004, in *Clusters of galaxies: probes of cosmological structure and galaxy evolution*, p.246, Eds. J.S. Mulchaey, A. Dressler, A. Oemler, Cambridge University press
Postman, M., Geller, M.J., 1984, ApJ, 281, 95
Postman, M., Lubin, L. M., Oke, J. B. 1998, AJ, 116, 560
Postman, M., Lubin, L. M., Oke, J. B. 2001, AJ, 122, 1125
Quilis, V., Moore, B., Bower, R., 2000, Science, 288, 1617
Richstone, D.O., 1976, ApJ, 204, 642
Ridley, M., 2003, *Nature via nurture - Genees, experience and what makes us human*, Harper Collins Publishers, New York
Sanchis, T., Mamon, G.A., Salvador-Sole, E., Solanes, J.M., 2004, A&A, 418, 393
Shankar, F., Salucci, P., Granato, G.L., De Zotti, G., Danese, L., 2004, MNRAS, 354, 1020
Shioya, Y. & Bekki, K. 2000, ApJL, 539, L29
Smail, I., Edge, A.C., Ellis, R. S., Blandford, R.D., 1998, MNRAS, 293, 124
Smail, I., Morrison, G., Gray, M. E., Owen, F. N., Ivison, R. J., Kneib, J.-P., Ellis, R. S. 1999, ApJ, 525, 609
Smith, G.P., Treu. T. et al., 2005, submitted (astro-ph 0403455)
Solanes, J.M., et al., 2001, ApJ, 548, 97
Toomre, A., Toomre, J., 1972, ApJ, 178, 623
Tran, K.-V.H., Franx, M., Illingworth, G.D., et al., 2003, ApJ, 599, 865
Tran, K.-V.H., Franx, M., Illingworth, G.D., et al., 2004, ApJ, 609, 683
Umeda, K., et al., 2004, ApJ, 601, 805
van Dokkum, P. G., Franx, M., Fabricant, D., Illingworth, G.D., Kelson, D. D. 2000, ApJ, 541, 95
van Gorkom, J., 2004, in *Clusters of galaxies: probes of cosmological structure and galaxy evolution*, p.306, Eds. J.S. Mulchaey, A. Dressler, A. Oemler, Cambridge University press
Vogt, N.P., Haynes, M.P., Giovanelli, R., Herter, T., 2004, AJ, 127, 3300

ACTIVE GALACTIC NUCLEI AND SURVEYS: THE VIEW FROM THE NEW X-RAY OBSERVATORIES

X. Barcons
Instituto de Física de Cantabria, CSIC-UC
E-39005 Santander, Spain
barcons@ifca.unican.es

Abstract I review some of the most important advances in our knowledge and understanding of the AGN phenomenon derived from the scientific exploitation of the new X-ray missions *Chandra* and *XMM-Newton*, with special emphasis on the survey content of these missions. In particular I discuss Fe line diagnostics of the inner accretion disk, the circumnuclear environment, challenges to the unified AGN model, obscured accretion and the origin of the X-ray background.

Keywords: X-rays: galaxies, Galaxies: active, Surveys: X-ray

1. Introduction

Active Galactic Nuclei (AGN) constitute a very important component of the Universe. Aside from the Big-Bang itself (as mapped by the Cosmic Microwave Background) and star formation (which dominates the infrared background), AGN are the most important energy generators in the Universe, and resposible for the peak in energy density of the extragalactic background radiation at around 30 keV. It is now known that the formation of the super-massive black holes residing in the centres of virtually all galaxies, is strongly linked to the formation of the galaxies themselves and to the formation of stars in these galaxies. Just what is the specific interplay between these phenomena is still not know.

The X-ray view of AGN

X-ray observations are a major handle to study the AGN phenomenon. In the standard model, the AGN central engine consists of a supermassive black hole (with mass in the range $10^{5-9}\,M_{\odot}$) surrounded by an accretion disk and a jet perpendicular to it. The viscosity of the accretion disk is resposible for both its heating and of transporting angular momentum outwards. Assuming Eddington limited accretion, the disk temperature scales like $T \propto M^{-1/4}$ with

J. C. del Toro Iniesta (eds.), The Many Scales in the Universe, 87–98.

the black hole mass M, peaking in the ultraviolet zone ($T \sim 10^5$ K) for an AGN-like black hole. Re-processing by energetic electrons in the disk corona, upscatters UV photons into the X-ray regime. These X-rays are the basic X-ray energy source in AGN.

The X-ray spectrum of a radio-quiet AGN or of a radio-loud AGN with the jets oriented away from the line of sight has several components. The basic disk coronal radiation is usually modeled as a power law with photon index $\Gamma \sim 2$. Whether that component rolls over at high energies or not, depends on whether the disk corona has a thermal or non-thermal structure. If this break at high energies is present (and there are evidences for that in some AGN), it happens at photon energies $E > 100$ keV. A reflection component, whose strength depends on the geometry of the X-ray emitting region and the accretion disk, is also seen in AGN. It consists of fluorescence emission lines, most notably an Fe Kα emission line, and a Compton reflection hump apparent above 20 keV resulting from the combination of photoelectric absorption at low energies and a drop at high energies produced by electron recoil. In addition to the reflection component, the tail of the quasi-thermal radiation from the accretion disk is also seen sometimes at soft X-ray energies.

Absorption, produced either by neutral or partially ionised material, is also often seen in AGN. In type-1 AGN, where we have a direct view of the Broad Line Region (BLR) clouds, we should also have a direct view of the primary X-ray source, and therefore little (if any) photoelectric absorption should be detected in X-ray spectra of these AGN. On the contrary, if type-2 AGN contain the same type of central engine, but with the line of sight to the BLR blocked by obscuring material, we expect significant photoelectric absorption in the X-ray spectrum. These are specific predictions of the AGN unified model (Antonucci 1993), which are on average seen to happen in X-ray observations of AGN.

New windows to the X-ray Universe opened by Chandra and XMM-Newton

The new X-ray observatories, NASA's *Chandra* and ESA's *XMM-Newton* have opened a new window into the X-ray Universe which has had a particularly deep impact into our knowledge and understanding of AGN. New windows opened by these two observatories include:

- The access to high photon energies. Although the USA-Japan mission *ASCA* was sensitive to photons of energy up to $\sim$ 8keV, *Chandra* and to a larger extent *XMM-Newton*, have increased substantially the sensitivity to hard X-rays (up to 12 keV) thanks to their much larger effective area and improved spatial resolution. Fe-line diagnostics and the discovery and study of absorbed AGN have become possible thanks to this new window.

- High-resolution imaging over a large field of view. *Chandra*, with its on-axis superb spatial resolution ($< 1''$) and *XMM-Newton* with its modest $\sim 15'' HEW$ resolution over a field of view as large as 30' in diameter, have revolutionised X-ray astronomy in many respects. *Chandra* has been able to detect X-ray jets in AGN, as well as to resolve a binary AGN in the centre of the same galaxy. *XMM-Newton* is a superb survey instrument at medium and faint fluxes, delivering at the same time X-ray spectra of the sources detected. Of course, *Chandra* is much more sensitive at the very faint soft (0.5-2 keV) X-ray flux limit, where it is photon counting limited up to exposures of many Msec, while *XMM-Newton* is confusion limited in the same band at around 100 ksec.

- The access to moderate resolution dispersive spectroscopy. Both *Chandra* and *XMM-Newton* are equipped with wavelength-dispersion grating spectrometers. The *Chandra* LETGS and HETGS spectrometers have about twice higher resolution (~ 400) than the *XMM-Newton* RGS (~ 200) and they cover also the high energy domain. The *XMM-Newton* RGS spectra are, however, of average higher signal to noise. These instruments have revealed the complex structure of the absorbing material that surrounds many AGN.

2. The inner disk: Fe line diagnostics

The Fe line emission from the AGN accretion disk comes from reflection of the primary (coronal) X-ray radiation. X-ray photons are photoelectrically absorbed by the K-shell electrons in heavy atoms/ions of the disk, and in some cases fluorescence (i.e., decay of an upper shell electron into the ground shell, with the corresponding emission of a monochromatic photon) is produced. Reflection from cold matter was studied in detail by George & Fabian (1991), where they found that the Fe Kα line would be the strongest one, due to the high value of its cosmic abundance times the fluorescence yield (as opposed to Auger effect). Indeed, only the Fe Kα emission line, at 6.4 keV for neutral Fe, is generally seen in X-ray spectra of AGN, with exceptions (e.g., the Circinus galaxy where Fe Kα, Fe Kβ and Ni Kα are detected, Molendi et al. 2003).

The strength and the profile of the line is modified by various processes happening in the inner accretion disk. Line profiles can be used to diagnose this environment.

The first effect to come into play is photoionisation by the central source. For low ionisation parameters ($\log \xi < 3$, where $\xi = L/(nR^2)$, L being the luminosity in ionising photons, n the particle density and R the distance to the ionising source), a number of fluorescence lines from a variety of elements (not fully ionised) should be seen (Ballantyne et al. 2001). For increasingly large ionisation parameters, most of these elements are fully ionised and only the

Fe K line survives. In addition the line centroid is shifted to higher energies: the He-like Fe Kα line occurs at 6.7 keV. In the limit of very high ionisation ($\log \xi \sim 4-5$) even Fe can be fully ionised and there is no trace of emission lines in the spectrum. This qualitatively explains the fact that Fe lines are commonly detected and strong in Seyfert 1 galaxies (Pounds et al. 1990), but are typically weaker and elusive in higher luminosity QSOs.

The most appealing effect on the Fe line profile is due to the presence of the very strong gravitational field produced by the black hole in the inner part of the accretion disk (see Fabian et al. 1989 for a comprehensive modeling of the line). Since the Fe atoms/ions sit in a rotating accretion disk, Newtonian kinematics predicts a double horned structure for the emission line (a blue-shifted and a red-shifted component). Beaming makes the blue horn brighter and the transverse Doppler effect mixes both components. Last, but not least, all photons must lose some energy to escape the deep gravitational potential well where they are emitted. This produces a red wing towards lower photon energies.

The specific shape of the line depends very strongly on the black hole spin parameter $a = J/(Mc^2)$, where J is the angular momentum. The basic idea is that reflecting Fe atoms can only sit in stable orbits, and the minimum radius (which drives the gravitational redshift) around a black hole is a decreasing function of a. For a non-rotating Schwarzschild black hole, the innermost stable orbit happens at a radius $3R_S$, where $R_S = 2GM/c^2$, while for a maximally rotating Kerr black hole ($a = GM/c^2$) the innermost stable circular orbit occurs at $R_S/2$, enabling a much larger gravitational redshift effect. Figure 1 illustrates the different line profiles in two cases.

The presence of relativistic Fe line profiles in AGN was first reported by Tanaka et al. (1995) using a long *ASCA* observation of the Seyfert 1 galaxy MCG-6-30-15. Further study by Iwasawa et al. (1996) suggested that the Fe line in the "low" state of that galaxy required an innermost radius of $0.6R_S$, implying a rapidly rotating black hole.

With *XMM-Newton* it is now known that the structure of the Fe line is more complex than in the case of MCG-6-30-15, with usually two components: an ionized disk component (with a likely relativistic profile) plus a cold reflection line. Mrk 205 is perhaps a prototypical case where the two components can be seen (Reeves et al. 2001), but often the broad relativistic line is weak. While other relativistic lines have been confirmed by *XMM-Newton* (e.g., MCG-5-23-16 Dewangan 2003, NGC3516 Turner et al. 2002, PG 1211+143 Pounds et al. 2003, IRAS 18325 Iwasawa et al. 2004) MCG-6-30-15 remains the best studied case. Vaughan & Fabian (2004) performed a detailed spectral fit of a long *XMM-Newton* exposure on that source, confirming the relativistic disk profile of the Fe emission line which, at the same time, can explain the presence of a strong Compton reflection hump in a simultaneous observation with *BeppoSAX* extending to 80 keV (Ballantyne et al. 2003).

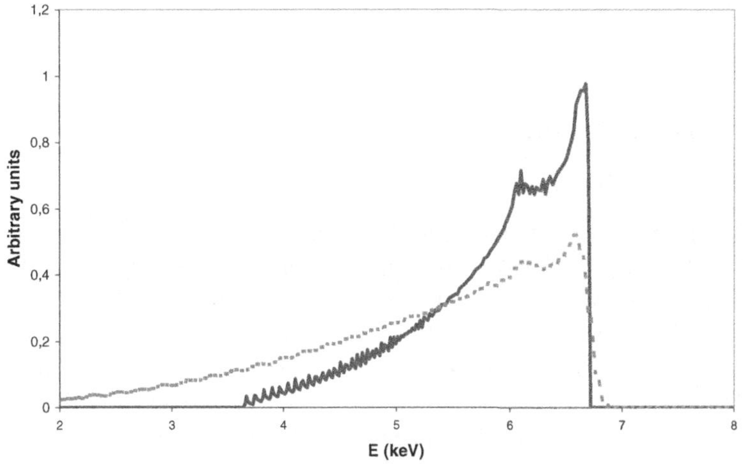

Figure 1. Examples of Fe line profiles for a non-rotating black hole (continuous line) and a close to maximally rotating black hole ($a = 0.998$, dotted line), where the disk emissivity is $\propto R^{-3}$ (R is the radius in the disk plane) and inclination 30°.

The presence of other relativistically broadened emission lines in the soft X-ray spectra of AGN is the subject of current debate. This was first claimed by Branduardi-Raymont et al. (2001) for MCG-6-30-15 and Mrk 766 based on *XMM-Newton* RGS spectroscopy. Lee et al. (2001), based on *Chandra* LETGS, claimed on the contrary that the soft X-ray spectrum could be explained by the presence of a dusty warm absorber. The arguments have continued (Sako et al. 2003, Mason et al. 2003, Turner et al. 2003) but no clear consensus about the reality of these lines is yet emerging.

What is certainly not yet fully understood is reverberation mapping. Reverberation can be used to define the geometry of the various componente responsible for different spectral features, by measuring time lags among them. Since the relativistic Fe emission line arises from reflection in the accretion disk, its intensity should respond to variations in the continuum, but no such effect is seen. In fact, *XMM-Newton* observations of MCG-6-30-15 show that the Fe line varies very little and in a manner uncorrelated to the continuum (Fabian & Vaughan 2003). Miniutti & Fabian (2004) suggest that strong light bending effects can explain this behaviour, if continuum variability arises from changes in the latitude of the X-ray source with respect to the plane of the disk. When the X-ray source is very close to the disk, we do not only detect radiation from the near side, but also from the back side thanks to the light bending effect. The geometry then becomes complicated and there is a regime where the reflection component is mostly independent on variations in the continuum. There is clearly much more work to be done on this.

3. The circumnuclear environment

Ionised absorbers

The presence of partially ionised material around type 1 AGN was noted early on by Halpern (1984) based on *Einstein IPC* data. *ASCA* represented a major leap forward, as it showed that at least 50% of Seyfert 1 AGN displayed some sort of signature related to ionized absorption, most often an OVII K-edge at 0.74 keV or an OVIII K-edge at 0.87 keV (Reynolds 1997, George et al. 1998). There is a systematic link between X-ray ionised absorbers and the so-called "associated" absorbers detected in the UV via the CIV $\lambda\lambda$ 1548,1550 doublet.

The new view of ionised absorbers has come thanks to the much improved spectral resolution delivered by dispersive spectrometers. Note that the imaging capabilities are of limited use here, as only the brightest AGN can be observed with the dispersive spectrographs and yield an acceptable S/N (a few tens of objects in total). One of the first examples studied was the red QSO IRAS 13349+2438 (Sako et al. 2001) which contains most of the general features detected in other objects. First, the presence of a number of absorption and a few emission lines, corresponding to (at least) two components appears to be the rule rather than the exception (see, e.g, Krongold et al. 2003 for a study of NGC 3783 with $\sim$ 100 features detected). The different components have different ionisation states and different velocities. Low ionisation components display invariably an Unresolved Transition Array (UTA) due to a large number of Fe M unresolved transitions. The shape of the UTA is similar to that of a weak absorption edge and therefore care has to be taken when analysing these low ionisation absorbers with CCD low spectral resolution data. The current *Chandra* and *XMM-Newton* dispersive spectrometers are at the edge of resolving the absorption features, whose turbulent velocities are of the order $\sim 300\,\mathrm{km\,s^{-1}}$. The general conclusion is that most the material appears to be outflowing from the AGN, and that its structure is complex both kinematically and in terms of ionisation.

Jets and Outflows

This is a new area in X-ray astronomy which has been opened by *Chandra* thanks to its superb angular resolution and sensitivity to low surface brightness objetcts. It is remarkable that one of the first targets to be observed by *Chandra* to demonstrate the sharpness of its point spread function (Pks 0637-752) showed a very clear jet coincident with the brightest of the two radiojets. Since then *Chandra* has shown that X-ray jets are common among AGN and have been detected and studied in the majority of nearby objects that show radio jets. An example of the science that can be done with these observations can be illustrated in the case of 3C273, where several knots in the X-ray jet have been resolved and matched to optical and radio emission (Marshall et al. 2001). The spectral energy distributions of each knot demonstrate that synchrotron is the dominant process in most cases, but that in others additional processes might be needed.

Binary AGN in galactic centres

Another important contribution due to the high-resolution imaging capability of *Chandra* has been the discovery of a binary AGN in the centre of the same galaxy (Komossa et al. 2002). Ironically, this has occurred in NGC 6240, a dusty star-forming galaxy which was only demonstrated to host an AGN thanks to X-ray observations and that might actually constitute a paradigm of the heavily obscured AGN population that is supposed to dominate the high redshift Universe.

The existence of binary supermassive black holes in the centres of galaxies has been reviewed by Gaskell (1996) based on optical data. Binary black hole mergers are expected to play a significant role in the triggering of the AGN phenomenon and perhaps in the growth of supermassive black holes themselves (in addition to accretion). The possible implications in the X-ray band, particularly in the shift of the blue edge of the Fe line due to disk precession, has been discussed by Torres et al. (2003), but this will have to await to future X-ray observatories to be observed.

4. Challenges to the unified AGN model

The simplest version of the AGN unified scheme predicts a direct relation between the optical spectroscopic properties and X-ray absorption. Maiolino (2001) compiled a sample of AGN selected by their [OIII] emission (a likely isotropic property in the unified model), where the tendency of increasing spectral type with increasing X-ray absorbing column was evident. There are, however, a number of exceptions along the lines of type 1 AGN displaying X-ray absorption and type 1.8, 1.9 or 2 AGN displaying little or no absorption at all. Studies with *Chandra* and *XMM-Newton* have both been instrumental in evidencing these discrepancies and they are also paving the way for an explanation.

Surveys conducted with *XMM-Newton* (Piconcelli et al. 2002, Mateos et al. 2004, Mateos et al. 2005), where detailed X-ray spectral information is obtained, have shown that about 10% of type 1 AGN have absorbing columns, typically below $10^{22}\,\mathrm{cm}^{-2}$. It is possible indeed that in these objects absorption happens so close to the central engine that sputtering destroys efficiently any dust and therefore no reddening is expected in the optical lines, although ionized absorption would be expected instead, as in the case of the (non-dusty) warm absorbers. Alternative possibilities to explain this fact include the contribution to absorption from the host galaxy or from some sort of weak Broad Absorption Line material.

The same surveys indicate that at least 40-80% of the type 2 AGN show absorption, the underlying power law being of similar shape to that of type 1 AGN. In the Lockman Hole observations, Mateos et al. (2005) report that 5 out of 28 type 2 AGN do not display any amount of photoelectric absorption. Explanations for this unusual behaviour, already detected in surveys with previous X-ray missions (Pappa et al. 2002, Panessa & Bassani 2002), were thought to be either due to the presence of a Compton-thick absorber (in which case only scattered radiation would be seen), to a dusty warm absorber (so X-ray absorption would only be detected in absorption edges but not in the overall shape) or to a genuine property of the BLR. Fortunately, the *XMM-Newton* data on some bright examples of these mismatched cases rules out the first two options (e.g. H1320+551, Barcons et al. 2003), as a Compton-thick AGN would produce a strong Fe line (not seen) and the high-quality X-ray spectra would be enough to detect absorption edges from any warm absorber. A further possibility would be that non-simultaneous optical and X-ray spectroscopy could map a perfectly standard but variable absorber. This has been recently ruled out in at least one case (Mrk 993, Corral et al. 2004), where simultaneous X-ray and optical spectroscopy shows that the type 1.8 character of this Seyfert galaxy is already imprinted on its BLR, rather than arising as a result of absorption.

5. X-ray surveys, obscured accretion and the X-ray background

The X-ray background, discovered in the very first astronomical X-ray observation (Giacconi et al. 1962), is now resolved to 94% in the 0.5-2 keV band (Moretti et al. 2003) and to 90% in the 2-10 keV band (Cowie et al. 2002). Above 50 keV the directly resolved fraction drops to $\sim 50\%$ (Worsely et al. 2004), probably due to the missing population of Compton-thick AGN at moderate redshift.

Currently accepted models for the cosmic X-ray background (XRB) require a mixture of absorbed and unabsorbed AGN to produce the right spectral shape (Comastri et al. 1995, Gilli et al. 2001, Ueda et al. 2003). The variety of

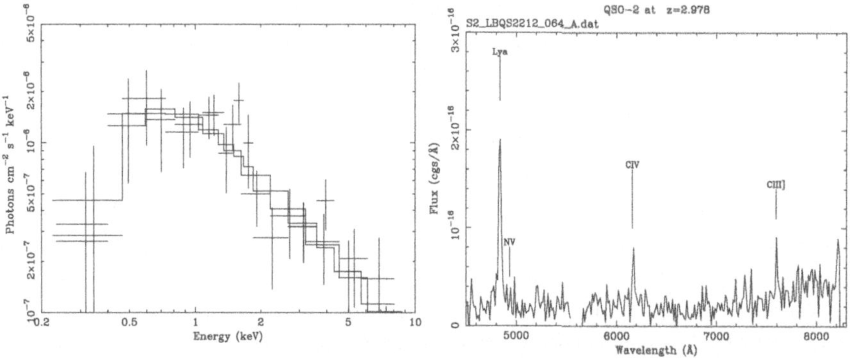

Figure 2. A type-2 QSO discovered by XMM-Newton at $z = 2.978$, with intrinsic 2-10 keV X-ray luminosity 4×10^{44} erg s^{-1} and absorbed by a column density $\sim 5 \times 10^{22}$ cm^{-2}. Left: XMM-Newton unfolded X-ray spectrum, right: AAT/2dF optical spectrum (obtained by Mat Page).

models proposed rely upon our knowledge of the AGN luminosity function, mostly based on unabsorbed AGN, its evolution, the fraction of absorbed to unabsorbed AGN (usually large $\sim$ 4) and the specific distribution of absorbing columns as well as any possible redshift evolution. The large number of parameters going into these ingredients means that detailed surveys at all flux levels are needed to constrain them, besides the overall constraint of the XRB spectral shape and intensity.

There are various generic predictions of these models. The first one is that a large fraction of accretion onto super-massive black holes occurs in obscured/absorbed objects. Since accretion is responsible for the growth of supermassive black holes, once they are formed, there should be a link between the X-ray background and the local density of supermassive black holes (Soltan 1982). For a standard 10% accretion efficiency, the local black hole density inferred from the kinematics around the centres of galaxies (which is $\sim 5 \times 10^5 M_\odot \,\mathrm{Mpc}^{-3}$ for $H_0 = 70 \,\mathrm{km\,s^{-1}\,Mpc^{-1}}$) is far too large to be accounted for only by unobscured AGN (QSOs and Seyfert 1 galaxies), and therefore obscured accretion is important (Marconi et al. 2004, Shankar et al. 2004).

The other prediction is the existence of genuine type 2 QSOs, the high luminosity version of Seyfert 2 galaxies. Indeed both radio-loud and radio-quiet type-2 QSOs are being found in medium and deep surveys, although it is unclear whether in large enough numbers as predicted by specific models (see Fig. 2 for an example).

Surveys of the X-ray sky are being conducted with the new observatories at various depths (Watson et al. 2001), from the Bright Source Survey (Della

Ceca et al. 2004), through various medium sensitivity surveys (Barcons et al. 2002, Kim et al. 2004a, Kim et al. 2004b, Green et al. 2004), and the deep surveys (Hasinger et al. 2001, Barger et al. 2003, Alexander et al. 2003). The picture emerging from all of these surveys is that the high-galactic latitude X-ray sky is dominated by AGN, and that the fraction of absorbed AGN appears to increase towards lower fluxes. In the *XMM-Newton* deep surveys, type 1 and type 2 AGN are already detected in similar numbers and in the deeper *Chandra* surveys a population of very faint starforming galaxies appears to become increasingly important. Another general property is that the deeper in X-rays, the redder the optical counterparts, with a significant fraction of Extremely Red Objects (EROs). The redshift distribution of the type 2 AGN, however, appears significantly different from that of type 1 AGNs, as it peaks at moderate redshifts ($z \sim 0.8$). This has changed our vision of the cosmic history of accretion onto supermassive black holes, as about 50% of the cosmic X-ray emissivity is now seen to arise at $z < 1$.

X-ray spectrosocpy of the most distant AGN, obtained with deep *XMM-Newton* observations of the Lockman Hole, reveal that at least 50% of the accretion occurs in obscured/absorbed objects (Mateos et al. 2005). The average stacked spectra of these AGN, exhibits a strong relativistic Fe line profile, implying rapid black hole rotation both in type 1 and type 2 AGN spectra (Streblyanska et al. 2004). This opens the possibility of measuring the redshift and characterising the accretion mode for each individual AGN using X-ray spectrosocpy with future high-throughput missions.

6. Outlook

Rather than summarizing what has already been said, I highlight a few questions that will be addressed in the future, most of them requiring the next generation of X-ray observatories (*XEUS/Constellation-X*). These include performing and understanding reverberation mapping on shorter timescales; test the properties of individual supermassive black holes powering high-z AGN and measure their mass, angular momentum and accretion rate; reach the "thermal limit" in X-ray spectroscopy so all lines due to circumnuclear material can be resolved; trace the cosmic evolution of obscured and unobscured accretion and finally relate the formation of supermassive black holes to galaxy formation itself.

Acknowledgments

I'm grateful to many colleagues of the *XMM-Newton* Survey Science Centre and of the Lockman Hole collaboration, for long-standing and fruitful collaborations on this subject. I'm especially indebted to Andy Fabian, Francisco Carrera, Guenther Hasinger and Mat Page for specific help with this presentation. Financial support for this work was provided by the Spanish MEC, under project ESP2003-00812.

References

Alexander, D.M. et al. (2003). AJ, 126, 539
Antonucci, R. (1993). ARAA, 31, 473
Ballantyne, D. et al. (2001). MNRAS, 327, 10
Ballantyne, D. et al. (2003). MNRAS, 342, 239
Barcons, X. et al. (2002). A&A, 382, 522
Barcons, X. et al. (2003). MNRAS, 339, 757
Barger, A.J. et al. (2003). AJ, 126, 632
Branduardi-Raymont, G. et al. (2001). A&A, 365, L140
Comastri, A. et al. (1995). A&A, 296, 1
Corral, A. et al. (2004). A&A,in the press (astro-ph/0410347)
Cowie, L.L. et al. (2002). ApJ, 566, L5
Della Ceca, R. et al. (2004). A&A, in the press (astro-ph/0407481)
Dewangan,G.C., et al. (2003). ApJ, 592, 52
Fabian, A.C. et al. (1989). MNRAS, 238, 729
Fabian, A.C. and Vaughan, S. (2003). MNRAS, 340, L28
Gaskell, C.M. (1996). In: *Jets from Stars and Galaxies*, ed. W. Kundt (Berlin: Springer), p. 165
George, I.M. and Fabian, A.C. (1991). MNRAS, 249, 352
George, I.M. et al. (1998). ApJS, 114, 73
Giacconi, R. et al. (1962). PRL, 9, 439
Gilli, R. et al. (2001). A&A, 366, 407
Green, P. et al. (2004). ApJS, 150, 43
Halpern, J.P. (1984). ApJ, 281, 90
Hasinger, G. et al. (2001). A&A, 365, L45
Iwasawa, K. et al. (1996). MNRAS, 282, 1038
Iwasawa, K. et al. (2004). MNRAS, 347, 411
Kim, D.-W. et al. (2004a). ApJS, 150, 19
Kim, D.-W. et al. (2004b). ApJ, 600, 59
Komossa, S. et al. (2002). ApJ, 582, L15
Krongold, Y. et al. (2003). ApJ, 597, 832
Lee, J.C. et al. (2001). ApJ, 554, L13
Maiolino, R. (2001). MmSAI, 72, 43
Marconi, A. et al. (2004). MNRAS, 351, 169
Marshall, H.L. et al. (2001). ApJ, 549, 167
Mason, K.O. et al. (2003). ApJ, 582, 95M
Mateos, S. et al. (2004). A&A, submitted
Mateos, S. et al. (2005). A&A, in preparation
Miniutti, G. and Fabian, A.C. (2004). MNRAS, 349, 1435
Molendi, S. et al. (2003). A&A, 343, L1
Moretti, A. et al. (2003). ApJ, 588, 696
Panessa, F. and Bassani, L. (2002). A&A, 394, 435
Pappa, I. et al. (2002). MNRAS, 336, 714
Piconcelli, E. et al. (2002).A&A, 394, 835
Pounds, K.A. et al. (1990). Nat, 344, 132
Pounds, K.A. et al. (2003). MNRAS, 345, 705

Reeves, J.N. et al. (2001). A&A, 365, L134
Reynolds, C.S. (1997). MNRAS, 286, 513
Sako, M. et al. (2001). A&A, 365, L168
Sako, M. et al. (2003). ApJ, 596, 114
Shankar, F. et al. (2004). MNRAS, in the press (astro-ph/0405585)
Soltan, A. (1982). MNRAS, 200, 115
Streblyanska, A. et al. (2004). A&A, in the press (astro-ph/0411340)
Tanaka, Y. et al. (1995). Nat, 375, 659
Torres, D. et al. (2003). ApJ, 596, L31
Turner, T.J. et al. (2002). ApJ, 574, L123
Turner, A.K. et al. (2003). MNRAS, 346, 833
Ueda, Y. et al. (2003). ApJ, 598, 886
Vaughan, S. and Fabian, A.C. (2004). MNRAS, 348, 1415
Watson, M.G. et al. (2001). A&A, 365, L51
Worsley, M.A. et al. (2004). MNRAS, 352, L28

STAR-FORMING COMPLEXES IN GALAXIES

Bruce G. Elmegreen
IBM Research Division, T.J. Watson Research Center
1101 Kitchawan Road, Yorktown Hts., NY, USA
bge@watson.ibm.com

Abstract Star complexes are the largest globular regions of star formation in galaxies. If there is a spiral density wave, nuclear ring, tidal arm, or other well-defined stellar structure, then gravitational instabilities in the gaseous component produce giant cloud complexes with a spacing of about three times the width. These gas complexes form star complexes, giving the familiar beads on a string of star formation along spiral arms, or nuclear hotspots in the case of a ring. Turbulence compression, supernovae, and self-gravitational contraction inside the giant clouds produce a nearly scale-free structure, including giant molecular clouds that form OB associations and molecular cloud cores that form clusters. Without stellar density waves or similar structures, random gravitational instabilities form flocculent spirals and these fragment into star complexes, OB associations and star clusters in the same way. In this case, all of the structure originates with gravitational instabilities and turbulence compression, but the usual concept of a star complex applies only to the largest globular object in the hierarchy, which has a size defined by the flocculent arm width or galaxy thickness. The largest coherent star-forming regions are the flocculent arms themselves. At the core of the hierarchy are the very dense clumps in which individual and binary stars form. The overall star formation rate in a galaxy appears to be regulated by gravitational collapse on large scales, giving the Kennicutt/Schmidt law scaling with density, but the efficiency factor in front of this scaling law depends on the fraction of the gas that is in a dense form. Turbulence compression probably contributes to this fraction, producing a universal efficiency on galactic scales and the observed star formation rate in disk systems. The CO version of the Schmidt law, recently derived by Heyer et al. (2004), follows from the turbulent hierarchy as well, as do the local efficiencies of star formation in OB associations and clusters. The efficiency of star formation increases with cloud density, and this is why most stars form in clusters that are initially self-bound.

Keywords: Star formation, Star clusters, Star complexes, Galactic structure

1. Introduction to star complexes

Star complexes are the largest coherent groupings of young stars in galaxies. Shapley (1931) first noted that they appear like "small irregular star clouds" up

J. C. del Toro Iniesta (eds.), The Many Scales in the Universe, 99–110.

to 400 pc in size in the Large Magellanic Cloud (LMC). McKibben Nail & Shapley (1953) later defined 12 "constellations" in the LMC. Baade (1963) considered the most active examples of star complexes and called them "superassociations," such as 30 Dor in the LMC and NGC 206 in M31.

The definition of star complexes was broadened by Efremov (1979). He considered complexes to be collections of Cepheid variables with similar periods and velocities, that is, within a narrow range of ages (e.g., $\sim$ 50 My) and moving as a group. This makes them distinct from "OB associations," which are collections of OB stars within a narrower range of ages ($\sim$ 10 My). Efremov noted that the OB associations in M31 defined by van den Bergh (1981) have an average size of 480 pc, which is much larger than local OB associations (80 pc). He concluded that the M31 associations are complexes, and that star complexes generally contain OB associations as sub-parts. Battinelli et al. (1996) quantitatively found stellar groupings in M31 and demonstrated this 2-component, or hierarchical, nature of associations and complexes.

Hierarchical structure in the LMC stars was first quantified by Feitzinger & Galinski (1987). Most recently, Maragoudaki et al. (1998) measured stellar groupings in the LMC using discrete magnitude limits in the U-band. They noted that the smaller groupings in the hierarchy are rounder. Gouliermis et al. (2000) did the same for B stars in LMC fields. Hierarchical structure as a general property of interstellar gas was discussed much earlier than this (see review in Scalo 1985). The observation that stellar groupings generally have the same type of hierarchical structure as the gas is not surprising since the stars form from the gas (e.g., see Bonnell, Bate & Vine 2003).

The definition of star complexes was broadened further by Elmegreen & Efremov (1996) to mean the largest "globular" scale in a hierarchy of star formation ranging from multiple stars to flocculent spiral arms. They showed that bigger scales evolve slower, with $t \propto L^{0.5}$, that all scales evolve on about a dynamical crossing time, and that the largest globular scale for star formation should be about the disk thickness (times $\sim \pi$), at which point the shear time becomes comparable to the crossing time. Bigger regions form the same way and are part of the hierarchy, but they look like flocculent spiral arms instead of roundish star complexes. Efremov & Elmegreen (1998) also found that the size-duration correlation for star formation is about the same as the size-crossing time correlation for molecular clouds, which suggests that turbulence regulates star formation on scales comparable to or smaller than the ambient ISM Jeans length.

The observation that star formation is somewhat scale-free, even up to $\sim$ 0.1 times galactic scales, received considerable support after the gas was found to be scale-free over similar lengths. Fractal structure in the gas on very small scales had been observed for a long time (e.g., Falgarone, Phillips & Walker 1991 and references therein), but the first observations of fractal structure in whole galaxies was by Westpfahl et al. (1999) and Stanimirovic et al. (1999). Westpfahl et al. found the fractal dimension for gas in M81 group galaxies using area-perimeter relations and box counting techniques on HI maps, while Stanimirovic et al. found that the power spectrum of HI emission from the entire Small Magellanic Clouds is a scale-free power law. A similar power-law was later found for the LMC (Elmegreen, Kim & Staveley-Smith 2001). In the LMC, much of the gas also resembles shells rather than blobs, and these shells are hierarchical too (*ibid.*). The exact relation between this shell structure, turbulence, and star formation is not clear yet (Wada, Spaans, & Kim 2000). Yamaguchi et al. (2001a,b) showed that some of the shells and other high-pressure sources trigger star formation directly in the LMC. Gouliermis et al. (2003) found that stellar systems line the edges of the supershells found by Kim et al. (1999). Thus some of the stellar hierarchy could be the result of high-pressure triggering in hierarchical shells.

The connection between the hierarchical structures of the gas and stars is emphasized further by the power spectra of optical emission from galaxies. Elmegreen et al. (2003a,b) found that power spectra of optical emission along azimuthal cuts through several galaxies have the same near-power law form as the power spectra of HI emission from the LMC. In one case, M81, a spiral density wave contributes to the power spectrum at the lowest wavenumber, but otherwise the power spectrum is the same as in a flocculent galaxy.

We might summarize these observations as follows: "Star complexes" are the largest globular regions of star formation in galaxies. They include associations of Cepheid variables, red supergiants, WR stars, HII regions, and OB-associations (e.g., Ivanov 2004). They could be the origin of moving stellar groups (Asiain et al. 1999). They are part of a continuum of star-formation scales between clusters and swing-amplified spirals. This continuum has at least two characteristics of turbulence: power-law power spectra and a velocity-size relation. We would like to know how star complexes form and how stars form in them. Is there any evidence also for star complexes with a characteristic scale, rather than an observationally selected scale among many scales?

2. Formation of star complexes

Most galaxies with star formation have at least a few regions with sizes comparable to the main stellar structures – spiral arms, tidal arms, resonance rings, etc.. Comparable sizes means comparable to the minor dimensions, e.g., the widths of the arms or rings. These regions are star complexes. They are probably the largest scale in a local hierarchy of scales beginning with some instability length and extending down to OB associations and clusters. If this is the case, then the complexes will have a characteristic length and mass.

Galaxies without stellar spiral waves or rings would not have a characteristic scale limited by these structures, so they could produce an even wider range of scales in the hierarchy of star formation. For example, the optical structures could range from flocculent arms, which are driven primarily by sheared gravitational instabilities, down to star complexes, OB associations, and clusters, which are fragments produced by self-gravity and turbulence (e.g., Huber & Pfenniger 2001). Recent CO observation of M33 (Engargiola et al. 2004), combined with stellar complex data (Ivanov 2004) and older HI observations (Deul & van der Hulst 1987) show star complexes associated with molecular and atomic gas. They also show that most molecular clouds are inside giant HI clouds, as observed locally (Grabelsky et al. 1987; Elmegreen & Elmegreen 1987) and in other galaxies (Lada et al. 1988). In addition, in M33, the CO cloud spin axes are correlated up to scales of $\sim$ 1 kpc, suggesting coherence on this scale. This is also the scale of the giant HI clouds. Thus star formation proceeds first by forming giant HI clouds, and then by forming molecular clouds inside of them and OB associations inside the molecular clouds. Each collection of OB associations, aged by $\sim 30-100$ My, is a star complex (Efremov 1995). The Gould's Belt region (e.g., Lallement et al. 2003) may be an example. The regular distribution of giant HI clouds along spiral arms has been known for many years, starting with the first HI observations of the Milky Way (McGee & Milton 1964) and proceeding through the 1970's and 1980's when HI was routinely mapped in nearby galaxies (e.g., Boulanger & Viallefond 1992). Regularity implies a characteristic scale, which is most likely the Jean length (Elmegreen & Elmegreen 1983; Kuno et al. 1995). These observations demonstrate that star complexes form in CO/HI cloud complexes. The stellar parts are hierarchically clumped, but still coherent up to $\sim$ 1 kpc in the main disks of galaxies, while the gaseous parts can extend for two or three times this distance. The primary objects formed by galactic processes are $\sim 10^7$ $M_\odot$ HI clouds, while giant molecular clouds (GMCs) are their fragments. This 10^7 $M_\odot$ mass is the Jeans mass in the ambient ISM, suggesting that the HI clouds and ultimately the star complexes form by gravitational instabilities. The formation of GMCs follows by a combination of self-gravitational contraction and turbulence compression inside the 10^7 $M_\odot$ clouds, but GMCs are not special, distinct objects. The CO/HI ratio depends primarily on self-shielding, not cloud formation processes. Low pressures, low metallicities, or high radiation fields imply low CO/HI ratios in each cloud. CO/HI varies with galactic radius or galaxy type because of variations in pressure, radiation field and metallicity, without any change in physical cloud structure or star formation properties (Elmegreen & Elmegreen 1987; Elmegreen 1993; Honma, Sofue, & Arimoto 1995; Engargiola et al. 2004).

There are 2 basic dynamical phases and 2 basic chemical phases for neutral clouds: atomic or molecular gas in a self-gravitating cloud, and atomic or molecular gas in a non-self-gravitating cloud. Usually the atomic phase dominates at low density, the molecular phase at high density, the non-self-gravitating phase at low column density, and the self-gravitating phase at high column density (Elmegreen 1995). The virial theorem also plays a role: considering the presence of some external pressure, both diffuse and self-gravitating clouds occur at low mass but only self gravitating clouds occur at high mass. In the general ISM, gravitating cloud complexes are evident mostly from knots in spiral arms, spurs, and the presence of star formation (e.g., Kim & Ostriker 2002). In accord with the virial theorem result above, the FCRAO Outer Galaxy Survey (Heyer, Carpenter & Snell 2001) shows that self-gravity is important only in the most massive CO clouds, $M > 10^4$ $M_\odot$. This is the same mass at which the virial theorem suggests a transition from both diffuse and self-gravitating clouds (at lower mass) to purely self-gravitating clouds (at higher mass), given a near-constant diffuse cloud density of $\sim$ 50 cm^{-3} (Elmegreen 1995). This mass limit should depend on pressure and the molecule detection threshold. Small self-gravitating cores are undoubtedly present inside these FCRAO clouds (because stars are forming), but they have a higher density threshold for detection and a smaller angular size, making their detection not as likely in large-scale CO surveys.

3. Characteristic size versus scale-free?

If a galaxy has a global spiral density wave in the stars, or if it has a stellar ring, then gaseous gravitational instabilities in these structures have a size and mass defined by the stellar geometry: i.e., the instability length is $\sim 3\times$ the spiral arm (or ring) width. Examples are the well-known "beads on a string of star formation" along spiral arms, and the "nuclear ring hotspots." If a galaxy has no spiral density wave, then the stars and gas become unstable together, forming multiple spiral arms or flocculent arms that are made of old stars, gas and star formation. The instability involved is the swing amplifier (Toomre 1981), usually enhanced by magnetic fields (Kim, Ostriker & Stone 2002, 2003). The instability should also drive turbulence, producing scale-free clouds and star formation as observed.

Stellar spirals define two characteristic scales: $2\pi G\Sigma/\kappa^2$, which is the Toomre (1964) length for the separation between spiral arms, and $2c^2/G\Sigma$, which is the Jeans length for pressure balance against self-gravity. Here, κ is the epicyclic frequency, Σ is the mass column density, and c is the gas velocity dispersion. The Jeans mass is $c^4/G^2\Sigma$. Inside spiral arms, the Jeans instability is one-dimensional, i.e., the collapse is parallel to the arms, and the characteristic length is usually about 3 times the arm width (Elmegreen & Elmegreen 1983; Bastien et al. 1991). The condition for rapid instability in this case is not the Toomre Q condition, but the 1-dimensional analog: $\pi G\mu/c^2 > 1$ for mass/length μ (Elmegreen 1994).

Wada & Norman (2001) modeled 2D hydrodynamics of galaxy disks without spiral density waves and found that gravitational instabilities drive turbulence, giving a log-normal density pdf that is typical for isothermal compressible turbulence (Vazquez-Semadeni 1994). Three-dimensional SPH models of galaxy disks without spiral density waves get the Schmidt/Kennicutt laws of star formation (Li et al. 2004).

4. Theory of the star formation rate

A sensible local SF law is (Elmegreen 2002b)

$$\mathrm{SFR/V} = \epsilon\rho(G\rho)^{1/2}, \tag{1}$$

which is the efficiency times the mass per unit volume, times the conversion rate from gas into stars. Kennicutt (1998) observes

$$\mathrm{SFR/Area} = 2.5 \times 10^{-4} \left(\Sigma/\mathrm{M}_{\odot}\ \mathrm{pc}^{-2}\right)^{1.4}\ \ \mathrm{M}_{\odot}\ \mathrm{kpc}^{-2}\ \mathrm{yr}^{-1} \sim 0.033\Sigma\Omega \tag{2}$$

for average mass column density Σ in the whole disk and rotation rate in the outer part Ω.

If we convert the global SFR/A into a local SFR/V using the local tidal density for gas, $\rho = 3\Omega^2/2\pi G$, a flat rotation curve, and an exponential disk with scale length $r_D/r_{edge} = 0.25$, then the average SFR/A converts to a local

$$\mathrm{SFR/V} = 0.012\rho(G\rho)^{1/2}. \tag{3}$$

Why is the efficiency $\epsilon = 0.012$, and is this the right star formation law? Boissier et al. (2003) compared the star formation rates versus radii in 16 galaxies with three simple expressions, finding factor of 3 variations around each law with no apparent cause, and no preferred law. Either we do not know the "right" star formation law, or additional processes give big variations around one of the assumed laws. By the way, all of the Boissier et al. laws, and that discussed by Hunter, Elmegreen, & Baker (1998) for dwarf irregular galaxies, are consistent with a local star formation rate proportional to the stellar surface density. Whether this is a cause or an effect of star formation is not clear. That is, one might expect the star formation rate to scale with the existing stellar surface density if star formation is building up that surface density and the exponential scale length does not change much with time. On the other hand, the relationship might also exist if background stars trigger star formation, as might be the case if supernova and HII regions directly compress the gas to trigger star formation or if these pressures indirectly compress the gas by driving supersonic turbulence.

Let us proceed with equation (3) and ask what determines the star formation efficiency, ϵ. Assume that stars form in dense cores where the efficiency is $\epsilon_c \sim 0.5$ and $\rho_c \sim 10^5$ cm^{-3}, giving

$$\mathrm{SFR/V}_{core} = \epsilon_c \rho_c \left(G\rho_c\right)^{1/2}. \tag{4}$$

Then make a conversion:

$$\mathrm{SFR/V}_{gal} = \mathrm{SFR/V}_{core} \times (V_{core}/V_{gal}) = \mathrm{SFR/V}_{core} \times (M_{core}/M_{gal}) \times \rho/\rho_c, \tag{5}$$

where $\rho_c = M_{core}/V_{core}$, $\rho = M_{gal}/V_{gal}$, M_{core} and V_{core} are the summed mass and volume of all cores, while M_{gal} and V_{gal} are the total gas mass and gas volume of the galaxy. Now substitute from equation 4 and set $\mathrm{SFR/V}_{gal}$ equal to the observed rate $0.012\rho(G\rho)^{1/2}$. Then we get the gas mass fraction in star-forming cores:

$$f_M \equiv (M_{core}/M_{gal}) \equiv 0.012/\epsilon_c \times (\rho/\rho_c)^{1/2} \sim 10^{-4}. \tag{6}$$

Thus the observed average efficiency of $\epsilon = 0.012$ (per dynamical time) requires 10^{-4} of the total ISM mass to be in star-forming cores if all regions evolve on a dynamical timescale (Elmegreen 2002b). This is the same mass fraction as in the Wada & Normal (2001) log-normal for $\rho/\rho_{ave} > 10^5$, as assumed above with $\rho_c \sim 10^5$ cm^{-3} and $\rho \sim 1$ cm^{-3}. This result depends on the density probability density function (pdf) for the ISM, which is not observed yet, and it assumes a log-normal form for this pdf. In fact, the high density portion may become a power-law after collapse starts (Klessen 2000). Nevertheless, the agreement between the simple theory and the Kennicutt (1998) star formation rate, which applies to essentially all late-type galaxy disks and their nuclear regions, suggests that something universal like turbulence helps partition the gas in a hierarchical fashion and that only the dense regions at the bottom of the hierarchy form stars.

A turbulent ISM has a small fraction of its mass at a high enough density to form stars. Most of the mass is either too low a density to form stars, or is not self-gravitating enough to resist turbulent disruption. For a turbulent medium, every structure forms on a local crossing time, but the progression to high density is not monotonic. The low-density clumps are destroyed easily and they are smashed and sheared into smaller pieces by transient pressure bursts. The progress toward high density is more like a random walk, with some interactions making denser regions and some making lower densities. Eventually the lucky ones that had a long succession of compressive interactions become dense enough and massive enough to be strongly self-gravitating at the typical pressure in the cloud. Then they presumably produce stars quickly. The delays from magnetic diffusion, disk formation, and turbulent energy dissipation are not nearly as time consuming as the fragmentation process on larger scales. Thus the largest scales control the overall rate. All that a microscopic delay might do is change the form of the density pdf, producing a bump at high density, for example, or a power-law instead of a log-normal, if the collapse slows down. If only turbulence is involved, though, the random walk in density produces a log-normal density pdf (Vazquez-Semadeni 1994). Our integration over the pdf for densities $\rho_c/\rho > 10^5$ involves an assumption that the collapse delays occur at higher densities, where the detailed shape of the pdf will not affect the integral under it if there is a steady flow toward higher density during the star formation process.

If we now denote the galactic average quantities by a subscript "0", then the efficiency at any density is given by:

$$\epsilon_0 \rho_0 (G\rho_0)^{1/2} = \epsilon(\rho)\rho(G\rho)^{1/2} f_V(\rho) = \epsilon_c \rho_c (G\rho_c)^{1/2} f_V(\rho_c). \qquad (7)$$

But $\rho f_V(\rho) = \rho_0 f_M(\rho)$, etc., so

$$\epsilon(\rho) = \epsilon_c \left(\rho_c/\rho\right)^{1/2} \left[f_M\left(\rho_c\right)/f_M\left(\rho\right)\right]. \qquad (8)$$

Here, $f_V(\rho)$ is the fraction of the volume having a density larger than ρ and $f_M(\rho)$ is the fraction of the mass having a density larger than ρ.

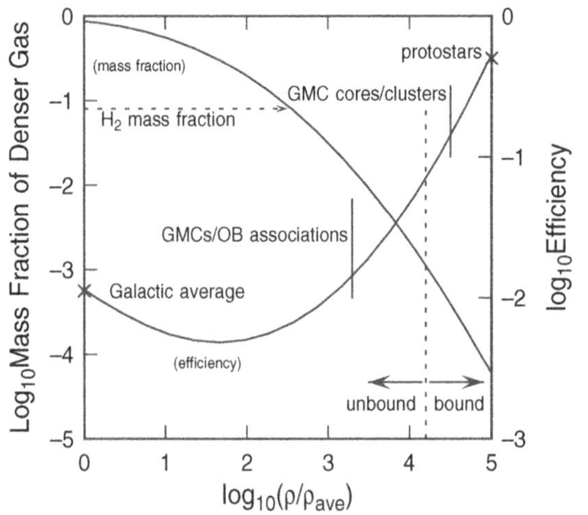

Figure 1. The mass fraction (downward-sloping curve) and efficiency of star formation as a function of density, derived from the log-normal density pdf in Wada & Norman (2001). Observed efficiencies in cloud cores and OB associations are shown as vertical lines. The increase of efficiency with density suggests that star formation proceeds in a hierarchical medium.

Figure 1 shows as a decreasing line the mass fraction, $f_M(\rho)$, versus density ρ using the log-normal found by Wada & Norman (2001) for a 2D disk with star formation, turbulence, and self-gravity. The scale for f_M is on the left-hand axis. The figure also shows as an increasing line the efficiency $\epsilon(\rho)$, using the right-hand axis. The efficiency increases with increasing average density because the hierarchical nature of clouds gives them a higher filling factor for dense gas at higher average density. The mass fraction decreases with density because only a small fraction of the matter is dense. Several commonly observed values for the efficiency are indicated: the average galactic value of 0.012, derived above, the range of $\sim 1-5\%$ for whole OB associations, and the range of $\sim 10-30\%$ for the cores of OB associations, where most stars actually form, producing clusters. Note that this efficiency is not the local efficiency where single stars form; that is assumed to be the constant value of $\epsilon_c = 0.5$ at $\rho/\rho_{ave} = 10^5$. Rather, it is the efficiency inside a cloud whose boundary density is ρ. OB associations form with an overall low efficiency because there is a lot of inactive gas at low density. Only the cores, and in them, only the small dense cores of these cores, form stars with high efficiency. This increase of efficiency with cloud density is commonly observed and easily explained for hierarchical stellar regions when all star formation occurs locally at the highest density. Note that star formation in simulations (e.g., Mac Low & Klessen 2004) proceeds over some prolonged time as does dense core formation, but the total time for this is still about the crossing time on the largest scale, making equation 1 appropriate.

Figure 1 highlights the boundary between bound and unbound stellar regions, which is where the average efficiency is greater than several tenths (Lada & Lada 2003). Bound regions have high densities and may therefore be identified with clusters rather than OB associations or any other part of the hierarchy on larger scales. The masses of the bound regions are not specified by this derivation but may be anything, always distributed as $dN/dM \sim M^{-2}$ for hierarchical gas structures (Fleck 1996; Elmegreen & Efremov 1997). This is the essential explanation for the formation of most stars in clusters. After ~ 10 My, most clusters dissolve and their stars fill in the region where they once clustered together. This is an OB association. In another $\sim 30-100$ My, these OB associations dissolve and fill in the region where they clustered together; this makes a star complex. Star complexes are so big that their dispersal is relatively slow and therefore accompanied by significant shear in normal galaxy disks. Thus there is no globular-shaped super-collection of star complexes, only flocculent spiral arms on larger scales.

Heyer et al. (2004) studied the star formation rate versus radius for molecular gas in M33. Both the SFR and the column density increase toward the center, with a mutual relationship

$$\mathrm{SFR} = 3.2\left(\Sigma_{H2}/\mathrm{M}_\odot\ \mathrm{pc}^{-2}\right)^{1.36} \mathrm{M}_\odot \mathrm{pc}^{-2}\ \mathrm{Gy}^{-1}. \tag{9}$$

This may be written as in Eq. (1), SFR $\sim 0.6\rho_{H2}(G\rho_{H2})^{1/2}$, assuming a disk thickness of 150 pc. The efficiency is higher for the distributed CO density ($\epsilon_{H2} \sim 0.6$) than it is for the total gas density ($\epsilon_0 \sim 0.012$) because the average density of H_2 spread around a disk is lower than the total density by $\rho_{H2}/\rho_0 = (\epsilon_0/\epsilon_{H2})^{2/3} = (0.012/0.6)^{2/3} = 0.08$ (i.e., 0.6 is not the ϵ inside a CO cloud). This is the M_{H2}/M_{gal} mass ratio actually obtained from Fig. 1 if the density of CO-emitting material is taken to be $\sim$ 300 cm^{-3} (use the left-hand axis). Thus the CO-Schmidt law follows from this hierarchical model too.

As mentioned briefly above, giant molecular clouds, OB associations, and star clusters all get their mass distribution functions from the structure of a compressibly turbulent medium. These mass functions are a property of fractals, sampled in various ways (Elmegreen 2002a; also see Elmegreen 2004). For a region sampled at low density, far from the peak, the mass function is approximately a power law with a shallow slope, ~ -1.5, as observed for GMCs. When the same region is sampled at a higher density, the mass function has a steeper slope, ~ -2, as observed for clusters. This explains how GMCs and clusters can both form from the same gas distribution and yet have slightly different mass functions. Note that this works because the the ratio of cluster mass to cloud-core mass is about constant when there is a threshold efficiency required for bound cluster formation.

5. Conclusions

Star complexes usually form in clouds that result from a gravitational instability in the ISM. If there is an imposed structure on the gas distribution, such as a stellar spiral arm or ring, then the clouds can line up along this structure with a semi-regular spacing, producing beads on a string in spiral arms, tidal arm star-forming clumps, collisional ring star formation, nuclear ring hotspots, dwarf galaxy hot spots, etc. The cloud typically has 10^7 M$_\odot$ of gas in the main galaxy disk, and it makes a 10^5 M$_\odot$ star complex over a $\sim 50 - 100$ My period. In galaxies with no imposed stellar structures, gaseous instabilities and turbulence compression still make giant clouds and their fragments, but the star complexes are usually selected to be the largest globular scale, excluding the flocculent arms themselves. In both cases, star complexes appear as groupings of intermediate-age stars, such as Cepheid variables and red supergiants.

For a turbulent self-gravitating medium, star formation should operate on about the local dynamical time over a wide range of scales. Turbulence structures the gas, placing only a small fraction of the mass at a high enough density to form stars. The resulting hierarchical structure is somewhat continuous from the scale of the disk thickness to individual pre-stellar condensations. Stars form in the turbulent structures, making a hierarchy of stellar complexes, associations, clusters, multiple stars, and binaries. The efficiency of star formation increases with density in such a hierarchical medium. Only high density regions have high enough efficiencies to form bound clusters.

Acknowledgments

Much of this work was done with the help of a National Science Foundation grant AST-0205097.

References

Asiain, R., Figueras, E., & Torra, J. 1999, A&A, 350, 434

Baade W. 1963, Evolution of Star and Galaxies, Cambridge: Harvard Univ. Press

Bastien, P., Bonnell, I., Martel, H., Arcoragi, J., & Benz, W. 1991, ApJ, 378, 255

Battinelli, P., Efremov, Y., & Magnier, E.A. 1996, A&A, 314, 51

Bonnell, I.A., Bate, M.R., & Vine, S.G. 2003, MNRAS, 343, 413

Bossier, S., Prantzos, N., Boselli, A., & Gavazzi, G. 2003, MNRAS, 346, 1215

Boulanger, F., & Viallefond, F. 1992, A&A, 266, 37

Duel, E.R., & van der Hulst, J.M. 1987, A&AS, 67, 509

Efremov, Yu.N. 1979, Sov. Astr. Lett., 5, 12

Efremov, Yu.N. 1995, AJ, 110, 2757

Efremov, Yu.N., & Elmegreen, B. G. 1998, MNRAS, 299, 588

Elmegreen, B.G. 1993, ApJ, 411, 170

Elmegreen, B.G. 1994, ApJ, 433, 39

Elmegreen, B.G. 1995, in The 7th Guo Shoujing Summer School on Astrophysics: Molecular Clouds and Star Formation, ed. C. Yuan and Hunhan You, Singapore: World Press, p. 149

Elmegreen, B.G. 2002a, ApJ, 564, 773

Elmegreen, B.G. 2002b, ApJ, 577, 206

Elmegreen, B.G. 2004, in Formation and Evolution of Massive Young Star Clusters, eds. H. Lamers, L. Smith, & A. Nota, ASP, PASP Conference series, in press, astroph/0405552

Elmegreen, B.G. & Efremov, Yu.N. 1996, ApJ, 466, 802

Elmegreen, B.G. & Efremov, Yu.N. 1997, ApJ, 480, 235

Elmegreen, B.G., & Elmegreen, D.M. 1983, MNRAS, 203, 31

Elmegreen, B.G., & Elmegreen, D.M. 1987, ApJ, 320, 182

Elmegreen, B. G., Elmegreen, D. M., & Leitner, S. N. 2003a, ApJ, 590, 271

Elmegreen, B. G., Leitner, S. N., Elmegreen, D. M., & Cuillandre, J.-C. 2003b, ApJ, 593, 333

Elmegreen, B. G., Kim, S., & Staveley-Smith, L. 2001, ApJ, 548, 749

Engargiola, G., Plambeck, R.L., Rosolowsky, E., & Blitz, L. 2003, ApJS, 149, 343

Falgarone, E., Phillips, T.G., & Walker, C.K. 1991, ApJ, 378, 186

Feitzinger, J. V., & Galinski, T. 1987, A&A, 179, 249

Fleck, R. C., Jr. 1996, ApJ, 458, 739

Gouliermis, D., Kontizas, M., Korakitis, R., Morgan, D.H., Kontizas, E., & Dapergolas, A. 2000, AJ, 119, 1737

Gouliermis, D., Kontizas, M., Kontizas, E., & Korakitis, R. 2003, A&A, 405, 111

Grabelsky, D. A., Cohen, R. S., Bronfman, L., Thaddeus, P., & May, J. 1987, ApJ, 315, 122

Heyer, M.H., Carpenter, J.M., & Snell, R.L. 2001, ApJ, 551, 852

Heyer, M.H., Corbelli, E., Schneider, S.E., & Young, J.S. 2004, ApJ, 602, 723

Honma, M., Sofue, Y., & Arimoto, N. 1995, A&A, 304, 1
Huber, D., & Pfenniger, D. 2001, A&A, 374, 465
Hunter, D.A., Elmegreen, B.G., & Baker, A.L. 1998, ApJ, 493, 595
Ivanov, G.R. 2004, astroph/0407043
Kennicutt, R.C. 1998, ApJ, 498, 541
Kim, S., Dopita, M.A., Staveley-Smith, L., Bessell, M.S. 1999, AJ, 118, 2797
Kim, W.-T., & Ostriker, E.C. 2002, ApJ, 570, 132
Kim, W.-T., Ostriker, E.C., Stone, J.M. 2002, ApJ, 581, 1080
Kim, W.-T., Ostriker, E.C., Stone, J. 2003, ApJ, 599, 1157
Klessen, R.S. 2000, ApJ, 535, 869
Kuno, N., Nakai, N., Handa, T., & Sofue, Y. 1995, PASJ, 47, 745
Lada, C.J., Margulis, M., Sofue, Y., Nakai, N., & Handa, T. 1988, ApJ, 328, 143
Lada, C.J., & Lada, E.A. 2003, ARAA, 41, 57
Lallement, R., Welsh, B. Y., Vergely, J. L., Crifo, F., Sfeir, D. 2003, A&A, 411, 447
Li, Y., Mac Low, M.-M., & Klessen, R.S. 2004, astroph/047247
Mac Low, M.-M., & Klessen, R. S. 2004, RvMP, 76, 125
Maragoudaki, F., Kontizas, M., Kontizas, E., Dapergolas, A., & Morgan, D.H. 1998, A&A, 338, L29
McGee, R.X., & Milton, J.A. 1964. Austr. J. Phys, 17, 128
McKibben Nail, V. & Shapley, H. 1953, Proc.Nat.Acad.Sciences, 39, 358
Scalo, J. 1985, in Protostars and Planets II, D.C. Black, & M.S. Matthews, Tucson: Univ. of Arizona, 201
Shapley, H. 1931, Harvard Bull. No. 884, 1
Stanimirovic, S., Staveley-Smith, L., Dickey, J.M., Sault, R.J., & Snowden, S.L. 1999, MNRAS, 302, 417
Toomre, A. 1964, ApJ, 139, 1217
Toomre, A. 1981, in The Structure and Evolution of Normal Galaxies, ed. S.M. Fall & D. Lynden-Bell, Cambridge, Cambridge University, p. 111.
van den Bergh, S. 1981, ApJS, 46, 79
Vázquez-Semadeni, E. 1994, ApJ, 423, 681
Wada, K., Spaans, M., & Kim, S. 2000, ApJ, 540, 797
Wada, K., & Norman, C. A. 2001, ApJ, 547, 172
Westpfahl, D. J., Coleman, P. H., Alexander, J., & Tongue, T. 1999, AJ, 117, 868
Yamaguchi, R., Mizuno, N., Onishi, T., Mizuno, A., & Fukui, Y. 2001a, ApJ, 553, 185
Yamaguchi, R., Mizuno, N., Onishi, T., Mizuno, A., Fukui, Y. 2001b, PASJ, 53, 959

STAR FORMATION AND INFRARED EMISSION IN GALAXIES

Nikolaos D. Kylafis and Angelos Misiriotis
University of Crete, Department of Physics
P.O. Box 2208, 71003 Heraklion, Crete, Greece
kylafis@physics.uoc.gr

Abstract The relationship between star formation and infrared emission in galaxies will be investigated. If galaxies were simple objects and young stars were completely covered with dust, then all the absorbed light of the young stars would be re-emitted in the infrared and from the infrared emission of galaxies we would infer the star formation rate (SFR) in them accurately. To show the complexities involved in real galaxies, we will use as a case study the late-type spiral galaxies. We will show that the heating of the dust is done mainly by the UV radiation of the young stars and therefore the infrared emission reveals the SFR in them. With a realistic model and its application to a number of galaxies, tight correlations are derived between SFR and total far infrared luminosity on one hand, and dust mass and 850 micron flux on the other. Other diagnostics of the SFR are examined and it is shown that there is consistency among them. Thus, the SFR for galaxies of all Hubble types has been determined as well as for interacting starburst galaxies. Combining different methods, the star-formation history of the universe has been determined and will be shown. Finally, some early results from the Spitzer Space Telescope will be presented.

Keywords: Star formation, Infrared emission, Galaxies

1. Introduction

We will start by asking the naive and rhetoric question of why should star formation and infrared emission in galaxies be connected.

If galaxies were simple objects, consisting of young stars surrounded by optically thick dust, then practically all the luminosity of the young stars would be absorbed by the dust and it would be re-emitted in the infrared (IR) part of the spectrum. Then, observation of the IR luminosity of a galaxy would give us, with the use of an initial mass function, the star formation rate (SFR) in the galaxy.

However, real galaxies are more complicated objects for the following reasons: 1) They consist mainly of old stars, which may also contribute to the

J. C. del Toro Iniesta (eds.), The Many Scales in the Universe, 111–119.

heating of the dust. 2) The dust is distributed throughout a galaxy and not around its stars. 3) The optical depth of the dust in a galaxy varies significantly with position in the galaxy and direction. Therefore, detailed modeling of the stars (young and old) and the dust in a galaxy is needed before trustworthy conclusions are drawn.

In what follows, we will treat late-type spiral galaxies as a case study. Then we will extend our discussion to all types of galaxies.

2. Model for late-type spiral galaxies

Modeling of the optical light

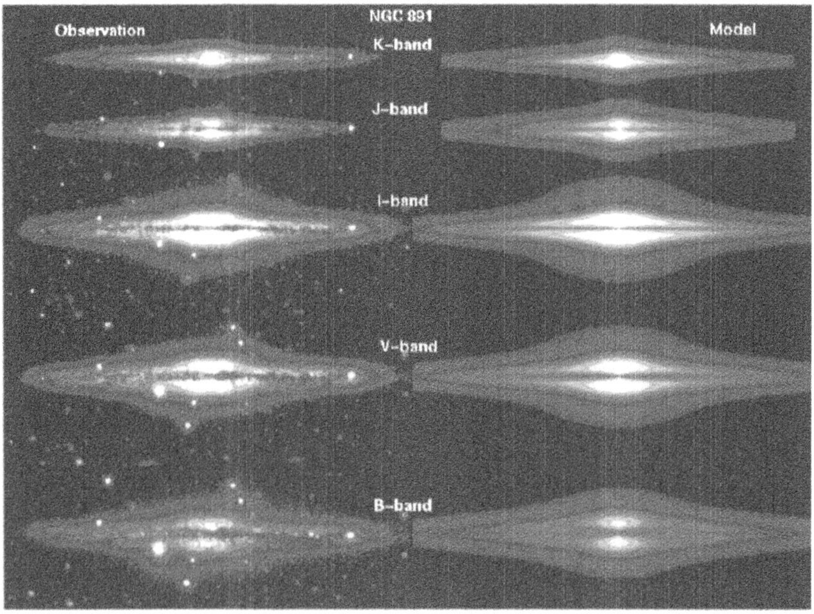

Figure 1. Images of NGC 891 in the K, J, I, V, and B bands and the corresponding models.

A successful model for late-type spiral galaxies has been applied to a number of galaxies by Xilouris et al. (1999). This model includes the old stellar population in the form of an exponential (both in radius and height) disk and a de Vaucouleurs (1953) spheroid.

The dust is distributed in another exponential disk (with scales different than those of the stars). Comparison of model images with optical and near infrared (NIR) images of galaxies reveals the total amount of dust (warm and cold)

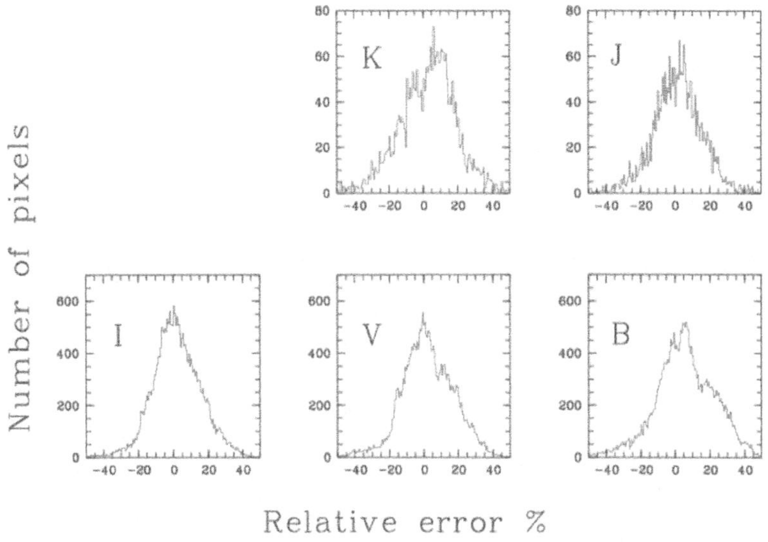

Figure 2. Histograms of the relative errors between the pixels of the K, J, I, V, and B band observations and the corresponding models.

in them and its distribution. Figure 1 shows NGC 891 in five bands and the corresponding model images (Xilouris et al. 1998). The agreement between the model and the observations is very good. This is seen more clearly in Fig. 2, where the residuals between the model and the observations are shown (Xilouris et al. 1998). The main conclusions of Xilouris et al. (1999) are:

1) The scale height of the dust is about half that of the stars.

2) The radial scale length of the dust is about 1.4 times that of the stars. Furthermore, the dust extends beyond the optical disk.

3) The average gas-to-dust ratio of seven spiral galaxies is about 400, i.e. comparable to that of our Galaxy.

4) The extinction coefficient in the optical and NIR parts of the spectrum is the same as in our Galaxy, indicating common dust properties among spiral galaxies.

5) The central, face-on optical depth in the B-band of all seven galaxies is less than one. This means that, *if* all the dust has been accounted for in these galaxies, then late-type spiral galaxies are transparent.

Similar conclusions have been reached by Alton et al. (1998) and Davies et al. (1999). The effects of spiral structure have been shown to be negligible (Misiriotis et al. 2000) and similarly for the effects of clumpiness in the dust (Misiriotis and Bianchi 2002).

Modeling of the infrared emission

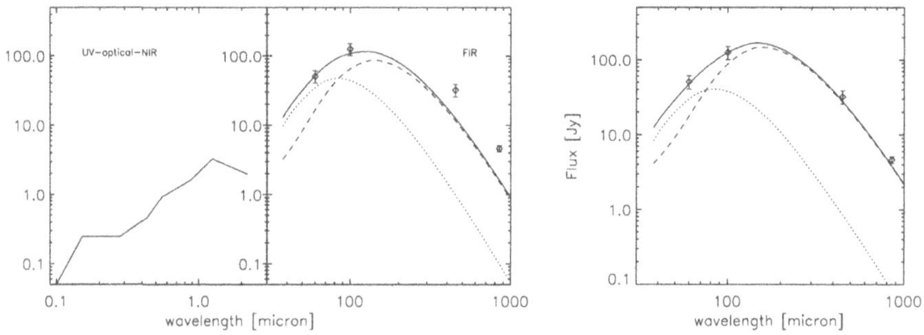

Figure 3. One- and two-dust-disk models of NGC 891.

In order to model the IR emission of late-type spiral galaxies, one has to take into account the heating of the dust not only by the old stars but also by the young ones. This was done by Popescu et al. (2000) for NGC 891 and by Misiriotis et al. (2001) for four more galaxies. The model utilizes the dust distribution derived from the optical images and assumes that the young stars (and therefore the UV emission) are distributed in an exponential disk with a small scale height (90 pc) and scale length equal to the B-band scale length of the old stellar population. Part of the UV luminosity is absorbed by local sources (HII complexes) and the rest is diffuse. The far infrared (FIR) emissivity of the dust was taken to be the same as the one thought appropriate for our Galaxy (Laor & Draine 1993).

It was found that model fluxes in the submillimeter (submm) part of the spectrum are significantly lower than the observed ones (Fig. 3, left). In order to account for the "missing" submm flux, Popescu et al. (2000) proposed that there is a second dust disk, not visible in the optical images, with scale height 90 pc (equal to that of the young stars). The mass in the second dust disk is about the same as that of the first. The model accounts (Fig. 3, right) not only for the spectral energy distribution (SED) of NGC 891 in the FIR/submm regime (Popescu et al. 2000), but also for its observed surface brightness at 170 and 200 μm (Popescu et al. 2004). Furthermore, the model showed that the heating of the dust is done mainly by the UV. Thus, the star formation rate in NGC 891 is directly connected to its infrared emission.

It is, however, possible that the FIR/submm dust emissivity value used for our Galaxy (e.g. Draine & Lee 1984) has been underestimated. Alton et al. (2004), using a simple but reliable model, studied the edge-on galaxies NGC 891, NGC 4013, and NGC 5907 and concluded that the emissivity at 850 μm is about four times the widely adopted value of Draine & Lee (1984). A detailed model by Dasyra et al. (2004) arrived at the same conclusion.

At the moment, the above degeneracy (i.e., more dust or higher emissivity) is not lifted, but the higher emissivity seems to be a real possibility (Dasyra et al. 2004).

3. Correlation between star formation and infrared emission

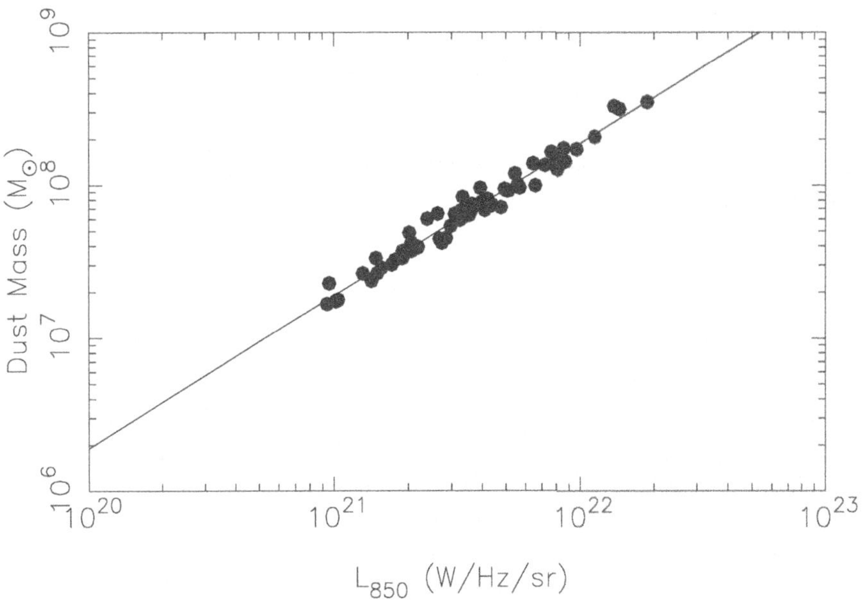

Figure 4. Dust mass as a function of luminosity at 850 μm. The solid line shows the best linear fit (in log-log space).

As we saw above, in order for one to fit the SED of a late-type spiral galaxy, a SFR must be assumed. The question then is: Can a correlation be found between SFR and infrared emission for a sample of late-type spiral galaxies?

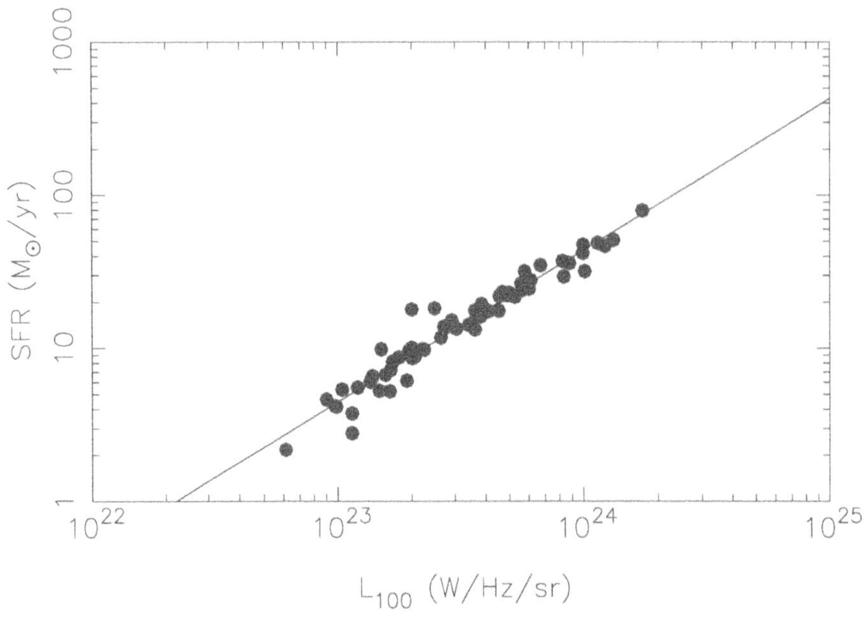

Figure 5. Star formation rate plotted as a function of L_{100}. The solid line is the best linear fit (in log-log space).

The answer is yes and it was given by Misiriotis et al. (2004). They studied 62 bright IRAS galaxies from the SCUBA Local Universe Galaxy Survey of Dunne et al. (2000). Figure 4 shows the derived dust mass of the galaxies as a function of their 850 μm luminosity. The correlation is linear and quite impressive. This is not surprising, because most of the dust in these galaxies (as in the galaxies studied by Xilouris et al. 1999) is cold ($\sim$ 15 K). In Fig. 5, the derived SFR is shown as a function of the 100 μm luminosity. Since the peak of the SED of these galaxies is near 100 μm, it is again not surprising that the correlation is linear and quite tight. In Fig. 6, the sought after correlation between SFR and IR luminosity is shown. The solid line is a linear fit to the data, while the dashed one is the correlation of Kennicutt (1998; see also Buat & Xu 1996). The dot-dashed line shows the Kennicutt correlation if one assumes that the UV to IR transformation efficiency is not 100% (as in starbursts) but 50%.

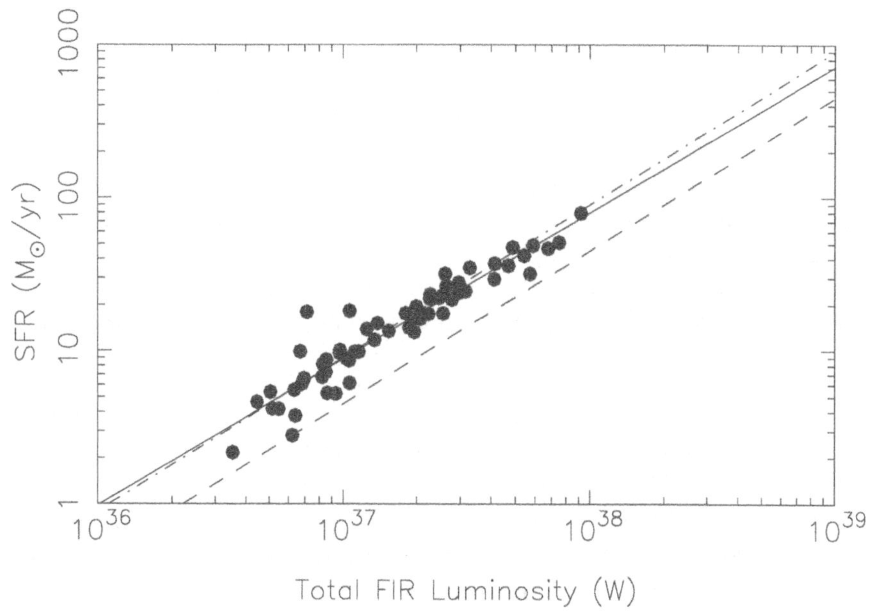

Figure 6. Star formation rate as a function of total FIR luminosity. The solid line is the best linear fit (in log-log space). The dashed line is drawn according to the SFR-L_{FIR} relation of Kennicutt (1998) for starburst galaxies. The dash-dotted line is Kennicut's relation if one 50% rather than 100% transformation efficiency of UV into FIR.

4. Other star formation diagnostics

Several other SFR diagnostics have been used over the years and the question is how well they agree among themselves. The answer now is quite well, though this was not the case a few years ago.

Kewley et al. (2002) showed that the SFR derived from the IR luminosity and that derived from Hα (properly corrected for extinction) are in excellent agreement *for all types of galaxies* (from elliptical, to spiral, to peculiar) and *for SFR ranging over four orders of magnitude*. The reader is also referred to the work of Buat et al. (2002), Hirashita et al. (2003), Panuzzo et al. (2003), and Flores et al. (2004).

The forbidden lines of [OII] have also been used as a measure of SFR. Kewley et al. (2004) showed that the SFR derived from [OII] and that derived from Hα are in excellent agreement for all types of galaxies and for four orders of magnitude of the SFR.

Not surprisingly, the SFR determined from UV observations is in excellent agreement with the SFR determined from the IR (Iglesias-Páramo et al. 2004).

On the other hand, what is somewhat surprising is the fact that the 1.4 GHz luminosity is a good indicator of the SFR. For an explanation of this, the reader is referred to Bressan et al. (2002), but the explanation is not widely accepted (Bell 2003; Pierini et al. 2003). Afonso et al. (2003) showed that the SFR determined from the 1.4 GHz luminosity is in good agreement with the SFR determined from Hα or [OII].

It is remarkable that even for interacting starburst galaxies there is a good correlation between the SFR derived from the Hα flux and the SFR derived from the FIR continuum (Dopita et al. 2002).

Last but not least, we want to mention the work of Förster Schreiber et al. (2004), who used a sample of galaxies containing a) quiescent spiral galaxies, b) spiral galaxies with active circumnuclear regions, c) starburst galaxies, d) LIRGs, and e) ULIRGs and showed that the monochromatic 15 μm continuum and the 5 - 8.5 μm emission constitute excellent indicators of SFR.

We believe that the above are more than convincing that the determination of the SFR in galaxies is a mature subject. Significant progress has also been made in the determination of the SFR history. The most recent work in this subject is that of Heavens et al. (2004), who did an analysis of the 'fossil record' of the current stellar populations of 96,545 nearby galaxies, from which they obtained a complete star-formation history. They found that the peak of star formation was at about five billion years ago, i.e. more recently than other studies had found. They also found that the bigger the stellar mass of the galaxy, the earlier the stars were formed, which indicates that high- and low-mass galaxies have very different histories.

5. Instead of a summary

Instead of a summary, we want to mention just three of the recently reported very exciting results from the Spitzer Space Telescope.

Appleton et al. (2004) reported that the FIR - radio correlation is valid to at least $z = 1$ and similarly for the mid-infrared - radio correlation.

Higdon et al. (2004; see also Charmandaris et al. 2004) reported that the redshift of high-z, faint galaxies can be determined from the 14 - 38 μm spectrum! This opens up tremendous possibilities for the determination of z of faint galaxies.

Houck et al. (2004) reported that the blue compact dwarf galaxy SBS 0335-052, which has very low metallicity ($Z \sim Z_{\odot}/41$), has a featureless mid-infrared spectrum with a peak at $\sim$ 28 μm! Taken at face value, this means that there is no cold dust in this galaxy.

The reader is referred to the Special Issue of ApJS (Volume 154) for additional exciting results from Spitzer.

References

Afonso, J., Hopkins, A., Mobasher, B.& Almeida, C. 2003, ApJ, 597, 269

Alton, P.B., Trewhella, M., Davies, J.I. et al. 1998, A&A, 507, 125

Alton, P.B., Xilouris, E.M., Misiriotis, A., Dasyra, K.M. & Dumke, M. 2004, A&A, 425, 109

Appleton, P.N., Fadda, D.T., Marleau, F.R. et al. 2004, ApJS, 154, 147

Bell, E.F. 2003, ApJ, 586, 794

Bressan, A., Silva, L. & Granato, G. L. 2002, A&A, 392, 377

Buat, V., Boselli, A., Gavazzi, G. & Bonfanti, C., 2002, A&A, 383, 801

Buat, V. & Xu, C. 1996, A&A, 306, 61

Charmandaris, V.; Uchida, K.I.; Weedman, D. et al., 2004, ApJS, 154, 142

Dasyra, K.M., Xilouris, E.M., Misiriotis, A. & Kylafis, N.D. 2004,A&A submitted

Davies, J.I., Alton, P., Trewhella, M., Evans, R. & Bianchi, S. 1999, MNRAS, 304, 495

de Vaucouleurs, G. 1953, MNRAS, 113, 134

Dopita, M.A., Pereira, M., Kewley, L.J. & Capaccioli, M. 2002, ApJS, 143, 47

Draine, B.T. & Lee, H.M. 1984, ApJ, 285, 89

Dunne, L., Eales, S., Edmunds, M., Ivison, R., Alexander, P. & Clements, D.L. 2000, MNRAS, 315, 115

Flores, H., Hammer, F., Elbaz, D. et al. 2004, A&A, 415, 885

Förster Schreiber, N. M., Roussel, H., Sauvage, M. & Charmandaris, V. 2004, A&A, 419, 501

Heavens, A., Panter, B., Jimenez, R. & Dunlop, J. 2004, Natur, 428, 625

Higdon, S.J.U., Weedman, D. & Higdon, J.L., et al. 2004, ApJS, 154, 174

Hirashita, H., Buat, V. & Inoue, A.K. 2003, A&A, 410, 83

Houck, J.R., Charmandaris, V. & Brandl, B.R. 2004, ApJS, 154, 211

Iglesias-Páramo, J., Buat, V., Donas, J., Boselli, A. & Milliard, B. 2004, A&A, 419, 109

Kennicutt, R.C., Jr. 1998, ARA&A, 36, 189

Kewley, L.J., Geller, M. J., Jansen, R.A. & Dopita, M.A. 2002, AJ, 124, 3135

Kewley, L.J., Geller, M.J. & Jansen, R.A. 2004, AJ, 127, 2002

Laor, A. & Draine, B.T. 1993, ApJ, 402, 441

Misiriotis, A., Bianchi, S. 2002, A&A, 384, 866

Misiriotis, A., Kylafis, N.D., Papamastorakis, J. & Xilouris, E.M. 2000, A&A, 353, 117

Misiriotis, A., Papadakis, I. E., Kylafis, N.D. & Papamastorakis, J. 2004, A&A, 417, 39

Misiriotis, A., Popescu, C.C., Tuffs, R. & Kylafis, N.D. 2001, A&A, 372, 775

Panuzzo, P., Bressan, A., Granato, G.L., Silva, L. & Danese, L. 2003, A&A, 409, 99

Pierini, D., Popescu, C.C., Tuffs, R.J. & Völk, H. J. 2003, A&A, 409, 907

Popescu, C.C., Misiriotis, A., Kylafis, N.D., Tuffs, R.J. & Fischera, J. 2000, A&A, 362, 138

Popescu, C.C., Tuffs, R.J., Kylafis, N.D. & Madore, B.F. 2004, A&A, 414, 45

Xilouris, E.M., Byun, Y.I., Kylafis, N.D., Paleologou, E.V. & Papamastorakis, J. 1999, A&A, 344, 868

Xilouris, E.M., Alton, P.B., Davies, J.I., Kylafis, N.D., Papamastorakis, J. & Trewhella, M. 1998, A&A, 331, 894

EVOLUTION OF THE MILKY WAY DISK

B. Nordström
Niels Bohr Institute
Juliane Maries Vej 30, DK – 2100 Copenhagen, Denmark
birgitta@astro.ku.dk

J. Andersen
Niels Bohr Institute
Juliane Maries Vej 30, DK – 2100 Copenhagen, Denmark
&
Nordic Optical Telescope Scientific Association
La Palma, Canarias, Spain
ja@astro.ku.dk

Abstract The Solar neighbourhood is where the physical basis for models of the evolution of spiral galaxy disks can be tested most stringently. A new survey has provided full space motions, metallicities, ages, and duplicity information for over 14,000 nearby F and G dwarfs. The sample is magnitude-limited, volume complete to 40 pc, and without significant bias as regards age, metallicity, and kinematics, and thus complements a variety of detailed spectroscopic studies of selected subsamples of stars. The results show that the local hydrodynamics of star formation and chemical enrichment are far more prominent in the disk than accounted for by traditional chemical evolution models.

Keywords: Galactic disk, Galactic evolution, Galactic dynamics, nearby stars.

1. Introduction

Galaxies are the building blocks of the universe, and an important fraction of them are spirals like the Milky Way. The most significant feature of a spiral galaxy is a *disk*, and we live right in the middle of the disk of the Milky Way itself. From this position we can study the stellar content of a typical galaxy disk more accurately and completely than in any other galaxy – best, of course, in our neighbourhood where the stars are bright and accurate data can be obtained efficiently.

But not all nearby stars took part in the evolution of our neighbourhood; many of them originate elsewhere and just happen to transit through our neigh-

J. C. del Toro Iniesta (eds.), The Many Scales in the Universe, 121–130.

bourhood right now. Thus, the velocities of the nearby stars are as important as their ages and chemical composition for our efforts to understand the chemical enrichment history of the Galaxy - and, further, of the universe.

Classical diagnostics for models of the evolution of the disk are the metallicity distribution function for long-lived stars, and the age-metallicity and age-velocity relations (AMR and AVR) for the Solar neighbourhood, and much observational effort has gone into populating these diagrams with the best possible data. More recently, additional insight has resulted from detailed studies of the evolution of individual element ratios (e.g., Edvardsson et al. 1993 or Fuhrmann 1998).

Ever since the pioneering work of Twarog (1980), the discussion has focused on the shape of the *mean* relation, assuming any scatter around it to be purely due to observational errors. From the mean trends, deductions were made on key features of the evolution (monolithic collapse or mergers, closed system or large-scale in- and outflow of gas; contributions and timescales of supernovae of different types, perturbation of stellar orbits by massive objects in the disk, etc.).

Two caveats must be observed in such work, however: *(i):* Statistical selection biases in the stellar samples, or in the computation of metallicity, age, etc., from the observations, may distort the mean relations and lead to wrong conclusions; and *(ii):* If the scatter of the data points cannot be explained by observational errors alone, the evolution is not described adequately by mean relations alone, and far more complex models are needed. The *Geneva-Copenhagen Survey of the Solar Neighbourhood* (Nordström et al. 2004) was conducted to address these two key issues, in tandem with the more detailed spectroscopic analysis of a small subsample of the stars by Edvardsson et al. (1993).

2. Stellar sample and observational data

Full details of our sample selection, observational material, and derivation of metallicities, ages, space motions, and Galactic orbits have been published (Nordstrom et al. 2004), so only a brief summary is given here.

Stellar sample

In order to study the evolution of the disk through its history, we need stars which are frequent, can become as old as the disk itself and whose ages can be determined observationally with useful accuracy, whose atmospheres reflect their initial chemical composition, and for which space motions can be determined efficiently. Only F and G dwarf stars fulfil these requirements, but the selection of stars must avoid introducing spurious correlations between age, metallicity, and kinematics, which could bias the conclusions of the analysis.

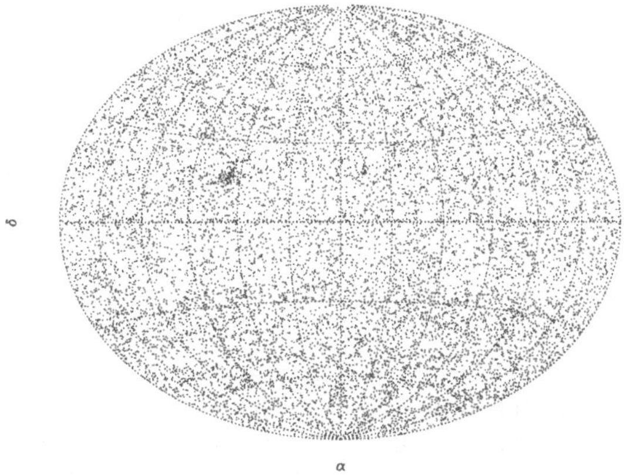

Figure 1. Distribution of the sample on the sky (note the Hyades cluster).

These considerations led us to define an initial sample of all HD stars brighter than apparent magnitude $V \simeq 8.5$ and within a generous range of spectral types. For all of these (∼30,000) stars Olsen (1994 and earlier papers) obtained $uvby\beta$ photometry, from which our final apparent-magnitude limited sample of 16,682 F and G dwarf stars was defined. In the southern hemisphere, spectral types extend to K2 V so as to include any old, metal-poor stars. The sample is shown projected on the sky and in the HR diagram in Figs. 2 - 2.

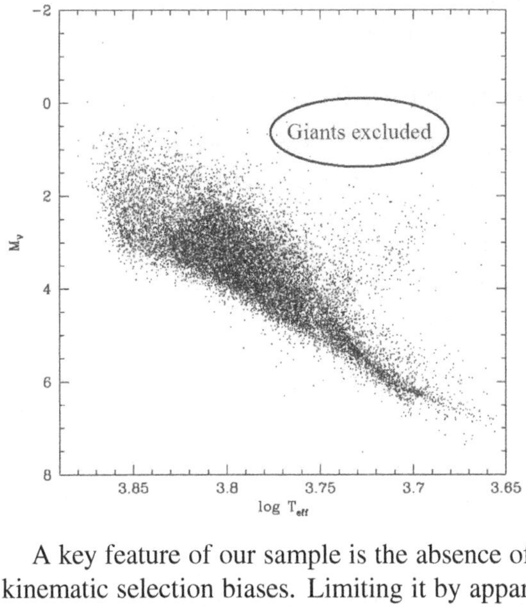

Figure 2. Distribution of the sample in the HR diagram.

A key feature of our sample is the absence of any significant metallicity or kinematic selection biases. Limiting it by apparent magnitude does introduce a strong absolute-magnitude (i.e. mass and/or age) bias, as young, intrinsically bright stars are included from a much larger volume than the fainter stars.

However, the sample is designed to be volume complete to a distance of 40 pc for all stellar types, while the brighter stars reach distances of 100-300 pc.

The new observational data

The $uvby\beta$ photometry also allows us to determine metallicities and ages for the stars. The missing information is then the space motions. Hipparcos (ESA 1997) did provide accurate distances and tangential velocities for $\sim$85% of our stars, but the *radial* component of their space velocities – the radial velocity – was missing. Filling this gap was the main motivation for the *Geneva-Copenhagen Survey*.

During more than 1,000 observing nights in both hemispheres, repeated radial-velocity observations were made of the $\sim$14,000 stars in the sample which also had Hipparcos data. The mean radial velocities, based on an average of 4 observations per star over 1-3 years, have typical mean errors of 0.25 km s^{-1}. The multiple observations also allow us to detect most of the spectroscopic binaries in the sample; if these, or the many visual binaries (also identified), are treated as single, the derived ages, metallicities, and space motions will be unreliable.

3. Derived astrophysical parameters

Here we briefly discuss the transformation from observational data to quantities that can be directly compared with models. Note in particular the new methodology introduced in the age computation.

Metallicities and effective temperatures

Overall metal abundances were derived from the m_1 index, using updated $uvby\beta$ calibrations based on accurate spectroscopic analyses. The resulting values of [Fe/H] have typical mean errors of 0.1 dex.

Effective temperatures were determined from the reddening-corrected $uvby\beta$ indices and the calibration by Alonso et al. (1996). Estimated errors from both photometry and calibration are about 150 K (0.01 dex in $\log T_{eff}$).

Distances, space motions, and galactic orbits

From the subset of stars with Hipparcos distances better than 3%, we find distances determined from the $uvby\beta$ photometry to be accurate to 13%. Accordingly, whenever the error of the Hipparcos parallax exceeds 13%, we use the photometric distance instead. Thus, our stars have absolute magnitudes with errors of 0.26 mag or less.

TYCHO-2 proper motions (Høg et al. 2000) exist for essentially all our stars and correspond to typical errors in the tangential velocities of only 0.7 km s^{-1};

typical radial-velocity errors are even smaller. The total uncertainty of the space motion is then dominated by the parallaxes, but still averages only ~1.5 km s^{-1} in each of *U*, *V*, and *W*.

From their present locations and space motions, we have integrated the Galactic orbits of all the stars backward in time for several orbital revolutions, assuming a smooth, axisymmetric gravitational potential satisfying the usual constraints (the rotation curve and mass density of the disk). Average orbital elements are given in the catalogue.

Ages

Computing ages free of systematic error and with reliably determined uncertainties for disk stars of all ages is a highly non-trivial task. The principle is shown in Fig. 3. Briefly, adopting the Padova stellar evolution models (Girardi et al. 2000, Salasnich et al. 2000) and assuming Gaussian errors in $\log T_{eff}$, M_V, and [Fe/H], we compute for every point in the three-dimensional 'HR cube' (Fig. 3a) the probability that the observed stars could be located there and have the age of the isochrone passing through that point. Integrating these probabilities over the whole isochrone set yields the likelihood distribution function (or 'G function') for the age, shown in Fig. 3b. The peak of the G function indicates the most probable age of the star, and its width at 0.6 of maximum gives the $\pm 1\sigma$ error limits for the age.

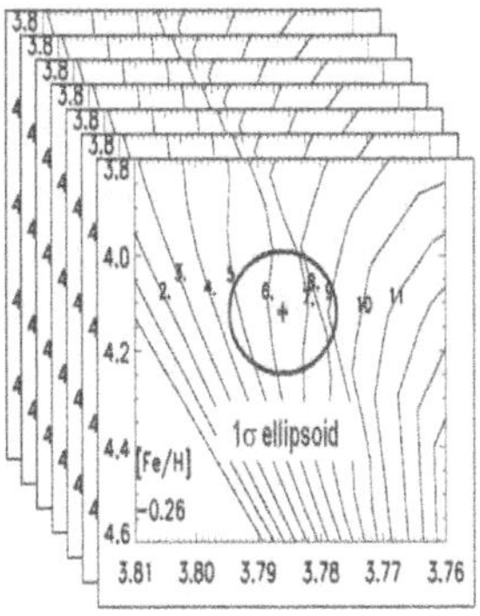

Figure 3a. The 'HR cube' and cuts in the 1-σ error ellipsoid.

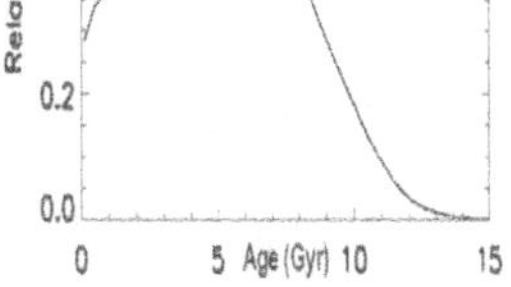

Figure 3b. The G-function resulting from integrating over the cube.

In the computations, we have accounted for the average α-enhancement of disk stars as a function of [Fe/H] (e.g. Edvardsson et al. 1993). We also checked the temperature scale of the models against the unevolved stars in our sample (i.e. those with $M_V >$ +5.5). Agreement is good for stars of Solar or very low metallicity, but we find – and correct - a significant temperature mismatch at intermediate metallicity. Neglecting them would have produced a spurious age-metallicity relation of conventional shape in the just metallicity range comprising most disk stars.

The varying speed of evolution of a star through the HR diagram, the higher frequency of low-mass stars (the IMF), and the metallicity distribution of disk stars tend to introduce biases in the resulting ages; our computational technique corrects for all these effects in a fully Bayesian manner (see Jørgensen & Lindegren 2005 for details). Simulations of artificial stellar samples show that we recover both the ages and their errors correctly, while traditional simplified methods lead to biased mean ages and substantially underestimated errors.

4. The key diagnostic relations

With this new data base, much larger and statistically better defined than ever before, we can reconsider the classical diagnostic relations for models of the evolution of the Solar neighbourhood. In particular, the $\sim$34% of the stars that are binary or multiple systems can, for the first time, be excluded from consideration when necessary.

Metallicity distribution of long-lived stars

The metallicity distribution of stars with lifetimes exceeding the age of the Milky Way disk is a record of the heavy-element production preceding the formation of those stars. The observed lack of the metal-poor dwarfs that, in a closed-box model, should have accompanied the massive stars that produced the heavy elements we see in later generations is commonly known as "The G dwarf problem" (Van den Bergh 1962). It was exhaustively rediscussed, based on our data, by Jørgensen (2000), who concluded that when kinematic selection biases are eliminated, the "G dwarf problem" is even more marked than before; the 'Solar circle' in the disk certainly did not evolve in isolation.

The age-velocity relation

The velocity dispersion of disk stars as a function of time records, on the one hand, the kinematics and shapes of the disk at various epochs, and on the other, the 'kinematic heating' of the disk as its stars are continuously scattered by local mass concentrations while orbiting the Galactic centre.

The age-velocity relation as determined from our sample is shown in Fig. 4. Its slope is large enough to rule out such known scattering agents as stationary spiral arms or giant molecular clouds, but may correspond to heating by stochastic spiral arms (De Simone et al. 2004). The heating effect is also large enough that no traces of very early galaxy mergers in the disk should survive today (Freeman & Bland-Hawthorn 2002), yet too small to explain the scatter

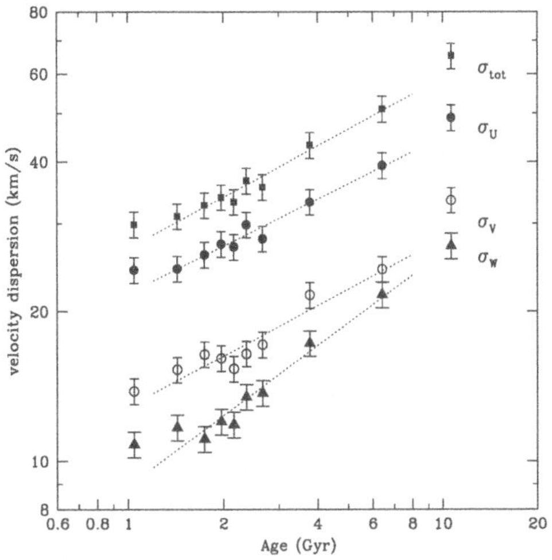

Figure 4. Velocity dispersions for the 2852 single stars with ages better than 25% as functions of age.

in the age-metallicity relation by migration of stellar orbits in the disk (Wielen et al. 1996).

The age-metallicity relation

The age-metallicity diagram summarises the heavy-element enrichment in stars over the lifetime of the disk, and the mean age-metallicity relation (AMR) is one of the most important classical constraints on models of the chemical evolution of galaxies.

Our volume-complete subsample of single stars within 40 pc is ideal for determining the AMR from an unbiased sample. The resulting diagram (Fig. 4a) is arguably our most far-reaching and controversial result: It shows *no* change in mean metallicity over the past $\sim$10 Gyr (large dots show means in bins of equal numbers of stars), but a scatter in [Fe/H] at all ages which greatly exceeds the observational errors.

The crucial differences between our results and such classical studies as Twarog (1980) or Edvardsson et al. (1993; Fig. 4b) are our colour cuts and age determination techniques. By limiting their samples to F and early G dwarfs, these earlier studies excluded *a priori* any old, metal-rich stars which might hide among the cooler stars that we deliberately sought to include. And such stars do in fact exist, as shown e.g. by the old, metal-rich open cluster NGC 6791 (Sandage et al. 2003). As a by-product, the cool dwarfs allowed us to detect temperature errors in the stellar models which, if uncorrected, would have produced a spurious correlation between metallicity and age (the more massive, evolved stars studied by Edvardsson et al. did not allow this test).

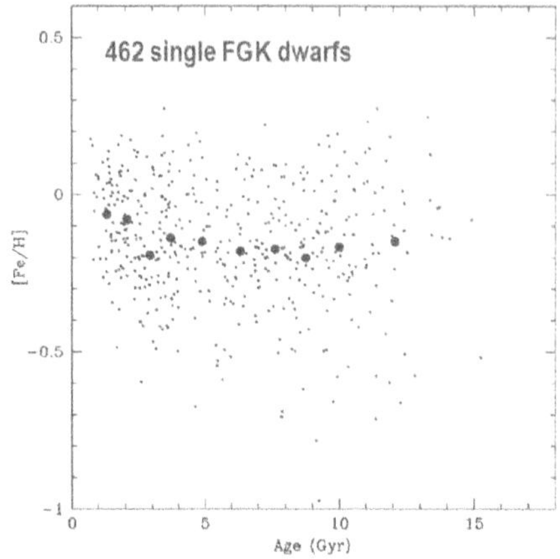

Figure 5a. Age-metallicity diagram for 462 single stars within 40 pc and with well-defined ages.

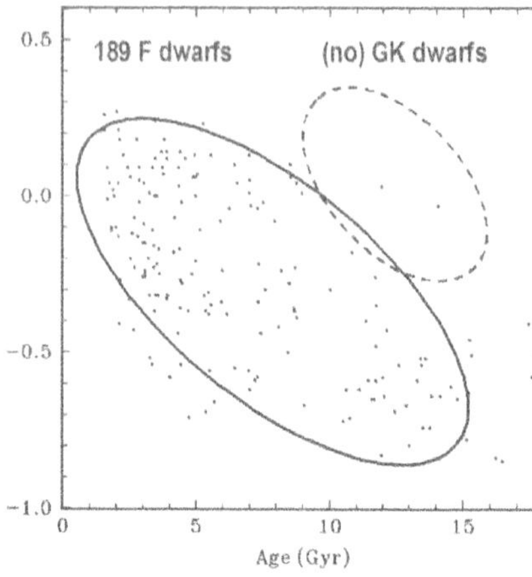

Figure 5b. The corresponding diagram from Edvardsson et al. (1993).

5. Discussion

The new data from the *Geneva-Copenhagen Survey* provide a much better basis for many classical investigations of the evolution of the Galactic disk than available before. Paradoxically, they also show the limits of the underlying paradigm: All classical models predict some form of gradual heavy-element enrichment over the life-time of the disk. Additional model parameters such as pre-enrichment or infall of metal-poor gas may flatten the relation, but still leave some slope and are poorly constrained from independent data.

However, the models confront another, more fundamental obstacle: If the range in metallicity seen at all ages greatly exceeds the change in the mean during the lifetime of the disk, and radial migration of stars cannot be invoked as an explanation, the assumptions of axial symmetry and approximate synchronisation of the chemical evolution in the disk must be abandoned. The real picture of the evolution of the disk of our Galaxy – and thus other spiral galaxies as well – seems to be a far more complex one, with far greater local variations in the rates of star formation and chemical enrichment than have been assumed so far.

Interestingly, this situation is reminiscent of what is found in QSO absorption line systems, where the mean metallicity is uncorrelated with redshift (i.e. time), but depends on the depth of the gravitational potential in which the absorbing clouds are located (Pettini 2004). One wonders if the underlying physics may be similar to that operating in the Milky Way disk.

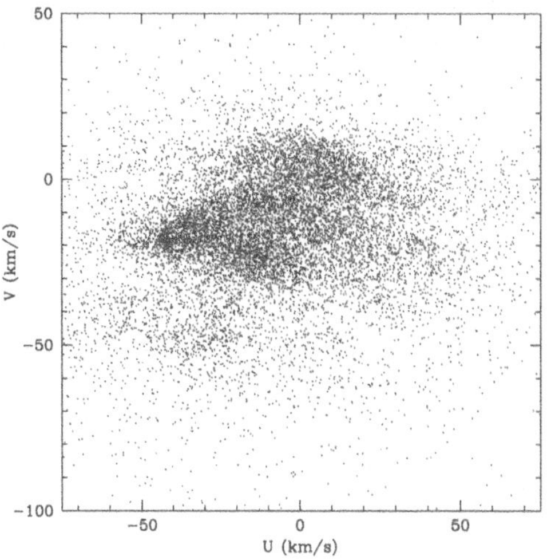

Figure 6. *U,V* diagram for all stars in the sample.

The demise of another popular paradigm is shown in Fig. 5, which bears little resemblance to the superposition of two Gaussian velocity distributions commonly assumed when assigning stars to the thin or thick disk (e.g. Bensby et al. 2003). The bands of stars crossing the diagram do not correspond to classical moving groups, as the stars show a wide range in both age and metallicity, but more likely result from kinematic focusing by spiral arms or a bar structure in the Galaxy. As a consequence, smooth, axisymmetric potentials cannot be used to pursue stellar orbits far back in time (see also Famaey et al. 2004).

6. Summary

Understanding the evolution of the Milky Way disk begins at home. While local in nature and planned on the basis of the classical paradigms for galactic chemical evolution models, the *Geneva-Copenhagen Survey of the Solar Neighbourhood* has highlighted a fundamental limitation in the classical models: The complex details of localised star formation, supernova explosions, and chemical enrichment must be taken into account, not only when studying the halo (e.g. Argast et al. 2000), but also for the disk. Twenty years of work have not made the picture simpler, but certainly much more realistic and interesting!

Acknowledgments

The Geneva-Copenhagen Survey was a joint effort with several colleagues from Geneva, Copenhagen, Lund, and Marseille observatories, the Harvard-Smithsonian Center for Astrophysics, and ESO; we thank them all most cordially. It was made possible by large grants of observing time and financial support from ESO, the Danish Natural Science Research Council, the Fonds

National Suisse pour la Recherche Scientifique, the Carlsberg Foundation, the Swedish Research Council, and the Nordic Academy for Advanced Study, all of which is gratefully acknowledged.

References

Alonso, A., Arribas, S. & Martinez-Roger, C. 1996, A&A, 313, 873

Argast, D., Samland, M., Gerhard, O.E. & Thielemann, F.K. 2000, A&A, 356, 873

Bensby, T., Feltzing, S., & Lundstrom, I. 2003, A&A, 410, 527

De Simone, R.A., Wu, X., & Tremaine, S. 2004, MNRAS, 350, 627

Edvardsson, B., Andersen, J., Gustafsson, B., Lambert, D.L., Nissen, P.E., & Tomkin, J. 1993, A&A, 275, 101

ESA 1997, The Hipparcos and Tycho Catalogues, ESA-SP 1200

Famaey, B., Jorissen, A., Luri, X., Mayor, M., Udry, S., Dejonghe, H. & Turon, C. 2004, A&A, 430, 165

Freeman, K.C., & Bland-Hawthorn, J. 2002, ARA&A, 40, 487

Fuhrmann, K. 1998, A&A, 338, 161

Girardi, L., Bressan, A., Bertelli, G., & Chiosi, C. 2000, A&AS, 141, 371

Høg, E., Fabricius, C., Makarov, V.V., Urban, S., Corbin, T., Wycoff, G., Bastian, U., Schwekendiek, P., & Wicenec, A. 2000, A&A, 355, L27

Jorgensen, B.R. 2000, A&A, 363, 947

Jorgensen, B.R., & Lindegren, L. 2005, A&A, in press

Nordstrom, B., Mayor, M., Andersen, J., Holmberg, J., Pont, F., Jorgensen, B.R., Olsen, E.H., Udry, S., & Mowlavi, N. 2004, A&A, 418, 989

Olsen, E.H. 1994, A&AS, 106, 257

Pettini, M. 2004, in *Cosmochemistry: The Melting Pot of the Elements.* Eds. C. Esteban, R. García Lopez, A. Herrero & F. Sánchez. Cambridge Univ. Press, 257

Salasnich, B., Girardi, L., Weiss, A., & Chiosi, C. 2000, A&A, 361, 1023

Sandage, A., Lubin, L.M., & VandenBerg, D.A. 2003, PASP, 115, 1187

Twarog, B.A. 1980, ApJ, 242, 242

van den Bergh, S. 1962, AJ, 67, 486

Wielen, R., Fuchs, D., & Dettbarn, C. 1996, A&A, 314, 438

MASSIVE STARS IN THE GALACTIC CENTER

F. Najarro
Instituto de Estructura de la Materia, CSIC
Serrano 121, 28006 Madrid, Spain
najarro@damir.iem.csic.es

Abstract We review quantitative spectroscopic studies of massive stars in the Galactic Center clusters. Thanks to the impressive evolution of IR detectors an the new generation of line blanketed models for the extended atmospheres of hot stars we are able to accurately derive the physical properties of the massive stars in these clusters. Our analysis of the Luminous Blue Variables (LBVs) in the Quintuplet cluster provides, for the first time, a direct estimate of α-elements and Fe chemical abundances in these objects. Preliminary results point to a slightly enhanced enrichment of α compared to Fe and suggest an initial mass function dominated by massive stars, as found for the Arches cluster. On the other hand, from our analysis of the Arches cluster, we introduce a new method to estimate metallicity in very young clusters based on the N abundance of WNL stars and the theory of evolution of massive stars. Results indicating solar metallicity are presented.

1. Introduction

The dramatic continuous progress in infrared detectors since the mid-eighties has opened the central parsecs of our galaxy to stellar spectroscopic studies. The detection of a He I emission line cluster (Krabbe et al. 1991) in the central parsec raised the question about the physical nature of its members and their role on the energetics of this region. This discovery triggered a substantial improvement in the atmospheric models for hot stars in the near-infrared, which theory was clearly lagging behind. For the first time, reliable values for luminosities, temperatures, mass-loss rates, helium abundances and ionizing photons were obtained for the brightest members of the He I cluster (Najarro et al. 1994, 1997), which solved the energy puzzle of the central cluster and placed important constraints on the theory of the evolution of massive stars.

Today, we now that the Galactic center is clearly a unique environment in the Galaxy concerning massive star formation. Only molecular clouds with higher densities than those found in the disk will survive the extreme tidal forces in

J. C. del Toro Iniesta (eds.), The Many Scales in the Universe, 131–144.

the center. These clouds will have temperatures about a factor of three higher and strong magnetic fields (Morris 1993, Morris & Serabyn 1996). With a current star formation rate being approximately 250 times higher than the mean rate in the Galaxy, this region is ideal to address crucial questions such as the universality of the initial mass function (IMF) or the maximum mass a star can possess (Figer 2004). In fact, the Galactic center hosts three dense and massive star clusters that have formed in the inner 30 pc within the past 5 Myr. The Central cluster and the Quintuplet and Arches clusters contain over 10% of the known massive stars in the Galaxy. Each cluster has a mass around $\sim 10^4 M_\odot$. The extreme youth (2-2.5Myr, Figer et al. 2002, Najarro et al. 2004) of the Arches cluster allows to address the above questions using photometry alone (Figer 2004), while the Central and Quintuplet clusters being twice older may have lost their most massive members into the supernova stage.

On the other hand, one of the fundamental aspects influencing stellar evolution, namely the metallicity, remains to be addressed. Indeed, the metallicity issue in the galactic center is still a source of controversy. Based on measurements of the gas-phase, Shields & Ferland (1994) obtained twice solar metallicity from Argon and Nitrogen emission lines while a solar abundance was derived for Neon. For the cool stars, and based on LTE-differential analysis with other cool supergiants, Carr, Sellgren & Balachandran (2000) and Ramirez et al. (2000) have obtained strong indications for a solar Fe abundance. Further, Maeda et al. (2002) have recently obtained four times solar abundance fitting the X-ray local emission around Sgr. A East. It is therefore crucial to obtain metallicity estimates from direct analysis of hot stars and confront them with those from the cool-star and gas-phase analyses. Spectroscopic studies of photospheres and winds of massive hot stars are ideal tracers of metal abundances because they provide the most recent information about the natal clouds and environments where these objects formed.

In this paper, we review progress in both infrared observations and quantitative infrared spectroscopy of massive stars in the Quintuplet and Arches clusters, which are allowing us to obtain, for the first time, direct abundance estimates of N, C, O, Si, Mg and Fe in the Galactic Center. We present preliminary results of detailed spectroscopic analyses of the "Pistol Star" and #362 in the Quintuplet cluster, pointing towards solar Fe metallicity and slightly α-elements enrichment in the cluster. We concentrate then on the much younger Arches cluster. We present a method to derive metallicity in such young clusters. The method is based on the study of WNL stars using their N surface abundances. We apply it to the WNLs in the Arches cluster, for which also solar abundance is obtained.

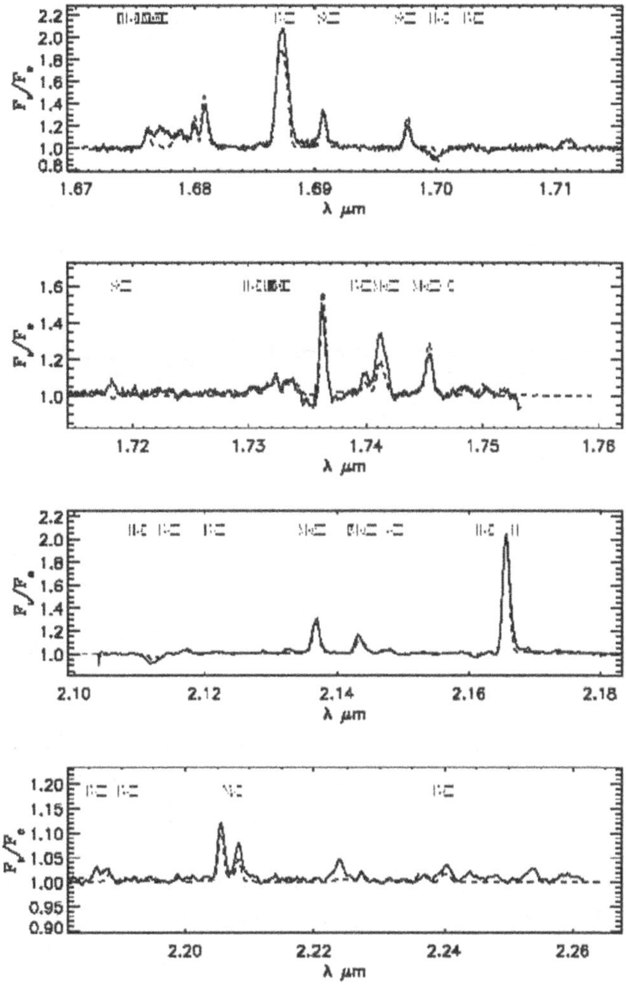

Figure 1. Impact of high S/N and mid-high spectral resolution and new blanketed models for early type stars in the infrared. UKIRT-CGS4 (R∼5000) H- and K-Band observations of the "Pistol Star" in the Quintuplet cluster and new model fits.

2. Improved observations and models

In the field of hot stars, high quality IR-spectra have been obtained during the last decade (Morris et al. 1996, Hanson, Conti & Rieke 1996, Figer, McLean & Najarro 1997, Blum et al. 1997). Most of these spectra were obtained with low-mid (R∼500-2000) resolution, which is enough to classify the stars but insufficient in most of the cases to perform accurate quantitative spectroscopic studies. Examples of mid-high resolution observations of early type

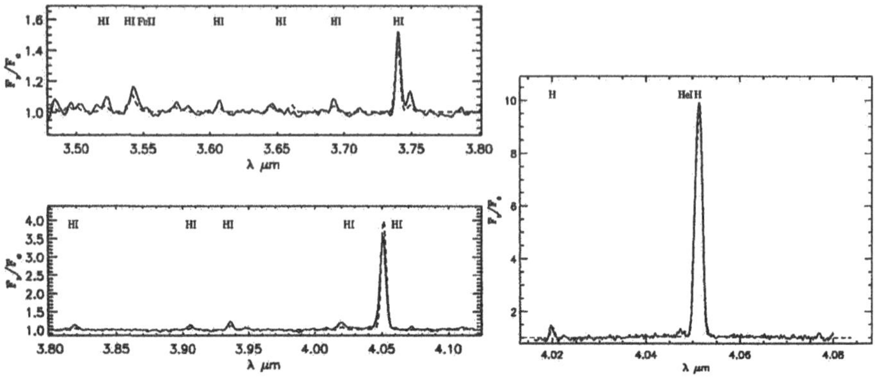

Figure 1. (continued)
L-Band and Brα observations of the "Pistol Star" in the Quintuplet cluster and new model fits.

stars with strong winds are shown in Figs. 1 and 2 which display the spectra at several IR bands of the two LBVs in the Quintuple cluster. The spectra of the "Pistol Star" and star #362 (Geballe et al. 2000) were obtained with UKIRT-CGS4 with a resolution of R$\sim$5000. The number of new observational constraints provided by the new spectroscopic data is striking when compared to previous low-mid observations from which we could gather information only from some H and He lines. Inspecting Figs. 1 and 2 we immediately note that the key diagnostic lines would be fully blurred at R$\sim$1000. The availability of high-quality IR spectroscopic data has been substantially improved with a new generation of IR-spectrographs on 8-m class telescopes (ISAAC, NIRSPEC, etc).

The new model (Hillier & Miller 1998a) is a line blanketing method based on the standard iterative, non-LTE method to solve the radiative transfer equation for the expanding atmospheres of early-type stars in spherical geometry, subject to the constraints of statistical and radiative equilibrium and steady state. The density structure is set by the mass-loss rate and the velocity field via the equation of continuity. We allow for the presence of clumping via a clumping law characterized by a volume filling factor f(r). New species O, Mg, Ca, Si, Na, Al, Fe, etc are included and the blanketing ensures that the effect of continua on lines and lines on the continua as well as overlapping lines are automatically handled (see Hillier & Miller 1998a, 1999 for a detailed discussion of the method). The new model is then prescribed by the stellar radius, R_*, the stellar luminosity, L_*, the mass-loss rate $\dot{M}$, the velocity field, $v(r)$, the volume filling factor f and the abundances of the element considered.

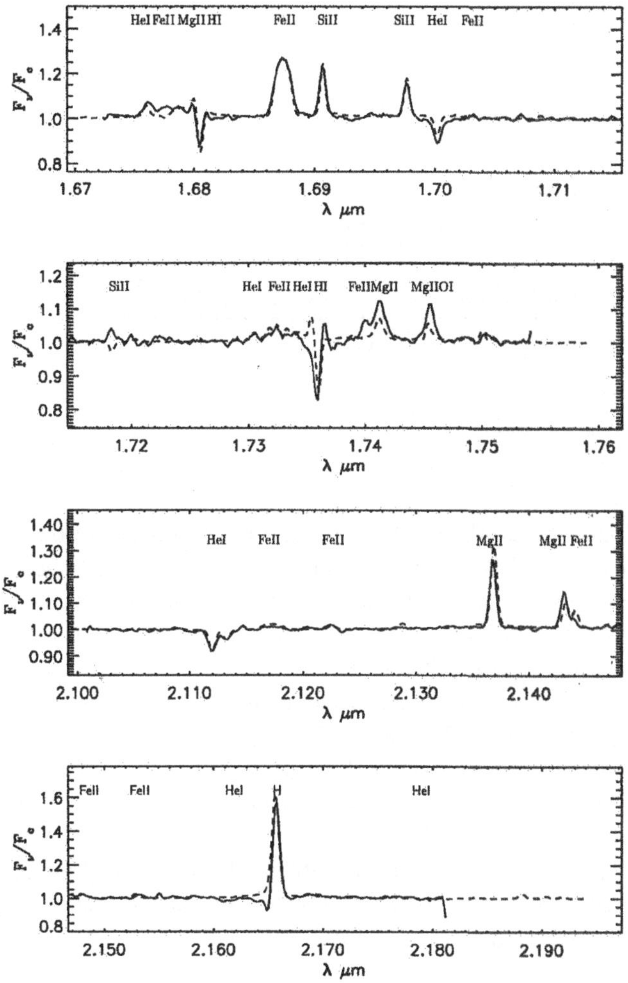

Figure 2. Impact of high S/N and mid-high spectral resolution and new blanketed models for early type stars in the infrared. UKIRT-CGS4 (R$\sim$5000) H- and K-Band observations of star #362 in the Quintuplet cluster and model fits.

3. The quintuplet cluster

The Quintuplet Cluster (Glass, Catchpole & Whitelock 1987, Figer et. al 1999a,b) contains a variety of massive stars, including WN, WC, WN9/Ofpe, LBV and less evolved blue-supergiants. The presence of such stars constrains the cluster age to be about 4Myr old, assuming coeval formation. The cluster provides enough ionizing flux ($\sim 10^{51}$ photons s^{-1}) to ionize the nearby "Sickle" H II region and enough luminosity ($\sim 10^{7.5} L_{\odot}$) to heat the nearby

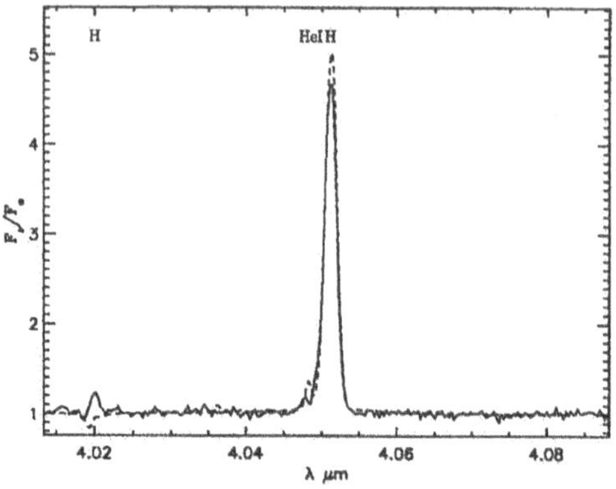

Figure 2. (continued)
Brα observations of star #362 in the Quintuplet cluster and model fits.

molecular cloud, M0.20-0.033. Its total mass is estimated to be $\sim 10^4 M_\odot$. The presence of two LBVs ("Pistol Star" & #362), with IR-spectra rich in metal lines (see Figs. 1 and 2), allows to obtain a direct estimate of the metallicity of the objects and hence constrain the metal enrichment history of the region. Furthermore, using the the Fe II, Si II & Mg II lines (Najarro et. al 2004, in prep.) we can measure the α-elements vs. Fe ratio and infer whether the initial mass function (IMF) is dominated by massive stars or is like in other clusters in the Galaxy with a steeper slope. If the IMF is dominated by massive stars, we should expect enhanced yields of α-elements compared to Fe through a higher than average SNII vs SNIa events.

To model the "Pistol Star" and star #362 we have assumed the atmosphere to be composed of H, He, C, N, O, Si and Fe. Below we briefly discuss preliminary results from a detailed spectroscopic analysis of both objects.

The "Pistol Star" and Star #362

The new blanketed models provide a significant improvement in our knowledge of the physical properties of the "Pistol Star" compared to the results we obtained in Figer et al. (1998) using non-blanketed models. The new model solves the dichotomy between the "high" and "low" luminosity (Teff) solutions in Figer et al. (1998) through the analysis of the metal lines (see the excellent fits in Fig. 1). The Si II, Mg II and Fe II lines "choose" the low luminosity model. We find a luminosity around $1.7510^6 L_\odot$ and a effective temperature of $T_{\rm eff} \sim 11000$K. This result, which reduces the previous estimate of the star luminosity by a factor of two shows the importance of the new generation of models. Given the $T_{\rm eff}$ and high wind density of the object we do find a degeneracy in the H/He ratio (see also Hillier et al. 1998b). In principle, fits of virtually equal quality may be obtained with H/He ratios varying from 10 to 0.05 by number. The only line that may help to break this degeneracy is the He I 2.112μm absorption line, which seems to favor H/He ratios between 3 and 0.05 by number. An important consequence of this degeneracy are the mass fractions derived for Fe, Si and Mg. Our models show that once the H/He ratio falls below one, the resulting metal abundances have to be scaled down. In other words, if H/He$\leq$1 then we will obtain only an upper limit on the metal abundances. We obtain solar iron abundance as upper limit for the "Pistol Star" (see line fits in Fig. 1). This estimate is rather robust as there is a large number of Fe II diagnostic lines. Further, if we assume that the object displays He enrichment consistent with an LBV evolutionary phase, H/He$\geq$1, we may conclude that the "Pistol Star" shows solar Fe abundance. The silicon abundance is obtained from the two Si II lines in the H band. These lines are extremely sensitive to the effective temperature of the star as well as to the transition zone between the star's photosphere and wind. Therefore, our Si$\approx$1.4Si$_\odot$ result should be regarded with some caution. Magnesium, on the other hand, provides more diagnostic lines through Mg II both in the H and

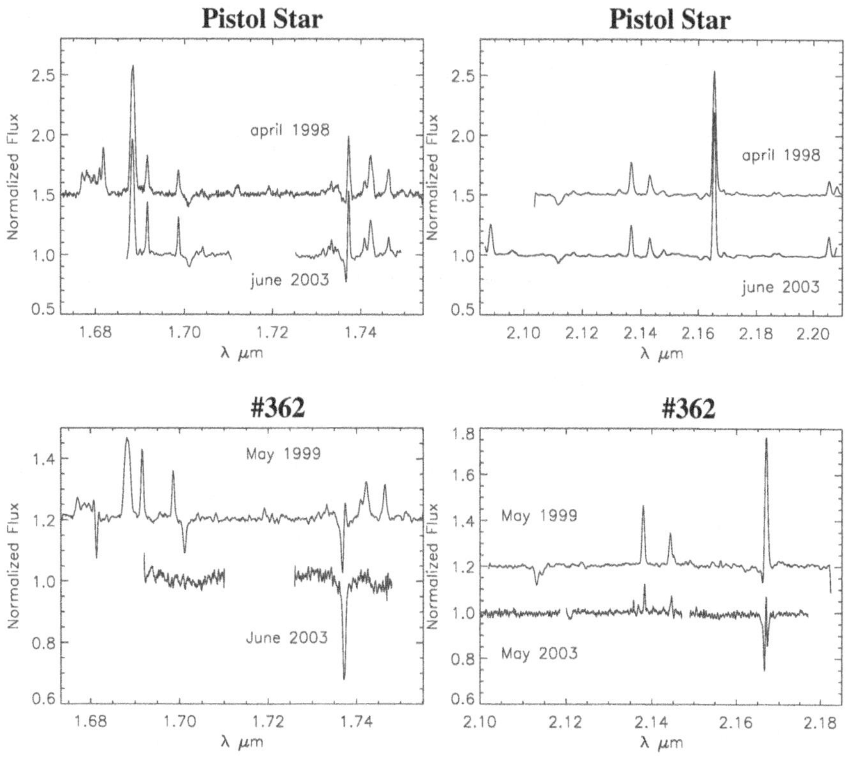

Figure 3. Variability of Pistol star and #362.

K-Band. We regard Mg$\approx$1.6Mg$_\odot$ as our current best estimate. This slightly α-elements vs Fe enrichment seems to favor the situation of an IMF dominated by massive stars in the GC. Finally, we may note that our result of solar Fe abundance agrees with previous estimates from differential analysis of cool stars in the GC (Carr et al. 2000, Ramirez et al. 2000).

From Fig. 2 we see that star #362 is nearly a twin of the "Pistol Star". In fact, our models show the same kind of degeneracy in the H/He ratio as for the "Pistol Star". We also obtain a luminosity around $1.710^6 L_\odot$ and an effective temperature of $T_{\rm eff}\sim$10500K for star #362. We derive a solar iron abundance as upper limit as well and obtain Si$\approx$1.8Si$_\odot$ and Mg$\approx$2.2Mg$_\odot$ in agreement with the slightly α-elements vs Fe enrichment found for the "Pistol Star".

Spectroscopic variability. One of the main characteristics of LBV is the change of their spectral type through their excursions to the red in the HR-diagram. Figure 3 displays H- and K-Band spectra of the "Pistol Star" and star #362 obtained in 1998 (CGS4-data presented in this paper) and 2003 (NIR-SPEC data). The spectra have been degraded to the same resolution. From Fig. 3 we see that while the "Pistol Star" has barely change its spectroscopic appearance (minor changes only in Brγ), star #362 shows a severe change in its spectral type. The He I lines have vanished, the Mg II lines have weakened drastically and the H-Bracket lines have gone into absorption. The 2003 data show that the star has started an excursion to the red in the HR-diagram, cooling down a bit and has expanded reducing significantly the stellar wind density. This result is consistent with photometric data (Figer priv. comm) which show an increase of the K magnitude of the star. The most important point of the 2003 data is given by the possibility to obtain, for the first time, a guess of the gravity of an LBV star, using the Bracket absorption lines (Najarro et al. in prep.).

4. The arches cluster

The Arches Cluster (Figer et al. 1999b, 2002) is the youngest and densest cluster at the Galactic center containing thousands of stars, including at least 160 O stars and around 10 WNLs (WN stars still showing H at their surface, Chiosi & Maeder 1986). The cluster is very young ($\leq$2.5Myr), and the only emission line stars present are WNLs and OIf$^+$ having infrared spectra dominated by H, He I, He II, N III lines. Some have weak C III/IV lines. The cluster gathers all requirements to study the high mass IMF slope and to estimate an upper mass cutoff: individual members can be resolved, large amount of mass in stars, young enough so that most massive members are not pre-supernovae and old enough for its stars to have emerged from their natal cocoons. Figure 4 (from Figer 2004) shows the mass function in the Arches cluster extended to very high masses as measured by Figer et al. (1999a). Its slope appears to

Table 1. WNL and O stars in the Arches Cluster. Object identifiers after Figer et al. (2002). Upper limits are quoted as (:). He/H is ratio by number and other abundances are mass fractions. η is the performance number and Q is the ionizing photons rate in photons/s

	Arches Object				
Parameter	#3	#4	#8	#10	#15
$R_*(R_\odot)$	43.5	39	48	48	29
$L_*(10^5 L_\odot)$	10.3	16.5	18.5	18.5	5.85
$T_{eff}(10^4 K)$	2.79	3.32	3.09	3.07	2.95
He/H	0.50	0.57	0.67	0.33	0.33
Z(He)	0.65	0.68	0.71	0.56	0.56
Z(N)	1.7	1.4	1.6	0.6:	0.6:
Z(C)	0.02:	0.03:	0.02:	0.08:	.15:
$\dot{M}(10^{-6} M_\odot\, yr^{-1})$	21.5	32.5	45.0	4.3	9.11
$v_\infty(km\ s^{-1})$	840	1400	1100	1000:	1000:
Clumping factor f	0.1	0.15	0.08	0.1	0.1
$\eta=\dot{M}v_\infty/(L_*/c)$	0.72	1.35	1.32	0.11	0.75
$LogQ_{H+}$	49.4	49.9	49.9	49.8	49.2
$LogQ_{He+}$	47.3	48.6	48.5	48.4	47.3

be shallow with respect to the Salpeter value favoring the formation of very massive stars. Further, we see that one might expect massive stars up to 500-1000$M_\odot$, yet none are seen beyond $\sim$120$M_\odot$. Figure 4, thus clearly suggests the existence of an upper mass cutoff at $\sim$150$M_\odot$.

Metallicity studies

The absence of late B-supergiants and LBVs prevents one from obtaining direct estimates of the important α-elements vs. Fe metallicity ratio as in the Quintuplet cluster or the Central parsec cluster. Being the youngest cluster at the Galactic center, any hint about its metallicity would constitute our "last-minute" picture of chemical enrichment of the central region in the Milky Way.

To analyze the stars in the Arches cluster we have assumed the atmosphere to be composed of H, He, C, N, O, Si and Fe. Observational constraints are provided by the K-Band spectra of the stars (see Fig. 5) and the narrow band HST/NICMOS photometry (filters $F_{\rm F110W}$, $F_{\rm F160W}$ & $F_{\rm F205W}$) and Pα equivalent width (filters $F_{\rm F187N}$ & $F_{\rm F190N}$). Object identifications are given according to Figer et al. (2002). Below we present the results of our analysis (Najarro et al. 2004).

The reduced spectra and model fits are shown in Fig. 5. The top three spectra correspond to some of the most luminous stars in the cluster. As described in Figer et al. (2002), these are nitrogen-rich Wolf-Rayet stars with thick and fast winds. The bottom two spectra in the figure correspond to slightly less evolved

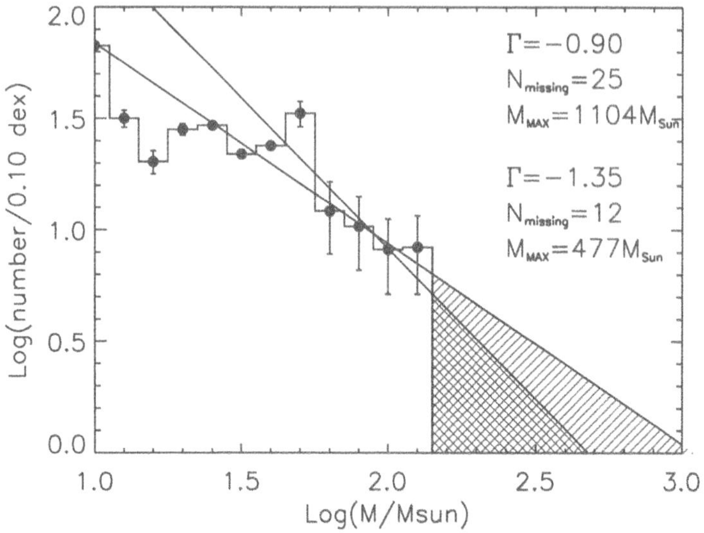

Figure 4. Mass function of the Arches Cluster (Figer 2004). Salpeter mass function is overploted for comparison purposes. The hatched regions demonstrate that one would expect a significant number of massive stars exceeding $120M_\odot$.

stars with the characteristic morphology of OIf^+ stars. The main diagnostic H and He lines in the observed K band portions (see Fig. 5) are Brγ, the He I lines at 2.058 μm, 2.060 μm, 2.112/3 μm, 2.160-66 μm and 2.181 μm and the He II lines at 2.165 μm, 2.189 μm. The main carbon lines which may be used to place upper limits on the carbon abundance are C IV 2.070 μm and C III 8-7 at 2.114 μm. Of concern are the N III 8-7 lines at 2.103 μm and 2.115 μm as well as the N III 5p ^{2}P–5s ^{2}S doublet at 2.247 μm and 2.251 μm. Figer et al. (1997) showed that these N III lines appear only for a narrow range of temperatures and wind densities, which occur in the WN9h (WNL) stage. The fairly distinct nature and energies of the multiplets involved in each of both N III line sets provide strong constraints for the determination of the nitrogen abundance. Thus, at the S/N of our spectra, our models show that the WNLs N III lines can easily track relative changes as low as 20% in the nitrogen abundance, and a 30% error should be regarded as a safe estimate, as shown in Fig. 6-*left*.

Individual stellar parameters of the WNL stars (objects #3,#4 & #8) are listed in Table 1, together with those derived for two OIf^+ stars (#10 & #15) in the cluster. It is important to note that given the extreme sensitivity of the He I and He II line profiles in this parameter domain to changes in effective temperature, we estimate our errors to be below 1000 K. For all objects, the relative strength between the H and He I lines is used to derive the He/H ratio while their absolute strengths provide $\dot{M}$. The stellar parameters displayed in

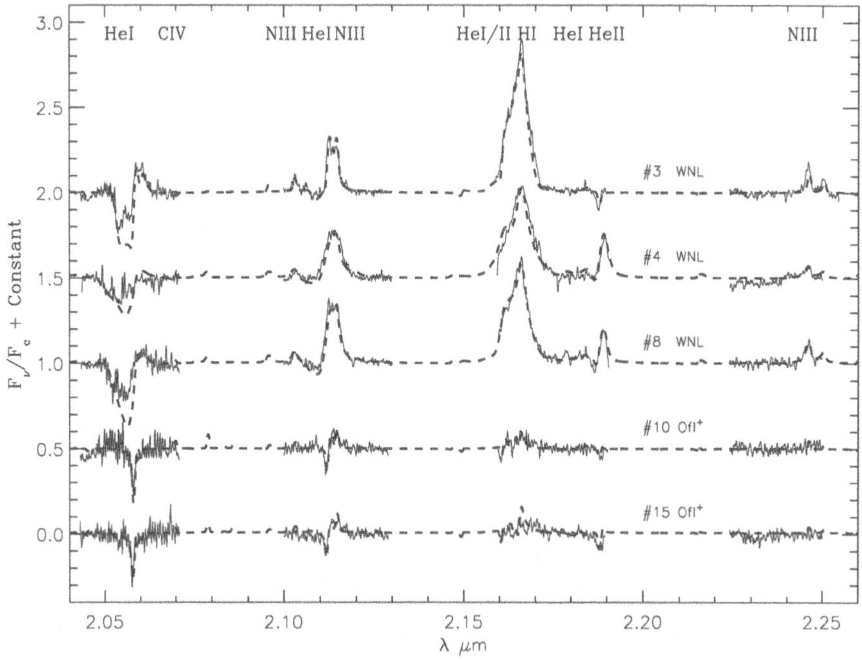

Figure 5. Observed spectra (solid) and model fits (dashed) for the three WNL and two OfI^+ Arches stars.

Table 1 ($\dot{M}$, He and N abundance) reflect the striking difference in morphology between the observed WNL and OIf^+ spectra. Of particular importance is the roughly same surface abundance fraction of N, Z(N), obtained for all three WNL objects ($\sim$1.6%) well above the upper limit found for the OIf^+ stars ($\sim$0.6%). WNL stars do not exhibit any primary diagnostic line in their K-Band spectra to estimate metallicity. However, the crucial rôle of Z(N) in determining metallicity from WNL stars can be immediately anticipated if we make use of the stellar evolution models for massive stars.

According to the evolutionary models by Schaller et al. (1992) and Charbonnel et al. (1993), a star entering the WNL phase still shows H at its surface together with strong enhancement of helium and nitrogen and strong depletion of carbon and oxygen as expected from processed CNO material. Further, once a massive star reaches this phase, it maintains a nearly constant Z(N) value throughout this period, and the amount achieved essentially depends only linearly on the original metallicity (see Fig. 6-*right*). Namely, the maximum Z(N) in the WNL phase is set by the initial content of CNO (sum of C, N and O mass fractions) present in the natal cloud. However, since we expect the CNO

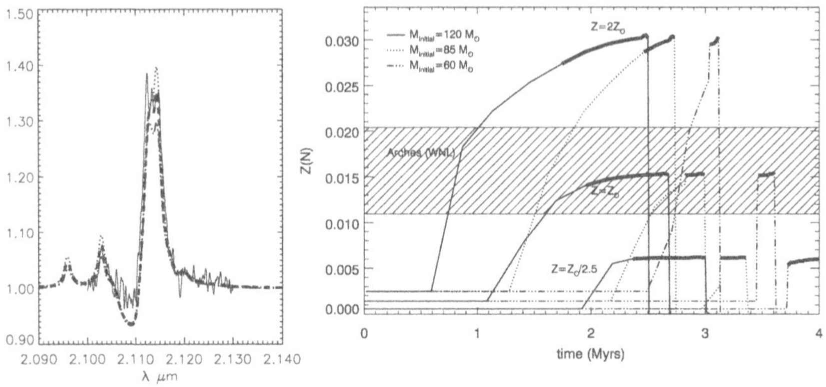

Figure 6. Left. Leverage of error estimates on N abundance. Shown is observed 2.10-2.13 μm region (solid) together with current best fit (dashed) and 30% enhanced (dotted) and 30% depleted (dashed-dotted) nitrogen mass fractions. Right. Nitrogen mass abundance versus time using Geneva models. The measurements require solar metallicity and an age of 2-2.5 Myr.

abundance in the natal cloud to scale as the rest of metals, the nitrogen surface abundance must trace the metallicity of the cluster. Further, this value is basically unaffected by the mass-loss rate assumed and the presence of stellar rotation during evolution (Meynet et al. 2004). Hence, once we clearly identify a star to be on the WNL evolutionary phase, we may confidently use its N surface abundance fraction to estimate metallicity. The He/H ratio, the upper limit on the very depleted carbon abundance and the T_{eff} and $\dot{M}$ values indicate that this is the case for objects #3, #4 & #8. The derived Z(N) ($\sim$1.6%) is the one expected for **solar** metallicity from the evolutionary models. Due to the linear behavior of Z(N) with metallicity, we make use of the uncertainties in the N abundance estimate to obtain an error of roughly 30% for the metallicity (see Fig. 6-*left*). The reliability of our method is demonstrated in Fig. 6-*right*, where we display the nitrogen mass fraction as a function of time for stars with initial masses of 60, 85, and 120 $M_{\odot}$, and metallicities equivalent to 2, 1, and 0.4 times solar, assuming the canonical mass-loss rates (Schaller et al. 1992). The cross-hatched region represents our estimates including the above quoted errors of the nitrogen mass fraction for the Arches WNLs. To test the validity of the nitrogen abundance method to trace metalliticy, we may compare our results with previous abundance estimates for these objects in regions of different metallicity. Indeed, Crowther (2000) found that the nitrogen surface abundance value reached during the WNL phase in the SMC (0.3%) reproduced very well the observed spectra of SK-41 while, Crowther et al. (1995) found surface abundances corresponding to solar metallicity in their analysis of galactic WNL stars.

Another important direct result from Fig. 6 and Table 1 is a tight estimate for the masses and ages of the objects. The luminosities of #4 & #8 are consistent with initial masses of $\sim$120 $M_\odot$ for these WNL stars while the luminosity of #15 seems to be consistent with masses 70-90$M_\odot$ for the O stars. Assuming these masses, equal initial abundances, and coeval formation, one finds that both the WNL and O stars require solar metallicity and an age of 2-2.5 *My* (see also Najarro et al. 2004).

Our result of solar metallicity for the Arches Cluster and Quintuplet Cluster runs counter to the trend in the disk (Rolleston et al. 2000, Smartt et al. 2001) but is consistent with the findings from cool star studies (Carr et al. 2000, Ramirez et al. 2000). This may imply that the ISM in the disk does not extend inward to the GC, or that the GC stars are forming out of an ISM that has an enrichment history that is distinctly different from that in the disk. Our result is more consistent with the values found for the bulge (Frogel et al. 1999, Felzing et al. 2000).

Acknowledgments

I would like to thank Don Figer, John Hillier, Rolf Kudritzki and Tom Geballe for invaluable discussions. F. N. acknowledges PNAYA-2003-02785-E and ESP2002-01627 grants and the Ramon y Cajal program.

References

Blum, R.D, Ramond, T.M., Conti, P.S., Figer, D.F., & Sellgren, K., 1997, AJ, 113, 1855
Carr, J.S., Sellgren, K., & Balachandran, S.C., 2000, ApJ, 530, 307
Charbonnel, C., Meynet, G., Maeder, A., Schaller, G., & Schaerer, D. 1993, A&As, 101, 415
Chiosi, C. & Maeder, A. 1986, Annual Review of A&A, 24, 329
Crowther Crowther, P. A. 2000, A&A, 356, 191
Crowther, P. A., Hillier, D. J., & Smith, L. J. 1995, A&A, 293, 403
Glass, I. S., Catchpole, R. M., & Whitelock, P. A. 1987, MNRAS, 227, 373
Geballe, T.R., Najarro, F., & Figer, D.F., 2000, ApJ, 530, L97
Feltzing, S., & Gilmore, G. 2000, A&A, 355, 949
Figer, D.F., McLean, I.S., & Najarro, F., 1997, ApJ, 486, 420
Figer, D. F., Najarro, F., Morris, M., McLean, I. S., Geballe, T. R., Ghez, A. M., & Langer, N. 1998, ApJ, 506, 384
Figer, D.F., McLean, I.S., & Morris, M., 1999a, ApJ, 514, 202
Figer, D.F., Kim, S.S., Morris, M., Serabyn, E., Rich, R.M., & McLean, I.S., 1999b, ApJ, 525, 750
Figer, D.F., Najarro, F., Gilmore, D., et al., 2002, Morris, M., Kim, S.S., ApJ, 581, 258
Figer, D.F., 2004, in proceedins of "IMF50"
Frogel, J.A., Tiede, G.P., & Kuchinski, L.E. 1999, AJ, 117, 2296
Hanson, M.M., Conti, P.S., & Rieke, M.J., 1996, ApJSS, 107, 281
Hillier, D.J., & Miller, D.L., 1998a, ApJ, 496, 407
Hillier, D.J., Crowther, P.A., Najarro, F., Fullerton, A.W., 1998b, A&A, 340, 483
Krabbe, A., Genzel, R., Drapatz, S., & Rotaciuc, V., 1991, ApJ, 382, L19
Maeda, Y., et. al., 2002, ApJ, 570, 671
Meynet, G., Maeder, A., Schaller, G., Schaerer, D., & Charbonnel, C. 1994, A&A Supp., 103, 97

Meynet, G.,& Maeder, A., 2004, A&A, 404, 975
Morris, M., 1993, ApJ, 408, 496
Morris, M. & Serabyn, E., 1996, ARA&A, 34, 645
Morris, P.W., Eenens, P.R.J., Hanson, M.M., Conti, P.S., & Blum, R.D, 1996, ApJ, 470, 597
Najarro, F., Hillier, D.J., Kudritzki, R.P., Krabbe, A., Genzel, R., Lutz, D., Drapatz, S., & Geballe, T.R., 1994, A&A, 285, 573
Najarro, F., Krabbe, A., Genzel, R., Lutz, D., Kudritzki, R.P., & Hillier, D.J., 1997, A&A, 325, 700
Najarro, F., Figer, D.F., Hillier, D.J., Kudritzki, R.P., 2004, ApJ, 611, L108
Ramirez, S.V., Sellgren, K., Carr, J.S., Balachandran, S.C., Blum, R., Terndrup, D.M., & Steed, A., 2000, ApJ, 537, 205
Rolleston, W.R., Smartt, S.J., Dufton, P.L., & Ryans, R.S.I., 2000, A&A, 363, 537
Schaller, G., Schaerer, D., Meynet, G., & Maeder, A., 1992, A&A, 96, 269
Shields, J.C., & Ferland, G.J., 1994, ApJ, 430, 236
Smartt, S.J., Venn, K.A., Dufton, P.L., Lennon, D.J., Rolleston, W.R., & Keenan, F.P., 2001, A&A, 367, 86

THE MASS SPECTRUM OF X-RAY BINARIES

Jorge Casares
Instituto de Astrofísica de Canarias
38200 La Laguna, Tenerife, Spain
jcv@iac.es

Abstract This review summarizes the observational constraints on the mass spectrum of compact objects in X-ray binaries. We currently have 20 X-ray binaries with confirmed black holes, based on dynamical information (i.e. mass in excess of 3 $M_\odot$). In two cases, V404 Cyg and GRS 1915+105, the black hole mass exceeds the maximum predicted by current Type Ib supernovae models and challenges black hole formation scenarios. The great majority of black hole binaries are members of the class of X-ray Transients, where long periods of quiescence enable spectroscopic studies of the faint donor stars. On the other hand, neutron star binaries are mostly found in persistent binaries, where reprocessed light from the accretion disc overwhelms the companion star and precludes mass estimates. New results, based on the detection of optical fluorescent lines from the donor star and X-ray burst oscillations, provide the best prospects for mass constraints of neutron stars in persistent X-ray binaries.

Keywords: Binaries: close, X-rays: binaries, Stars: neutron, black holes

1. Introduction

Building the mass distribution of compact objects is a fundamental experiment in modern astrophysics which can only be done in X-ray binaries. The interest of this research is two-fold: set constraints on the equation of state of nuclear matter (hereafter EOS) and test models of supernovae explosions and close binary evolution. X-ray binaries are interacting binaries where a normal star transfers matter onto a compact object, a black hole (BH) or neutron star (NS). Matter is accelerated in the strong gravitational field of the compact star and heated up to $\sim 10^7$ K before being accreted. This is the canonical model, first proposed by Shklovskii (1967) to explain the new X-ray sources detected in the 60's and 70's by X-ray satellites such as UHURU or Einstein. In the 80's and 90's, a new generation of higher sensitivity satellites, such as Ginga, Rosat and XTE, revealed a population of several hundred X-ray binaries in the Galaxy with X-ray luminosities in the range 10^{36} – 10^{39} ergs s^{-1} and displaying rapid variability, down to the kilohertz regime. And nowadays, the

J. C. del Toro Iniesta (eds.), The Many Scales in the Universe, 145–154.

large collecting area and high angular resolution of satellites such as Chandra and XMM-Newton allow us to resolve the X-ray binary population in nearby galaxies, obtaining the X-ray luminosity function and proving that they are responsible for typically a third of the galaxy's total X-ray emission.

X-ray binaries are classically divided into two populations, based on the nature of the optical (companion) stars: high mass X-ray binaries (HMXBs) and low mass X-ray binaries (LMXBs). In HMXBs the companion is a hot luminous O-B supergiant whose bolometric luminosity completely dominates the energy spectrum. These binaries are relatively young (short lived) with estimated ages in the range $10^7 - 10^8$ yrs. There are about ~ 100 such binaries in the Galaxy, strongly concentrated towards the spiral arms and, hence, they are considered as good tracers of the star formation rate (Grimm, Gilfanov & Sunyaev 2002). Neutron stars in HMXBs are recognized by the presence of regular X-ray pulsations as a consequence of their strong magnetic fields.

On the other hand, LMXBs contain low-mass companion stars of spectral types (mainly) K-M. These binaries are very compact, with orbital periods clustering around 4-8 hours and concentrated in location towards the galactic bulge. They are associated with the old stellar population and show a large spread in galactic latitude, interpreted as a signature of kick velocities received when the supernova explosion formed the compact star. Neutron stars in LMXBs are also revealed by the exhibition of sporadic X-ray bursts i.e. thermonuclear eruptions of matter slowly accreted over the surface. There are ~ 200 X-ray active or "persistent" sources in the Galaxy, and about 10^3 "transients", which only show X-ray activity occasionally. The reason for this transient behaviour is the mass transfer rate from the donor star $\dot{M}_2$, which is driven by the binary/donor evolution. In X-ray transients, $\dot{M}_2$ is lower than a critical value $\sim 10^{-9}$ $M_{\odot}yr^{-1}$ and this triggers thermal-viscous instabilities in the accretion disc. This causes enhanced mass transfer episodes onto the compact object, the so-called "outbursts", with recurrence times of a few decades (King 1999). In the interim, these systems remain in the "quiescent" state, with typical X-ray luminosities below $\sim 10^{32}$ ergs s^{-1}. It is during these periods when we can attempt to detect the faint low-mass star and extract dynamical information. The many studies performed on quiescent X-ray transients have demonstrated that BHs outnumber NS by more than 70%, and hence the transients provide excellent hunting grounds for BHs. More details about the optical and X-ray properties of galactic X-ray binaries can be found in Charles & Coe (2004), McClintock & Remillard (2004) and Psaltis (2004).

2. Establishing black holes

As opposed to NS, which often show X-ray bursts or pulses, the best observational evidence for a BH is still its mass function $f(M_x)$. This equation relates the masses of the compact object M_x, the companion star M_c (or alternatively the binary mass ratio $q = M_\mathrm{c}/M_\mathrm{x}$) and the binary inclination angle i with two quantities to be extracted directly from the radial velocity curve of the companion star: the orbital period P_orb and the radial velocity semi-amplitude K:

$$f(M) = \frac{K^3 P_\mathrm{orb}}{2\pi G} = \frac{M_\mathrm{x}^3 \sin^3 i}{(M_\mathrm{x} + M_\mathrm{c})^2} = \frac{M_\mathrm{x} \sin^3 i}{(1+q)^2} \quad (1)$$

Since $M_\mathrm{c} > 0$ and $0 < i < 90°$ it is straightforward to show that $f(M)$ is a lower limit on M_x. Therefore, a mass function larger than 3 $M_\odot$, i.e. the maximum gravitational mass of a NS (Rhoades & Ruffini 1974), is taken as a secure proof for the existence of a BH, independently of the actual values of i and q.

Table 1 presents an updated list of confirmed BHs based on this simple dynamical argument. We currently have 20 BHs, with orbital periods ranging from 4.1 hours to 33.5 days. There are 17 transient LMXBs and only 3 persistent systems, the HMXBs Cyg X-1 and two sources from the Large Magellanic Clouds: LMC X-1 and LMC X-3. Note that, in addition to the three HMXBs, six transient LMXBs have mass functions < 3 $M_\odot$. However, we have solid constraints on the inclination and the companion's mass for these binaries which result in M_x > 3 $M_\odot$. The last one added to the list, BW Cir, contains an evolved G5 donor star in a 2.5 day orbit. Its optical luminosity places the binary at a distance $\geq$ 27 kpc and makes BW Cir the furthest BH binary in the Galaxy yet. Remarkably, its large systemic velocity (103 km s^{-1}) is in good agreement with the projected velocity of the Galactic differential rotation at that distance (see more details in Casares et al. 2004a).

In addition to the mass function, one obviously needs a knowledge of the inclination and the mass ratio to fully determine the masses of the two stars. In the case of a BH binary, we face a "single-line spectroscopic binary" problem and all the information must be extracted from the optical companion. However, there is a complete solution to the problem which involves: (i) the determination of the binary mass ratio through the rotational broadening $V_\mathrm{rot} \sin\, i$ of the companion's absorption lines and (ii) the determination of the inclination angle by fitting synthetic models to ellipsoidal lightcurves. This is the classic method to derive masses. Further details of these techniques and the systematics involved can be found in several review papers e.g. Casares (2001), Charles & Coe (2004).

Following this prescription, we currently have 15 reliable BH masses which are listed in the last column of Table 1, updated from the compilations in Orosz (2003) and Charles & Coe (2004). Figure 1 plots these masses with their 1σ errorbars. BH masses spread between 4 and 14 $M_\odot$, with typical uncertainties in the range 5-33%. Also in Fig. 1 we show 33 well determined NS masses,

Table 1. Confirmed black holes and mass determinations

System	P_{orb} *(days)*	*f(M)* $(M_\odot)$	*Donor Spect. Type*	*Classification*	M_x $(M_\odot)$
GRS 1915+105	33.5	9.5 ± 3.0	K/M III	LMXB/Transient	14 ± 4
V404 Cyg	6.470	6.08 ± 0.06	K0 IV	"	12 ± 2
Cyg X-1	5.600	0.244 ± 0.005	09.7 Iab	HMXB/Persistent	10 ± 3
LMC X-1	4.229	0.14 ± 0.05	07 III	"	>4
XTE J1819-254	2.816	3.13 ± 0.13	B9 III	LMXB/Transient	7.1 ± 0.3
GRO J1655-40	2.620	2.73 ± 0.09	F3/5 IV	"	6.3 ± 0.3
BW Cir [a]	2.545	5.75 ± 0.30	G5 IV	"	> 7.8
GX 339-4	1.754	5.8 ± 0.5	–	"	
LMC X-3	1.704	2.3 ± 0.3	B3 V	HMXB/Persistent	7.6 ± 1.3
XTE J1550-564	1.542	6.86 ± 0.71	G8/K8 IV	LMXB/Transient	9.6 ± 1.2
4U 1543-475	1.125	0.25 ± 0.01	A2 V	"	9.4 ± 1.0
H1705-250	0.520	4.86 ± 0.13	K3/7 V	"	6 ± 2
GS 1124-684	0.433	3.01 ± 0.15	K3/5 V	"	7.0 ± 0.6
XTE J1859+226 [b]	0.382	7.4 ± 1.1	–	"	
GS2000+250	0.345	5.01 ± 0.12	K3/7 V	"	7.5 ± 0.3
A0620-003	0.325	2.72 ± 0.06	K4 V	"	11 ± 2
XTE J1650-500	0.321	2.73 ± 0.56	K4 V	"	
GRS 1009-45	0.283	3.17 ± 0.12	K7/M0 V	"	5.2 ± 0.6
GRO J0422+32	0.212	1.19 ± 0.02	M2 V	"	4 ± 1
XTE J1118+480 [c]	0.171	6.3 ± 0.2	K5/M0 V	"	6.8 ± 0.4

[a] The 1-year alias period at 2.564 days is equally significant. In this case the BH would be even strengthen with $f(M_x) = 6.60 \pm 0.36$ $M_\odot$ (see Casares et al. 2004a).
[b] Period is uncertain, with another possibility at 0.319 days (see Zurita et al. 2002). This would drop the mass function to $f(M_x) = 6.18$ $M_\odot$.
[c] Mass function updated after Torres et al. (2004).

extracted from Stairs (2004) and Lattimer & Prakash (2004), also with 1σ uncertainties. The most precise NS masses have been measured in a group of binary radio-pulsars. They are descendants of HMXBs, composed of two young pulsars whose orbits are known to great accuracy from pulse time delays. Relativistic effects lead to NS mass determinations with exquisite accuracy and they display a normal distribution centered at the canonical value of 1.35 $M_\odot$ with a very small dispersion of ± 0.04 $M_\odot$. Dynamical masses are also available from pulsing NS in seven HMXBs, six of which are eclipsing. However, the uncertainties are much larger because the radial velocities of the optical stars are distorted by non-Keplerian perturbations, caused by their strong winds. A few measurements also exist for LMXBs and binary millisecond pulsars (BMPs). BMPs are descendants of LMXBs, composed of a millisecond NS (spun up by accretion) and a detached white dwarf. Both LMXBs and BMPs are potential sites for massive NS because binary evolution predicts accretion of several tenths of a solar mass during their lifetimes.

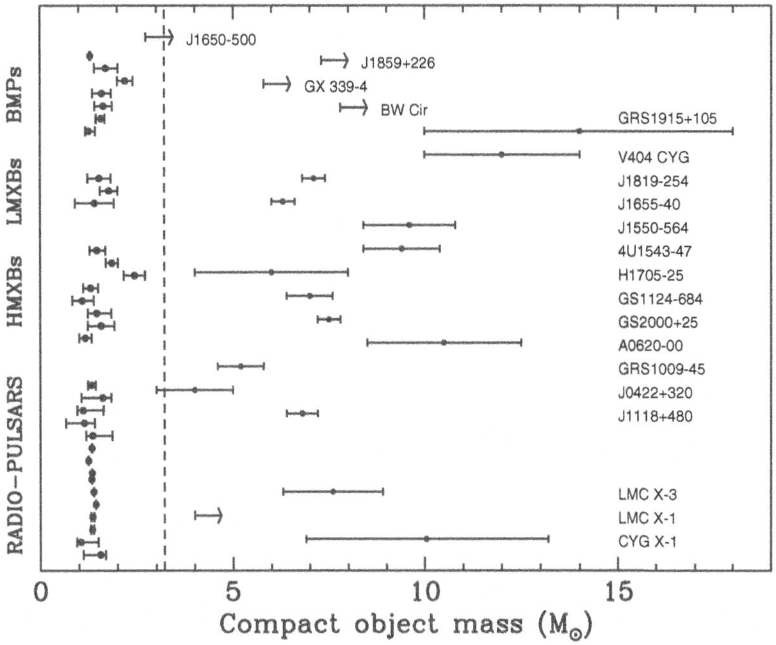

Figure 1. Mass distribution of compact objects in X-ray binaries. The arrows indicate lower limits to BH masses. The dotted vertical line marks the maximum mass allowed for a stable NS.

A fundamental result for the understanding of nuclear matter would be to find a NS more massive than 1.6 $M_\odot$ since this would rule out soft EOS (Brown & Bethe 1994). Currently the best candidates are found in the LMXB Cyg X-2 (1.78 $\pm$ 0.23 $M_\odot$: Casares, Charles & Kuulkers 1998, Orosz & Kuulkers 1999), the BMP J0751+1807 (2.2 $\pm$ 0.2 $M_\odot$: Nice, Splaver & Stairs 2004) and the HMXBs 4U 1700-37 (2.44 $\pm$ 0.27 $M_\odot$: Clark et al. 2002) and Vela X-1 (1.86 $\pm$ 0.16 $M_\odot$: Barziv et al. 2001). The latter case, however, is less secure since the radial velocity curve of the companion is affected by systematic excursions which prevents a confirmation of the mass estimate (see Barziv et al. 2001). In the case of J0751+1807 and 4U 1700-37, soft EOS are ruled out even at the 95% level.

Very recently, large NS masses have also been reported for 2S0921-630 (=V395 Car) by two different groups: 2.0–4.3 $M_\odot$ (Shahbaz et al. 2004) and 1.9–2.9 $M_\odot$ (Jonker et al. 2004). However, the compact object in this LMXB has never shown any evidence for X-ray bursts or pulsations, so it could be a low-mass BH. This is also the situation for the compact object in 4U 1700-37, since it has never shown any NS signature.

Figure 2 presents the histogram of compact object masses compared to the theoretical distribution of remnants computed in Fryer & Kalogera (2001) for

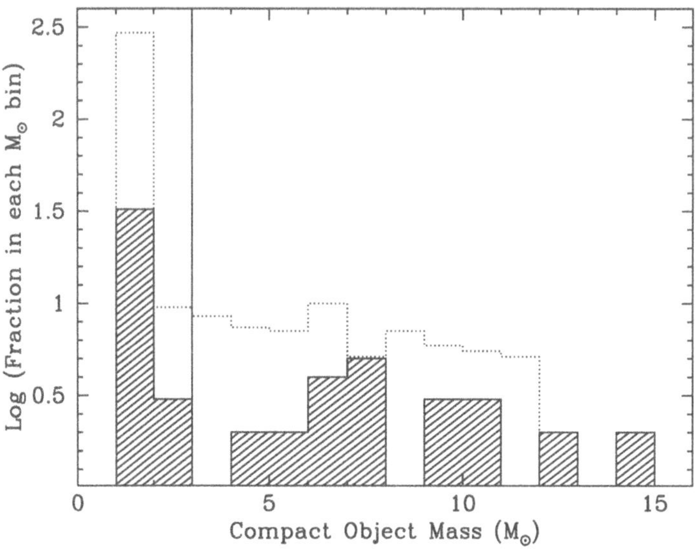

Figure 2. Observed mass distribution of compact objects in X-ray binaries (shaded histogram), compared to the theoretical distribution computed in Fryer & Kalogera (2001) for "Case C + Winds" scenario (dotted line). The model distribution has been re-scaled for clarity.

the case of binary interaction under Case C mass transfer (i.e. Common Envelope evolution after core helium ignition) and mass loss through winds in the Wolf-Rayet phase. This is the most realistic scenario since evolution through Case B mass transfer (i.e. Common Envelope and H envelope removal before core helium ignition) fails to produce BH remnants $> 3\ M_{\odot}$ with the current Wolf-Rayet mass loss rates (Woosley, Langer & Weaver 1995). The model predicts a BH mass cut at 12 $M_{\odot}$ which is difficult to reconcile with the high masses measured in V404 Cyg ($12 \pm 2\ M_{\odot}$) and GRS 1915+105 ($14 \pm 4\ M_{\odot}$). Despite the large uncertainties in the masses, these two X-ray binaries seem to pose a challenge to BH formation theories and, in particular, suggest that mass loss rates in the WR phase are overestimated.

Our histogram also shows a shortage of objects at 3–4 $M_{\odot}$, not predicted by the model distribution. If the gap is eventually confirmed, it could strongly restrict the supernova explosion energy since, as explained in Fryer & Kalogera (2001), it can be reproduced by a step function dependence with the progenitor's mass. However, selection effects could also be playing a role here since low mass BHs are likely to show up as persistent X-ray sources, where dynamical masses are difficult to obtain. The case of 2S0921-630 could well be an example. Obviously we are limited by low number statistics in the observed distribution of compact remnants. Clearly more X-ray transient discoveries and lower uncertainties in the mass determinations are required before these issues can be addressed and the form of the distribution can be used to constrain supernova models and X-ray binary evolution.

3. Mass determination in persistent LMXBs

So far we have been dealing with mass determination in quiescent X-ray binaries. Dynamical studies in persistent LMXBs are, on the other hand, hampered by the huge optical luminosity of the accretion disc. This is driven by reprocessing of the powerful (Eddington limited) X-ray luminosity and completely swamps the spectroscopic features of the faint companion stars.

New prospects for mass determination have been opened by the discovery of high-excitation emission lines arising from the donor star in Sco X-1 (Steeghs & Casares 2002). The most prominent are found in the core of the Bowen blend, namely the triplets NIII λ4634-40 and CIII λ4647-50. In particular, the NIII lines are powered by fluorescence resonance which requires seed photons of HeII Ly-α. These narrow components move in phase with each other and are not resolved (FWHM=50 km s^{-1} i.e. the instrumental resolution), an indication that the reprocessing region is very localized (Fig. 3). The extreme narrowness rules out the accretion flow or the hot spot and points to the companion star as the reprocessing site.

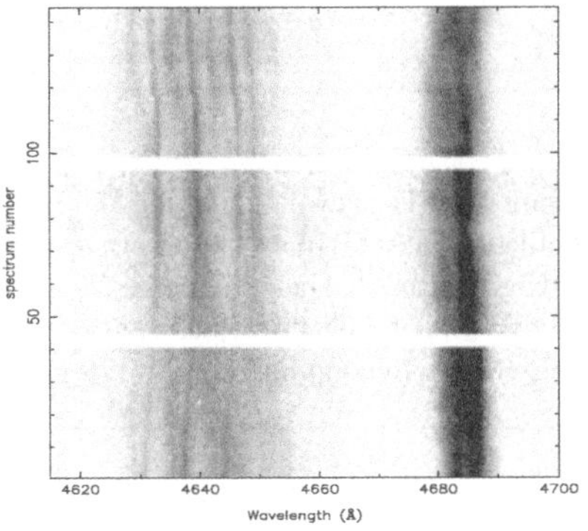

Figure 3. Trailed spectra of the narrow CIII+NIII Bowen emission lines and HeII λ4686 in Sco X-1. After Steeghs & Casares (2002).

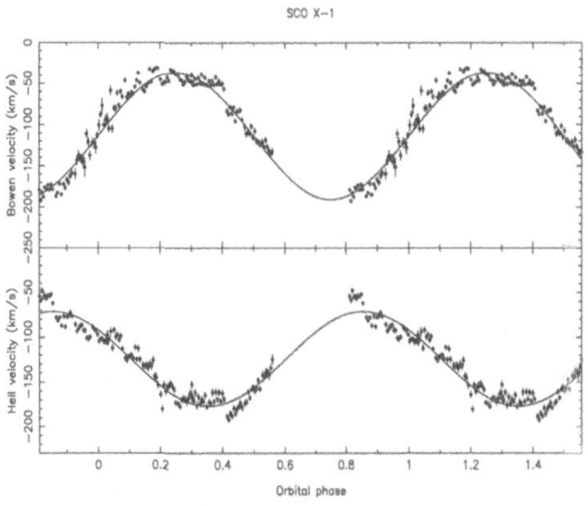

Figure 4. Radial velocity curve of the sharp CIII+NIII Bowen lines (top) and the wings of HeII λ4686 (bottom) in Sco X-1.

The radial velocity curve of the donor star can be extracted through a combined multigaussian fit to the three CIII/NIII lines (top panel of Fig. 4) or using more sophisticated Doppler tomography techniques (see e.g. Casares et al. 2003). Furthermore, the velocities are in antiphase with the wings of the HeII λ4686 emission, which approximately trace the motion of the compact star (bottom panel of Fig. 4). This work represents the first detection of the companion star in Sco X-1 and opens a new window for extracting dynamical information and deriving mass functions in the population of ~20 LMXBs with established optical counterparts.

Follow-up campaigns, using the AAT, NTT and VLT telescopes, have enabled us to extend this analysis to other fainter LMXBs leading to the detection of the secondary stars in 2A1822-371, MXB1636-536, MXB1735-444 and XTE J1814-338 and the determination of their orbital velocities, which lie in the range 200-300 km s^{-1} (see Casares et al. 2003, Casares et al. 2004b). In addition, the application of this technique to the BH candidate GX339-4 during its 2002 outburst provided the first determination of its mass function and hence dynamical proof that it is a BH (Hynes et al. 2003).

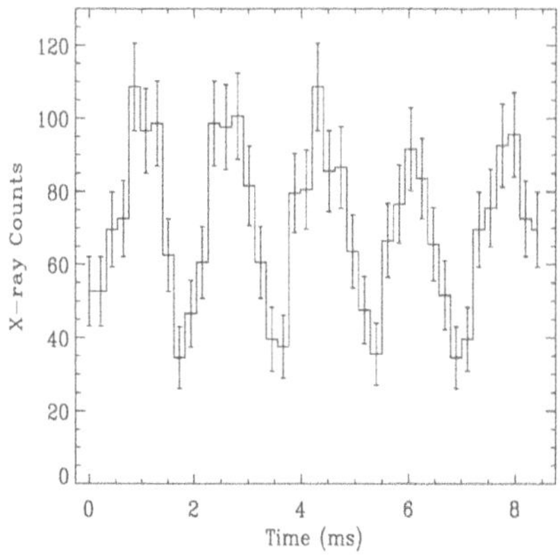

Figure 5. Millisecond pulses detected during the rise of an X-ray burst in MXB1636-536 (after Strohmayer et al. 1998).

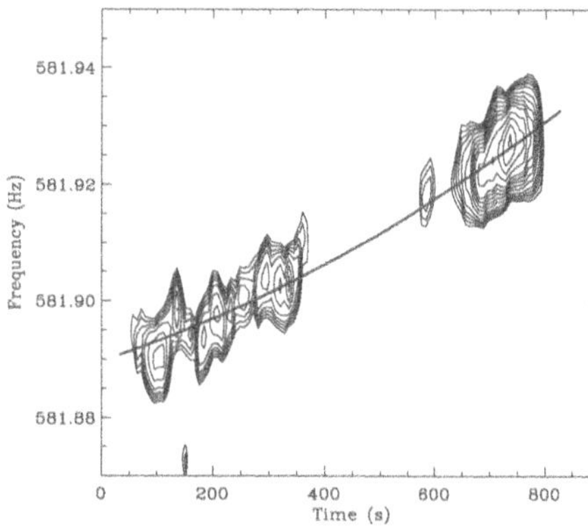

Figure 6. Time evolution of the 580 Hz oscillation in MXB1636-536 during a superburst episode. The solid curve shows the best orbital model fit which constrains the NS velocity amplitude to 90–175 km s^{-1}.

LMXBs are considered to be the progenitors of BMPs because they provide a mechanism for spinning up NS to millisecond periods, through the sustained accretion of matter with high angular momentum during their long active lives. Despite intensive efforts over 2 decades, the detection of millisecond pulsations in LMXBs proved elusive but the advent of XTE changed things dramatically with the discovery of: (i) persistent pulses in 5 transient LMXBs with spin periods in the range 185-435 Hz (ii) nearly coherent oscillations during X-ray bursts in 13 LMXBs. Figure 5 presents an example of a train of oscillations, with a frequency of 580 Hz, detected during an X-ray burst in MXB1636-536. In two LMXBs, SAX J1808-3658 (Chakrabarty et al. 2003) and XTE J1814-338 (Strohmayer et al. 2003), burst oscillations were detected in addition to persistent pulses and they showed identical frequencies. This confirmed that burst oscillations are indeed modulated with the spin of the NS.

Moreover, burst oscillations were detected in an 800 s interval during a superburst in MXB1636-536. The oscillations showed a clear frequency drift which was attributed to the orbital Doppler shift of the NS (see Fig. 6). A circular orbit model was fitted to the data and used to constrain the projected NS velocity to between 90 and 175 km s^{-1} (Strohmayer & Markwardt 2002). This is a remarkable result which shows that burst oscillations can be used to trace NS orbits and, in combination with information provided by the Bowen fluorescent lines, turn persistent LMXBs into double-lined spectroscopic binaries.

4. Conclusions

In the past 15 years the field of X-ray binaries has experienced significant progress with the discovery of 16 new BHs and 6 millisecond pulsars in LMXBs. Reliable mass determinations have been provided for 15 BHs and 33 NS which are starting to reveal the mass spectrum of compact remnants. Two X-ray binaries, V404 Cyg and GRS 1915+105, contain BHs too massive to be explained by current theoretical modelling. However, better statistics are required in order to use the observed distribution to constrain fundamental parameters of X-ray binary evolution and supernova models, such as the mass-loss rate in the W-R phase, the explosion energy dependence on progenitor mass, the amount of fall back or details of the Commom Envelope phase. The discovery of fluorescence emission from the companion star, together with X-ray burst oscillations has opened the door to derive NS masses in persistent X-ray binaries. This new technique, which will benefit from new instrumentation on large telescopes (e.g. OSIRIS on GTC), will likely provide further evidence for the existence of massive NS in LMXBs.

Note added. Two new NS spin periods were discovered when finishing writing this contribution: burst oscillations at 45 Hz in the persistent LMXB EXO 0748-676 (Villarreal & Strohmayer 2004) and persistent pulsations at 598.9 Hz in the new transient LMXB IGR J00291+5934 (Markwardt, Swank & Strohmayer 2004), discovered by INTEGRAL in December 2004.

Acknowledgments

The would like to acknowledge helpful comments from my colleagues Phil Charles, Danny Steeghs and Tariq Shahbaz. I'm also grateful for support from the Spanish MCYT grant AYA2002-0036 and the programme Ramon y Cajal.

References

Barziv, O., Kaper, L., van Kerkwijk, M., Telting, J.H. & van Paradijs, J. 2001, A&A, 377, 925

Brown, G.E. & Bethe, H.A. 1994, ApJ, 423, 659

Casares, J. 2001, in *Binary Stars: Selected Topics on Observations and Physical Processes*, Eds. F.C. Lazaro & M.J. Arevalo, LNP 563, p. 277

Casares, J., Charles, P.A. & Kuulkers, E. 1998, ApJ, 493, L39

Casares, J., Steeghs, D., Hynes, R.I, Charles, P.A. & O'Brien, K. 2003, ApJ, 590, 1041

Casares, J., Zurita, C., Shahbaz, T., Charles, P.A. & Fender, R. P. 2004, ApJ, 613, L133

Casares, J., Steeghs, D., Hynes, R.I., Charles, P.A., Cornelisse, R. & O'Brien, K. 2004, Rev. Mex. AA, 20, 21

Chakrabarty, D., Morgan, E.H., Muno, M.P., Galloway, D.K., Wijnands, R., van der Klis, M. & Markwardt, C.B. 2003, Nature, 424, 42

Charles, P.A. & Coe, M.J. 2004, in *Compact Stellar X-ray Sources*, Eds. W.H.G. Lewin & M. van der Klis, CUP (astro-ph/0308020)

Clark, J.S, Goodwin, S.P., Crowther, P.A., Kaper, L., Fairbairn, M., Langer, N. & Brocksopp, C. 2002, A&A, 392, 909

Fryer, C.L. & Kalogera, V. 2001, ApJ, 554, 548

Grimm, H.J., Gilfanov, M. & Sunyaev, R. 2002, A&A, 391, 923

Hynes, R.I., Steeghs, D., Casares, J., Charles, P.A. & O'Brien, K. 2003, ApJ, 583, L95

Jonker, P.G., Steeghs, D., Nelemans, G. & van der Klis, M. 2004, MNRAS, tmp, 583

King, A.R. 1999, Phys. Rev., 311, 337

J.M. Lattimer & M. Prakash 2004, Science, 304, 536

Markwardt, C.B., Swank, J.H. & Strohmayer. T.E. 2004, ATel, 353

McClintock, J.E. & Remillard, R.A. 2004, in Compact Stellar X-ray Sources, Eds. W.H.G. Lewin & M. van der Klis, CUP (astro-ph/0306213)

Nice, D.J., Splaver, E.M. & Stairs, I.H. 2004, in *IAU Symp. No 218 "Young Neutron Stars and Their Environments"*, Eds. F. Camilo & B.M. Gaensler (astro-ph/0311296)

Orosz, J.A. 2003, in *Proc. IAU Symp. No. 212 "A Massive Star Odyssey, from Main Sequence to Supernova"*, Eds. K.A. van der Hucht, A. Herrero & C. Esteban, p. 365

Orosz, J.A. & Kuulkers, E. 1999, MNRAS, 305, 1320

Psaltis, D. 2004, in *Compact Stellar X-ray Sources*, Eds. W.H.G. Lewin & M. van der Klis, CUP (astro-ph/0410536)

Rhoades, C.E. & Ruffini, R. 1974, Phys. Rev. Lett., 32, 324

Shahbaz, T., Casares, J., Watson, C.A., Charles, P.A., Hynes, R.I, Shih, S.C. & Steeghs, D. 2004, ApJ, 616, L123

Shklovskii, I.S. 1967, Astron. Zhur., 44, 930

Stairs, I.H. 2004, Science, 304, 547

Steeghs, D. & Casares, J. 2002, ApJ, 568, 273

Strohmayer, T.E., Zhang, W., Swank, J.H., White, N.E. & Lapidus, I. 1998, ApJ, 498, L135

Strohmayer, T.E. & Markwardt, C.B. 2002, ApJ, 577, 337

Strohmayer, T.E., Markwardt, C.B., Swank, J.H. & Zand, J.J.M.in't 2003, ApJ, 596, 67

Torres, M.A.P., Callanan, P.J., Garcia, M.R., Zhao, P., Laycock, S. & Kong, K.H. 2004, ApJ, 612, 1026

Villarreal, A.R. & Strohmayer, T.E. 2004, ApJ, 614, L121

Woosley, S.E., Langer, N. & Weaver, T.A. 1995, ApJ, 448, 315

Zurita, C. et al. 2002, MNRAS, 334, 999

MASSIVE STARS IN STARBURSTS

Radio Studies in Messier 82

Alan Pedlar
Jodrell Bank Observatory, University of Manchester
Nr. Macclesfield, Cheshire SK11 9DL, UK
ap@jb.man.ac.uk

Abstract Strong radio and infrared emission, coupled with disturbed optical morphology, indicates a high starformation rate in the central region of Messier 82 consistent with a starburst. Unfortunately this region is optically obscured and hence can be directly studied mainly via radio and infra-red measurements. High resolution radio observations of this region reveal approximately 60 compact sources, of which 15 are consistent with compact HII regions and most of the remaining 45 appear to be supernova remnants (SNR). In both cases massive stars are required, in the former to provide ionisation for the HII regions, and in the latter as progenitors to the supernovae. OH masers will also be discussed briefly as additional tracers of massive stars. Many of the SNR show clear shell structures at MERLIN resolution ($\sim$50mas=0.7pc) and 5 have been resolved on mas scales using VLBI. VLBI measurements of at least one of the SNR shows expansion of $\sim$10000 km s^{-1} and further MERLIN and VLBI measurements are underway to measure more expansion velocities. We estimate the ages of the remnants to be typically a few hundred years, leading to supernova rates consistent with those derived from other indicators of the starformation rate in M82.

Keywords: Galaxies, Starbursts, Star-formation, Supernova remnants

1. Introduction

Starburst galaxies contain regions with a significantly enhanced starformation rate, and estimates of the available gas present usually show that this rate can only be sustained for a small fraction of the galactic lifetime–hence the term ‘starburst’. The starburst usually has a size of order a kiloparsec, although in some cases merger events result in more extensive starbursts, whereas others can show intense starformation rates over regions of only a few hundred parsecs.

J. C. del Toro Iniesta (eds.), The Many Scales in the Universe, 155–164.

The best observational evidence for a starburst is probably the intense far-infrared (FIR) emission which is thought mostly to be dust heated by young, hot stars. The radio continuum flux density of a galaxy is also a good indicator of enhanced starformation, although there is sometimes contamination of this emission with radio emission produced by an active galactic nucleus (AGN). In nearby starburst galaxies it is usually possible to make detailed studies of the radio emission from both supernova remnants and HII regions, and hence infer the properties of the associated O stars, as well as distinguishing any AGN activity in the starburst.

2. Estimating starformation rates in messier 82

Messier 82 is one of the nearest, and certainly most studied starburst galaxy. However the dust associated with molecular clouds, coupled with its almost edge-on orientation, result in high levels of extinction (20-30M_v; Mattila & Meikle 2001) over much of the central region which contains the starburst. Consequently although inferences can be made concerning the starformation history from optical and UV studies (de Grijs et al. 2001), most of the region containing the current starburst is only accessible in the radio or infra-red. Consequently estimates of the global starformation rate cannot easily be made using optical or UV measurements. Fortunately the radio method (Cram et al. 1998) is not affected by dust extinction and hence the starformation rate can be estimated from the equation

$$SFR(M \geq 5M_o) = \frac{L_{1.4}}{4.0 \times 10^{21}} \ \mathrm{M_o yr^{-1}}$$

where $L_{1.4}$ is the total 1.4GHz luminosity of the galaxy (Cram et al. 1998). Note that to avoid complications with the initial mass function (IMF) of low mass stars, this relation gives the starformation rate for stars with mass $> 5M_o$. Hence as the total flux density of M82 at 1.4GHz is ~8.5Jy and using a distance of 3.2Mpc, then SFR($M \geq 5M_o$) for M82 can be estimated to be $\sim$ 2.5$\mathrm{M_o}$ yr^{-1}. Also the radio thermal free-free contribution is reasonably well defined in M82 (Carlstrom & Kronberg 1991) and can be used to estimate a starformation rate (Pedlar 2001) consistent with the above value. Finally the starformation rate derived from the total FIR luminosity also gives a value of SFR($M \geq 5M_o$) close to 2 M_o yr^{-1}.

3. Indicators of massive stars

Given the relatively high starformation rate in the central kpc of M82, there is no doubt that it contains large numbers of massive stars. Although the distribution of these stars cannot be observed directly because of extinction, their

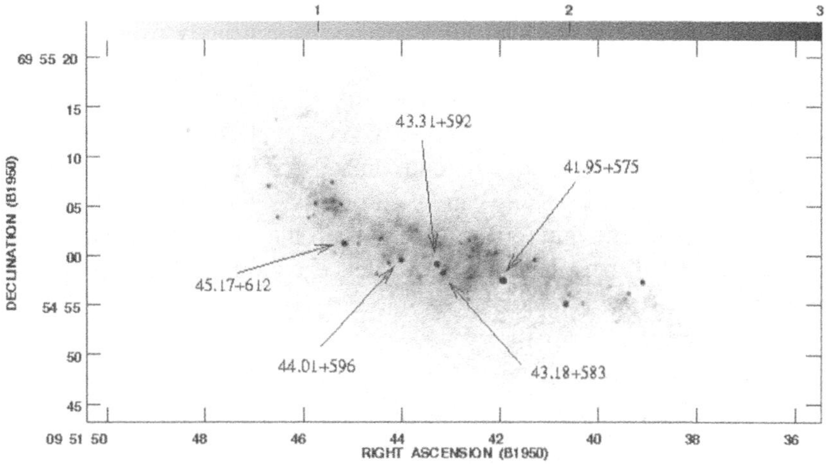

Figure 1. A 20cm combined MERLIN and VLA image of the M82 starburst. At this wavelength most of the compact sources are supernova remnants, and 5 of the objects which have been studied using VLBI are labeled (Pedlar et al. 1999).

presence can be inferred in the radio by observations of supernovae & supernova remnants, compact HII regions and OH masers.

Supernova Remnants

Although radio supernova remnant studies have been carried out on several nearby starburst galaxies, by far the most extensive have been carried out in the Messier 82 starburst. Compact sources were first noted in M82 by Kronberg & Wilkinson (1975), but it was not until the advent of MERLIN (Unger et al. 1984) and the VLA (Kronberg et al. 1985) that ~50 compact radio sources were separated from the extended background. This is clearly seen in the 20cm combined MERLIN and VLA image (Fig. 1) which shows both the diffuse background and the compact sources. Although initially it was thought that some of the compact sources could be radio supernovae, the lack of significant variability of most of the sources implied that they were supernova remnants. This was subsequently confirmed by 5GHz MERLIN observations (Muxlow et al. 1994) in which all the sources were shown to be extended with sizes of typically a few parsecs and many had the shell or partial shell structures typical of remnants.

If we assume that the SNR are all type II, then the distribution of SNR represents the distribution of massive stars ($> 8M_o$) approximately 10^7 years ago. Hence if the region of star-formation changed significantly over this period then the SNR would show a different distribution from more recent indicators of starformation such as compact HII regions(see below). As there is no significant evidence for such a change (McDonald et al. 2002), then it is reasonable to assume the starformation rate is approximately constant and hence, in principle the type II supernova rate can be derived directly from the current star formation rate and the initial mass function, on the assumption that all stars with masses $> 8M_o$ become supernovae. Hence assuming a Miller-Scalo IMF the supernova rate ν_{sn} is related to the star formation rate via

$$SFR(M \geq 5M_o) = 25\nu_{sn}\ \mathrm{M_o yr^{-1}}.$$

Thus if $SFR(M \geq 5M_o)$ is $\sim 2 - 2.5\mathrm{M_o yr^{-1}}$ as deduced in the previous section, this implies a supernova rate of $\sim 0.08 - 0.1\mathrm{yr}^{-1}$ in M82.

Several methods can be used to estimate the supernova rate directly from the radio parameters of the M82 supernova remnants (Muxlow et al. 1994, Pedlar 2001). The simplest method is to assume that all the M82 remnants which are brighter and smaller than the Cassiopeia A remnant in our galaxy are younger than 330 years. Alternatively if a constant expansion velocity is assumed (e.g. 5000-10000 km s^{-1}) this gives the ages of the remnants and hence the supernova rate (Muxlow et al. 1994). Finally van Buren & Greenhouse (1994) estimated the supernova rates from remnant luminosities. These methods give supernova rates ranging from 0.05 to 0.1 yr^{-1} - consistent with the starformation rates derived above. It is possible that not all supernovae produce radio emitting remnants, and if so this would imply an even larger supernova rates.

Compact HII regions

The majority of the more luminous compact sources in M82 have a steep spectrum, consistent with synchrotron emission from supernova remnants (e.g. Allen & Kronberg 1998). Although from these initial spectral studies it appeared that only one compact source had a spectrum consistent with thermal emission, the sample only included about half (~26) of the known compact sources and as they were selected at relatively low frequencies hence had a bias towards non-thermal sources.

McDonald et al. (2002) determined the two point (5 and 15 GHz) spectra of 46 the compact sources in M82 by using the extended VLA (ie including the Pie-Town Antenna) at 15GHz. This resulted in an angular resolution of 80mas (=1.2pc) which can be directly compared to MERLIN 5GHz observations with 50mas resolution. These observations showed that 16 of the sources had inverted spectra and brightness temperatures consistent with optically thick compact HII regions with emission measures of $\sim 10^7$ pc cm^{-6}. The distri-

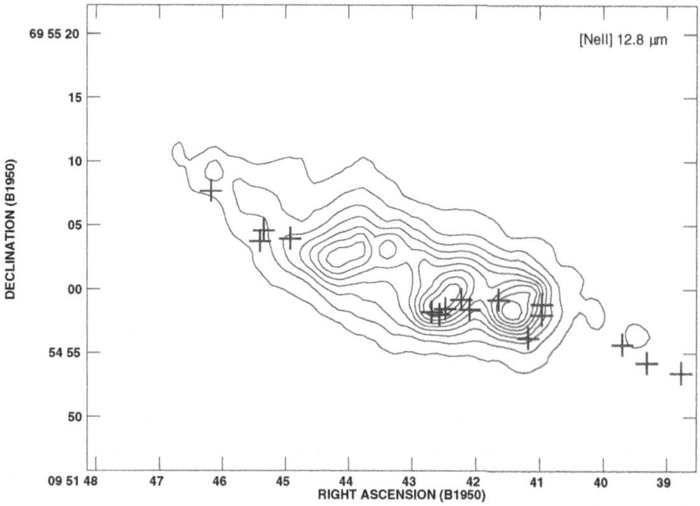

Figure 2. The position of the compact HII regions (crosses) reported by McDonald et al. 2002, superimposed on the diffuse ionised gas delineated by [NeII], Achtermann & Lacey 1995).

bution of these compact HII regions, superimposed on diffuse ionised gas inferred from [NeII] measurements (Achtermann & Lacey 1995), are shown in Fig. 2. From the size and emission measures of the compact HII regions it is possible to calculate the Lyman continuum flux required to ionize them. Thus, by assuming that the Lyman continuum output of an O5 star is $\sim 5 \times 10^{49}$ photon s^{-1} , it is possible to derive an equivalent number of O5 stars required to ionise the HII region. The number of O5 stars required to ionize these regions varies from a few to several hundred (McDonald et al. 2002). Therefore, these regions may be similar to those found in other galaxies such as NGC 2146 (Tarchi et al.) and NGC 5253 (Turner et al. 2000), which were deemed to be due to the precursors of 'super star clusters' or 'proto globular clusters'. Hence these compact HII regions are tracing the current position of clusters of massive stars in the M82 starburst.

OH Masers

Main line (1665 & 1667 MHz) OH masers are often indicators of massive star formation. Not only can they seen to be associated with starforming regions in our galaxy, but also OH megamasers are associated with galaxies with extremely high starformation rates. Our recent study of OH in M82 (Pedlar et al. 2003) has revealed at least 10 regions of strong maser emission in the M82 starburst (Fig. 3). Six of these regions are spatially coincident with the

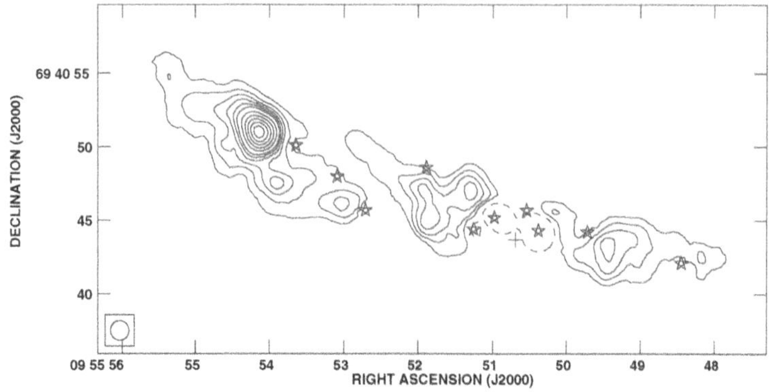

Figure 3. A contour map showing the integrated 1667MHz absorption seen against the starburst in M82. The contours of absorption are separated by units of 500 Jy beam^{-1} Hz. The stars mark the positions of the main line OH masers and the cross marks the position of the most luminous, compact radio source 41.95+575. The dashed contours show emission from the stronger masers (from Pedlar et al. 2003).

compact HII regions discussed in the previous section. In addition there is close positional agreement between the four H_2O masers reported by Baudry & Brouillet (1996) and four of the OH masers.

The association of these masers with the compact HII regions confirms their being associated with regions young star-formation as seen in galactic studies. From Figure 3 it can be seen that most of the main-line masers are situated at the edges of, rather than the centers of, the OH absorbing clouds, implying that starformation in taking place at the periphery, rather than within, the molecular clouds.

Several of the masers are in the region of free-free absorption detected at 73cm (Wills et al. 1997), initially interpreted as an ~100pc diameter HII region. This is the same region in which Weiss et al. (2001) reported a bubble of CO emission expanding at 45 km s^{-1} and Matsushita et al. (2000) imaged a shell of CO emission of radius ~100pc. Our OH data is also consistent with an expanding shell, although it is the OH masers which delineate the blue-shifted part of the shell, which could imply that starformation has been induced along the edge of the shell.

4. The age of the supernova remnants in M82

Recently some doubts have been raised concerning the evolution of the supernova remnants in M82. Kronberg et al. (2000), in a study of the variability of 24 remnants, claim, using statistical arguments, that in a large fraction of the remnants the radio luminosity is decaying at less than 0.1% per year. If true this could suggest that such remnants are over a thousand years old. Also recent theoretical studies by Chevalier & Fransson (2001) have suggested that the high pressure of the interstellar medium in M82 would result in the supernova remnants expanding at only 500km s^{-1}, hence as the observed sizes are typically a few parsecs, this again would imply ages of thousands of years. Both these results appear to be inconsistent with the supernova rates of $\sim$0.1 yr^{-1} derived above, which require that the ages of typical remnants are a few hundred years. In view of the implications of these results we will discuss the assumptions of such models, together with new observational data which gives a direct measurement of the expansion velocity of the remnants.

Interstellar medium pressures in M82

Chevalier & Fransson (2001) assume that the pressure in the starburst region of M82 is 10^7 $cm^{-3}K$ and from this derive the low (500 km s^{-1}) SNR expansion velocities. While there is no doubt that part of the starburst must be at this pressure (e.g. the compact HII regions McDonald et al. 2002), it seems likely that much of the starburst is not in static pressure equilibrium and that a range of pressures are present.

This can be demonstrated as, if the photoionised component in M82 has a temperature of $\sim 10^4$K, the pressure assumed by Chevalier & Fransson would require that the density of this gas be $\sim 10^3$ cm^{-3}. However the free-free absorption measurements taken against the individual remnants show emission measures of $\sim 5 \times 10^5$ pc cm^{-6} (Wills et al. 1997), and hence densities of $\sim 10^3$ cm^{-3} would imply ionised gas path lengths of only $\sim$0.5pc. Given that the extent of the starburst region is at least 500pc, unless the ionised gas is contrived to occupy the 0.1% of the volume of the starburst in the immediate vicinity of each remnant, it is difficult to see how the ionised component can provide sufficient pressure to slow the SNR expansion to 500 km s^{-1}.

Chevalier& Fransson suggest that the SNR may be confined by the interclump medium of molecular clouds in M82 and note that many of the remnants are in the line of sight to regions of strong CO emission. This is, however, not conclusive evidence that the remnants are embedded within the clouds, and often the HI absorption spectra measured directly against the remnants (Wills et al. 1998) do not support this interpretation. Furthermore studies by Weiss et al. 2001 have shown that the molecular gas in the starforming regions of M82 has a kinetic temperatures of $\sim$150K and densities of $\sim 10^3$ cm^{-3} which corresponds to a pressure two orders of magnitude less than that assumed by Chevalier & Fransson.

Clearly the ISM of M82 is complex, and to assign a single pressure to the starburst region may be unrealistic. Hence it is likely that many of the SNR in M82 are embedded in regions with pressures significantly lower than 10^7 cm^{-3} K and hence have expansion velocities much greater than 500 km s^{-1}.

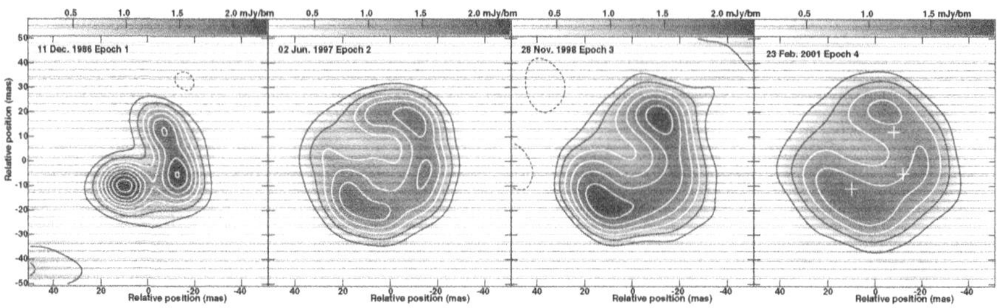

Figure 4. 20cm VLBI images of the M82 supernova remnant 43.31+592 plotted at epochs 1986, 1997, 1998 and 2001. The images have been convolved to a common angular resolution of 15mas.(Adapted from Riley et al. 2003).

Measuring expansion velocities

Now that we can resolve the compact sources in M82 (Muxlow et al. 1994), in principle it is a relatively simple matter to measure their expansion velocities by measuring their size over a period of time, although such studies are limited to the brightest sources by current instrumental sensitivity (see Fig. 1). The first expansion measurements were made by Pedlar et al. (1999) on the compact sources 41.95+575 & 43.31+592. The brightest, most compact source (41.95+575) was shown to have an elongated structure and the expansion velocity along the major axis was constrained to < 4000 km s^{-1}. Subsequent global VLBI measurements McDonald et al. 2001 showed it to have an expansion velocity of ~ 2000 km s^{-1}. This source is, however, rather anomalous on account of its high radio luminosity and its rapid decay. Also the global VLBI structure is highly elongated and more like a bipolar structure than a shell - unlike a typical SNR (McDonald et al. 2001).

The source 43.31+592 is much more typical an SNR of shows a well-defined shell structure (Fig. 4). By comparing data observed in 1986 and 1997, Pedlar et al. (1999) deduced an expansion velocity of $\sim$ 10000km s^{-1}. Subsequent Global VLBI measurements McDonald et al. 2001 Riley et al. 2003 have confirmed this result and shown its expansion to be consistent with no deceleration (Fig. 5), implying an age of about 40 years. As the SNR seems to be in free expansion we can set a simple constraint of < 2000 atoms cm^{-3} to the external density by assuming that the mass of gas swept up is less than the mass ejected. The expansion velocity of this source clearly exceeds the 500km s^{-1} predicted by Chevalier & Fransson. Although this could be accounted for if this particular SNR was in a region with relatively low external pressure, such an explanation seems rather contrived.

In fact recent MERLIN observations (Muxlow et al. 2003) have independently confirmed the above expansion velocities and also measured the expan-

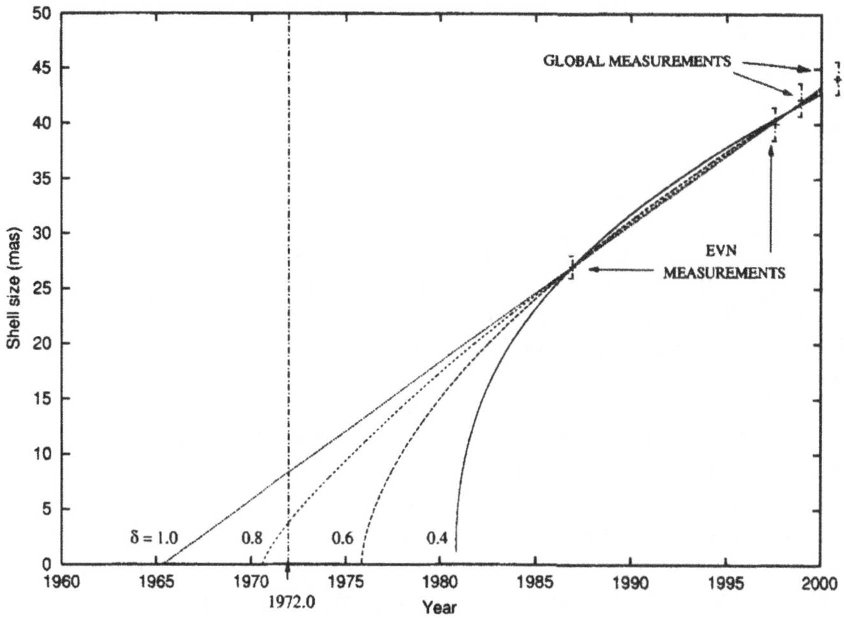

Figure 5. The size of 43.31+592 plotted as a function of age including the 2001.1 measurement. The figure also shows different expansion models. Adapted from McDonald et al. (2001).

sion velocities of 2 more remnants. One of the remnants has a velocity close to 10000 km s^{-1}, whereas the other is relatively low at 2000 km s^{-1}. We now have 4 expansion measurements and find little evidence for ~500 km s^{-1} expansion velocities. Hence we see no reason to revise the remnant ages or supernova rates that we have inferred previously for M82.

5. Summary

The starformation rates in M82 inferred from infrared, radio and mm observations are consistent with a supernova rate of one every 10 or 20 years. This is consistent with a rate derived from the size-diameter relation observed for SNR in M82, and with the expansion velocities of some of the brighter SNR.

Higher sensitivity observations using e-MERLIN, combined with an increasing time baseline, will enable the expansion velocities of many more remnants in M82 to be determined and hence constrain both the supernova rate and the parameters of the interstellar medium in M82.

Acknowledgments

The author wishes to thank Tom Muxlow, Rob Beswick, Jon Riley and Megan Argo for assistance with this review.

References

Achtermann, J.M. & Lacey, J.H. 1995, ApJ, 439, 163

Allen, M.L. & Kronberg, P.P. 1998, ApJ, 502,, 218

Baudry, A. & Brouillet, N. 1996, A&A, 316, 188

Carlstrom, J.E. & Kronberg, P.P. 1991, ApJ, 366, 422

Chevalier, R.A. & Fransson, C. 2001, ApJ, 558, 27

Cram, L., Hopkins, A., Mobasher, B. et al. 1998, ApJ, 507, 155

de Grijs, R., O'Connell, R.W. & Gallagher, J.S. 2001, AJ, 121, 768

Kronberg, P.P., Biermann, P. & Schwab, F.R. 1985, ApJ, 291, 693

Kronberg, P.P., Sramek, R.A., Birk et, G.T. et al. 2000, ApJ, 535, 706

Kronberg, P.P. & Wilkinson, P.N. 1975, ApJ, 200, 430

Matsushita, S. et al. 2000, ApJ, 545, 107

Mattila, S. & Meikle, W.P.S. 2001, MNRAS, 324, 325

Mcdonald, A.R., Muxlow, T.W.B., Pedlar, A. et al. 2001, MNRAS, 322, 100

McDonald, A.R., Muxlow, T.W.B., Wills, K.A. et al. 2002, MNRAS 334, 912

Muxlow, T.W.B., Pedlar, A., Wilkinson, P.N. et al. 1994, MNRAS, 266, 455

Muxlow, T.W.B. et al. 2003, in *Proc. IAU colloquium 192*

Pedlar, A. 2001, in: *Proc. IAU symposium 205*, Ed. R. Schilizzi, ASP Conf. Ser., p 366

Pedlar, A., Muxlow, T.W.B., Garrett, M.A. et al. 1999, MNRAS, 307, 761

Pedlar, A. et al. 2003 in *The neutral ISM in Starburst Galaxies*, Marstrand

Riley, J. et al. 2003, in *IAU Colloquium 192*

Tarchi, T. et al. 2004, MNRAS, 351, 339

Turner, J., Beck, S.C. & Ho, P.T.P 2000, ApJ, 532, 109

Unger, S.W., Pedlar, A., Axon, D.J. et al. 1984, MNRAS, 211, 783

van Buren, D. & Greenhouse, M.A. 1994, ApJ, 431, 640

Weiss, A., Neininger, N., Huettemeister, S. et al. 2001, A&A, 365, 571

Wills, K.A., Pedlar, A., Muxlow, T.W.B. et al. 1997, MNRAS, 291, 517

Wills, K.A., Pedlar, A., & Muxlow, T.W.B. 1998, MNRAS, 298, 347

THE HOT CONTENT OF PLANETARY NEBULAE

Martín A. Guerrero
Instituto de Astrofísica de Andalucía, CSIC
Apdo. 3004, Granada 18080, Spain
mar@iaa.es

You-Hua Chu and Robert A. Gruendl
Department of Astronomy, University of Illinois
USA
chu@astro.uiuc.edu, gruendl@astro.uiuc.edu

Abstract Planetary nebulae (PNe) are the latest addition to the zoo of X-ray sources. Besides the photospheric emission of hot central stars of PNe, *Chandra* and *XMM-Newton* observations have detected diffuse X-ray emission from shocked fast winds in PN interiors, providing support to the interacting-stellar-winds formation models of PNe. New types of X-ray sources in PNe have also been found: (1) diffuse X-ray sources from bow-shocks of fast collimated outflows impinging on the nebular envelope, and (2) hard ($\gg$0.5 keV) X-ray point sources at the PN central stars originating from instability shocks in the fast stellar wind itself or from a low-mass companion's coronal activity. X-ray observations of PNe offer a unique opportunity to assess the dynamic effects of fast stellar winds and collimated outflows in PN formation and shaping. The distribution and physical conditions of hot gas in PNe help us gain insight in similar astrophysical situations, e.g., bubbles and superbubbles blown by massive stars.

Keywords: ISM: planetary nebulae: general, ISM: planetary nebulae: individual: NGC 2392, NGC 6543, NGC 7009, BD+30°3639, Hen 3-1475, Stars: winds, outflows

1. Sources of X-ray emission from planetary nebulae

A planetary nebula (PN) consists of the stellar material ejected by a star with initial mass $\leq 8-10$ $M_\odot$. As such a star evolves off the asymptotic giant branch (AGB), its copious mass loss strips off the stellar envelope and exposes the hot stellar core. A PN emerges when the stellar UV radiation ionizes the ejected stellar material, causing it to emit in the optical.

J. C. del Toro Iniesta (eds.), The Many Scales in the Universe, 165–174.

PNe and their central stars have long been studied in the optical, UV, radio and infrared wavelenghts. In recent years, PNe have also been recognized as X-ray-emitting sources. Four different sources of X-ray emission can be hosted by PNe:

- Photospheric emission from hot central stars of PNe.
  The central stars of PNe can exhibit very high stellar effective temperatures, 100,000–200,000 K. The photospheric emission of such hot stars extends to the X-ray wavelength range, but only at energies $\ll$0.5 keV (Guerrero, Chu, & Gruendl 2000).
- Emission from shock-heated gas in the central cavity of PNe.
  In the interacting-stellar-winds model of PNe (Kwok, Purton, & Fitzgerald 1978), the fast wind (1000–4000 km s^{-1}) emanating from the central star sweeps up the slow AGB wind to form a sharp nebular shell. The interior structure of a PN would be similar to that of a wind-blown bubble (e.g., Weaver et al. 1977). The central cavity of a PN is expected to be filled with shocked fast wind at 10^7–10^8 K, but this hot gas is too tenuous to produce appreciable X-ray emission. Dynamic or evaporative mixing of cool nebular material into the hot gas at their interface produces optimal conditions for soft X-ray emission, which will show a limb-brightened morphology within the nebular shell.
- Emission from bow-shocks of fast outflows.
  Fast collimated outflows or jets that occur in the proto-PN phase, near the end of the AGB phase (Sahai & Trauger 1998; Bujarrabal et al. 2001) may also produce hot X-ray-emitting gas. When outflows with velocities $\geq$300 km s^{-1} initially impinge on the AGB wind, bow shocks and X-ray emission can be produced. The prolonged action of collimated outflows may bore through the AGB wind and form extended cavities filled by shocked hot gas that emits X-rays.
- Coronal activity of a dwarf late companion.
  The coronal activity of a late-type dwarf companion of the central star of a PN may produce detectable X-ray emission. A point X-ray source may be detected at the central star of a PN, but unlike the photospheric emission of a hot central star ($\ll$0.5 keV), the coronal emission from a late companion is substantial at $\sim$1 keV.

The X-ray emission from PNe and their central stars provides us with invaluable information to better understand the evolution of the central stars of PNe, the formation and evolution of PNe, and the role of binarity in their shaping. The photospheric X-ray emission from PN central stars is indicative of their chemical composition and surface temperature. The diffuse X-ray emission from shock-heated gas allows us to assess how fast stellar wind and collimated

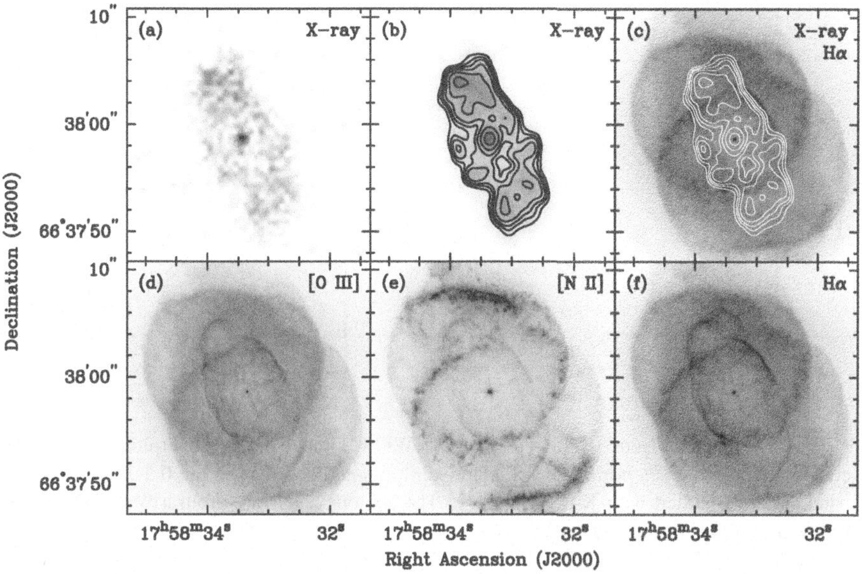

Figure 1. Image of NGC 6543, the Cat's Eye Nebula. *(a) Chandra* ACIS-S raw image in the 0.2-1.5 keV energy band; *(b)* the same X-ray image, but adaptively smoothed and overplotted with contours; *(c) HST* WFPC2 Hα image overplotted with X-ray contours; *(d,e,f) HST* WFPC2 image in the [O III] λ5007, [N II] λ6583, and Hα lines, respectively. The greyscales are selected to best reveal the surface brightness variations. This figure is adapted from Chu et al. (2001).

outflows interact with the AGB wind and transfer energy and momentum to the PN envelope. It is also important to detect low-mass binary companions to assess their implications in the PN shaping.

2. X-Ray observations of PNe

X-ray emission from PNe was reported in the mid-1980s using observations made with *Einstein* and *EXOSAT*; these X-ray sources were all soft and associated with the photospheric emission from their hot central stars (see Guerrero, Chu, & Gruendl 2000 for a complete review). In recent years, the *Chandra* and *XMM-Newton* X-ray Observatories have provided us with a great deal of information on the hottest component of PNe. Here we review the first detections of diffuse X-ray emission from PNe and the discovery of hard X-ray emission from their central stars.

Diffuse X-ray Emission from PNe

In the 1990s, *ROSAT* made useful observations of more than 60 PNe. These *ROSAT* observations showed for the first time hints of emission from the hot gas in PN interiors, but the evidence was not very convincing because of the

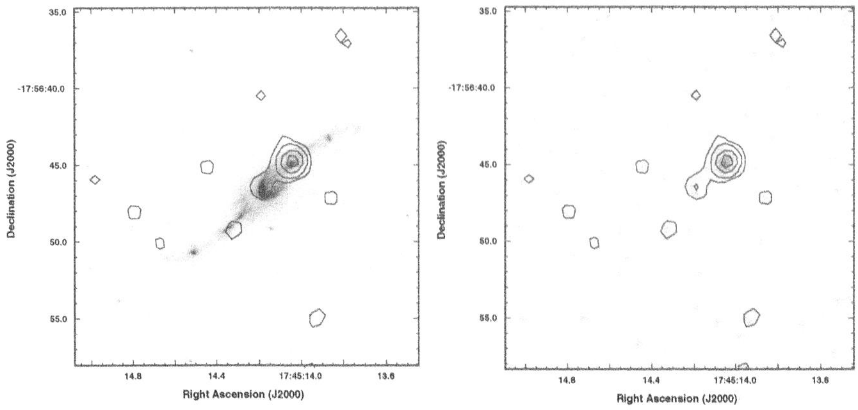

Figure 2. Image of Hen 3-1475 in *(right) HST* WFPC2 image in [N II] overplotted with X-ray contours, and *(left) Chandra* ACIS-S adaptive filter smoothed map in the 0.2-2.0 keV energy band also overplotted with X-ray contours. The contour levels are arbitrarily chosen to best follow the surface brightness variations.

limited angular resolution and low S/N ratios; only three PNe (BD+30°3639, A 30, and NGC 6543) show marginally extended X-ray emission, while two other PNe (NGC 7009 and NGC 7293) show a point-source at the central star with a hard, >0.5 keV, X-ray component not expected from the stellar photosphere (Guerrero, Chu, & Gruendl 2000).

The launch of modern X-ray observatories, *Chandra* and *XMM-Newton*, has made it possible to observe PNe with unprecedented sensitivity and angular resolution. To date, *Chandra* and *XMM-Newton* have unambiguously resolved diffuse X-ray emission in 9 PNe and proto-PNe. Some of these observations have produced exquisite X-ray images and useful spectra, for example, BD+30°3639 (Kastner et al. 2000), NGC 6543 (Chu et al. 2001), and NGC 7009 (Guerrero, Chu, & Gruendl 2002). The distribution of diffuse X-ray emission relative to the nebular shell is illustrated in Figure 1 for one of the most well-resolved PNe, NGC 6543, the Cat's Eye Nebula. The X-ray contours overplotted on the Hα image of this PN show that the diffuse X-ray emission is well bounded by the sharp, bright Hα filaments that outline the elliptical shell with two outward extensions along its major axis (Chu et al. 2001). With the exception of Hen 3-1475 (see below), the diffuse X-ray emission from PNe is confined within the innermost nebular shell of elliptical PNe or inside closed lobes of bipolar PNe, consistent with the expectation for shocked fast wind in the interacting-stellar-winds model.

As for Hen 3-1475, the *Chandra* ACIS-S observations have provided the only clear detection of X-ray emission from the bow-shock of a collimated outflow in a proto-PN or PN (Sahai et al. 2003). The bulk of X-ray emission

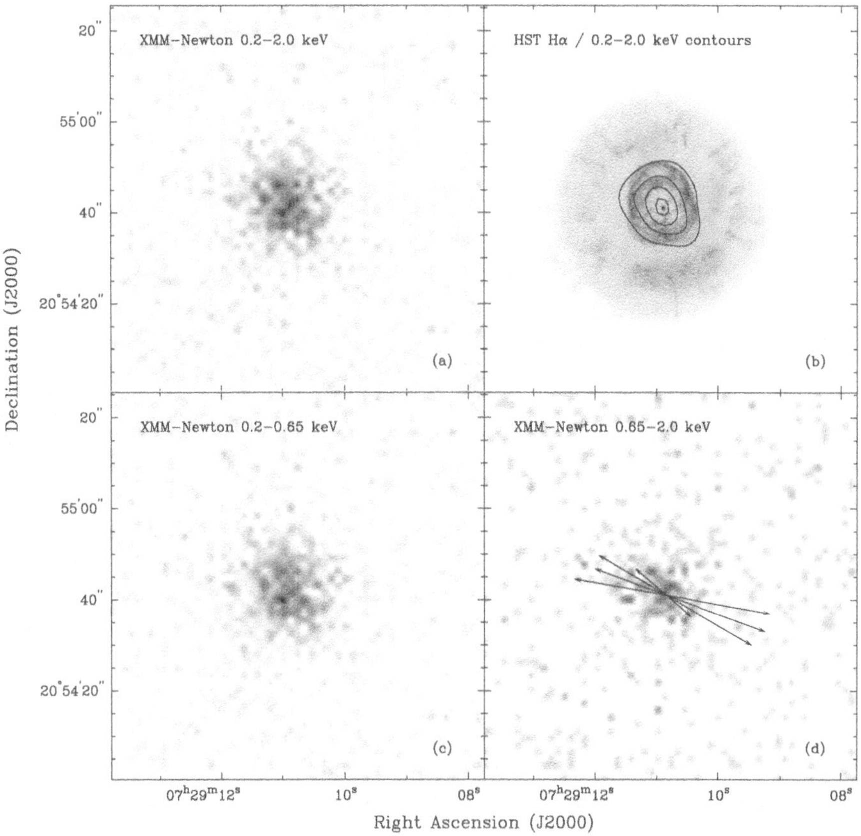

Figure 3. *XMM-Newton* and *HST* Hα images of the Eskimo Nebula. Panel *(a)* displays the *XMM-Newton* EPIC raw image in the 0.2-2.0 keV band. Panel *(b)* shows the *HST* Hα image overplotted by the 25%, 50%, 75%, and 95% X-ray contours. Panels *(c)* and *(d)* display *XMM-Newton* EPIC raw images in the 0.2-0.65 and 0.65-2.0 keV energy bands, respectively. The arrows in panel *(d)* mark the location of the fast collimated outflow as derived from high-dispersion echelle spectroscopy (Gieseking, Becker, & Solf 1985). This figure is adapted from Guerrero et al. (2005).

detected by *Chandra* is located at the tip of the NW bow-shock seen in the *HST* images (Fig. 2) where an abrupt change in velocity of the fast collimated outflow emanating from its core has been identified (Riera 2004). At the SE counterpart of the NW bow-shock, only a tentative detection of X-ray emission is obtained.

The existence of an X-ray-emitting collimated outflow has also been proposed for Mz 3 (Kastner et al. 2003). Recent *XMM-Newton* observations of NGC 2392, the Eskimo Nebula, have provided evidence for diffuse X-ray emission possibly associated with another fast outflow (Guerrero et al. 2005). The

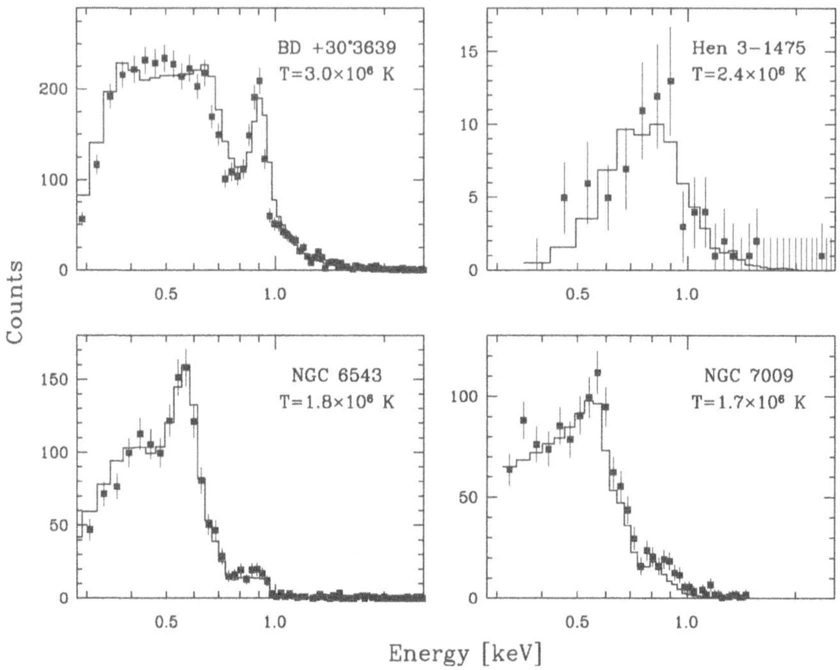

Figure 4. *Chandra* ACIS-S spectra of BD+30°3639, Hen 3-1475, and NGC 6543, and *XMM-Newton* EPIC/pn spectrum of NGC 7009. The histogram overlaid on each spectrum corresponds to the best-fit model. The temperature of the best-fit model is marked in each panel.

Eskimo Nebula is known for the exceptionally large expansion velocity of its inner shell, ~90 km s^{-1}, and the existence of a fast bipolar outflow with a line-of-sight expansion velocity approaching 200 km s^{-1}. The diffuse X-ray emission from NGC 2392 shows noticeably different spatial distributions between the 0.2–0.65 keV and 0.65–2.0 keV bands (Figure 3). The soft X-ray emission is closely outlined by the inner shell, while the harder X-ray emission is roughly aligned with the fast bipolar outflow, thus suggesting that the diffuse X-ray emission is mostly confined within its inner shell, but some of the harder X-ray emission in the 0.65-2.0 keV band may be produced by the interaction of the fast bipolar outflow with nebular material.

The X-ray spectra of four PNe with diffuse X-ray emission are presented in Figure 4. The X-ray emission from PNe is soft, peaking at energies <1.0 keV. The spectral shape is dominated by emission lines of N VII, O III, and Ne IX indicative of thin-plasma emission. Spectral fits using a thin-plasma emission model give plasma temperatures of 1–3$\times 10^6$ K and suggest chemical enrichment of nitrogen and neon. In the case of Hen 3-1475, the hot gas

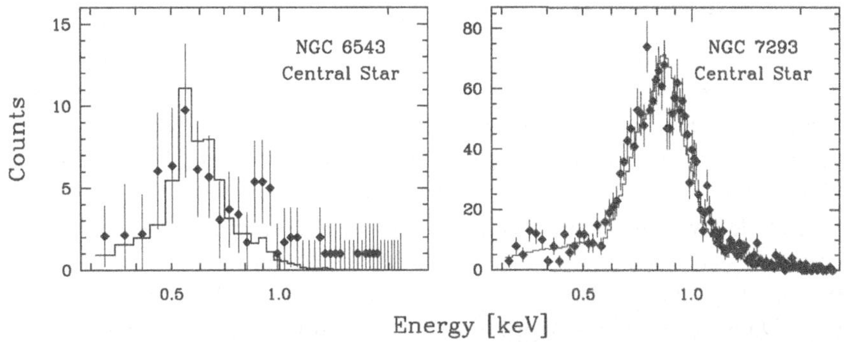

Figure 5. *Chandra* ACIS-S spectra of the central stars of NGC 6543 and NGC 7293. For comparison, the spectrum of a thin plasma emission model with temperature of 2×10^6 K is overplotted on the X-ray spectrum of NGC 6543. The best-fit model with temperature $\sim7\times10^6$ K is overplotted on the X-ray spectrum of NGC 7293.

temperature corresponds to the post-shock temperature for a shock velocity of $\sim$400 km s^{-1}.

The X-ray luminosities of PNe derived from these spectral fits range from 3×10^{31} ergs s^{-1} to 1×10^{33} ergs s^{-1}. The younger PNe (BD+30°3639, Mz 3, and NGC 7027) have systematically higher X-ray luminosities and temperatures than the more evolved PNe (NGC 2392, NGC 6543 and NGC 7009). All *Chandra* and *XMM-Newton* detections of diffuse X-ray emission are associated with PNe with a sharp-shell morphology indicative of highly pressurized hot gas, while observations of evolved, old PNe have rendered non detections of diffuse X-ray emission. Similarly, PNe with collimated outflows of moderate velocity ($<$200 km s^{-1}) do not possess detectable X-ray emission.

Hard X-ray Emission from PN Central Stars

The unprecedented *Chandra* resolution has made possible the detection of hard X-ray point sources at the central stars of Mz 3, NGC 6543, NGC 7293, and possibly Hen 3-1475 (Guerrero et al. 2001; Kastner et al. 2003). Figure 5 shows the X-ray spectra of the central stars of NGC 6543 and NGC 7293. These spectra suggest thin plasma emission at temperatures up to a few $\times10^6$ K and with X-ray luminosities $\sim10^{29}$ ergs s^{-1}.

Several possible origins of these point sources exist. For the central star of NGC 7293, the Helix Nebula, the temporal variations of X-rays and the Hα line profile suggest the presence of an unseen dMe companion with coronal activity (Gruendl et al. 2001; Guerrero et al. 2001). In other cases (e.g., NGC 6543), shocks in the fast stellar wind may be the origin of X-ray emission.

3. The conduction layers in PNe

The exciting X-ray detection of hot gas in PNe is undermined, however, by an outstanding problem related to the content of hot gas in PNe that needs further investigation. The observed X-ray luminosities of PNe are 1–2 orders of magnitude lower than the expectations of bubble models. For example, the observed L_X of NGC 6543, 1×10^{32} ergs s^{-1} (Chu et al. 2001), is an order of magnitude lower than that modeled by Zhekov & Perinotto (1998). Similar discrepancies between observed and modeled L_X have been seen in Wolf-Rayet bubbles (e.g., García-Segura, Langer, & Mac Low 1996) and superbubbles (e.g., Dunne et al. 2003), suggesting that this problem may be associated with thermal conduction per se. To understand the production and evolution of hot gas in PNe as well as their low X-ray luminosities, it is necessary to study the physics at the conduction layer between the hot gas in the PN interior and the cool nebular shell.

The hot interior gas and the cool nebular shell form a contact discontinuity where heat conduction (Spitzer 1962) is expected to occur. The resulting mass evaporation from the dense nebular shell into the hot interior lowers the temperature and raises the density of the hot gas (Weaver et al. 1977), significantly increasing the X-ray emissivity. Thermal conduction has been assumed in theoretical models but not constrained empirically through observation because interfaces at a few $\times 10^5$ K require difficult UV observations. Traditionally interfaces have been studied using spectral lines of C IV, N V, and O VI; however, these species can be photoionized by hot stars (especially PN central stars with effective temperatures >100,000 K) and the detection of these narrow absorption lines can be hampered by the stellar P Cygni profile and confused by interstellar absorption lines.

We have chosen to study the interface layer in the PN NGC 6543 using O VI emission because the physical properties of its interior hot gas have been established by *Chandra* observations (Chu et al. 2001) and its central star is too cool to produce O VI by photoionization ($T_{\rm eff} \sim 50,000$ K, Zweigle et al. 1997). *Far Ultraviolet Spectroscopic Explorer* (*FUSE*) observations of NGC 6543 have been made and O VI emission is indeed detected (Figure 6). By assuming pressure balance between the hot interior gas and the conduction layer, we find that the O VI emission is consistent with the level of X-ray emission (Gruendl, Chu, & Guerrero 2004).

4. Summary

Chandra and *XMM-Newton* observations of PNe have detected X-ray emission from hot gas in the central cavity (the so-called 'hot bubble') of PNe, bow-shocks associated with fast, $\simeq$1,000 km s^{-1}, collimated outflows, and point sources at the central stars. These observations have provided a wealth

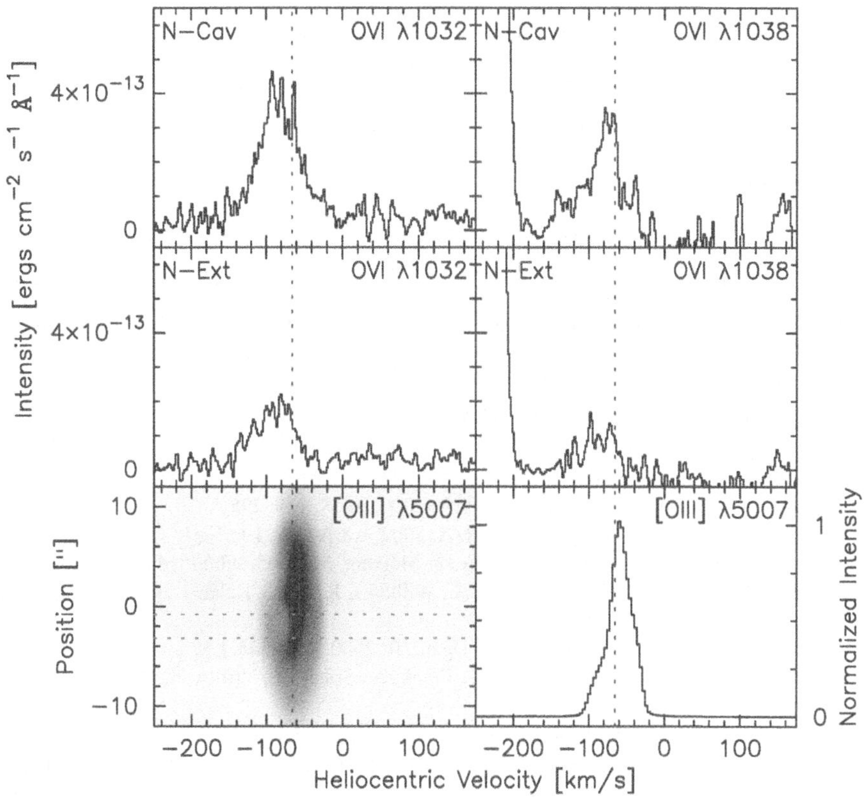

Figure 6. *FUSE* spectra for the O VI λλ1032,1038 lines from the northern region of the central cavity (top), and the northern extension (middle) of NGC 6543. For comparison the bottom panels show [O III] λ5007 nebular emission from a long-slit spectroscopic observation 4.5″ north of the central star. The bottom left panel shows the echellogram and the bottom right shows the line profile extracted from the position marked by the horizontal dashed lines on the echellogram. The systemic velocity of NGC 6543 is marked by a dashed vertical line in each panel. This figure is adapted from Gruendl, Chu, & Guerrero (2004).

of information on the distribution and physical conditions of hot gas in PNe, allowing us to investigate the physical structure of PNe as a whole and how collimated outflows transfer energy to the nebular envelope.

The emerging picture from these X-ray observations reveals that young PNe with a sharp shell morphology contain significant amounts of hot gas in their interiors. This gas is highly pressurized and drives the nebular expansion. In a short time-scale, however, the amount of X-ray-emitting hot gas diminishes drastically and the diffuse X-ray emission is no longer detectable. The action of the fast stellar wind in the formation and evolution of PNe appears to last only a short fraction of the PN lifetime.

Another important result from these observations is the detection of diffuse X-ray emission associated with collimated outflows in Hen 3-1475 and possibly NGC 2392 and Mz 3. Only outflows faster than $\sim$300 km s^{-1} can produce post-shock gas at temperatures greater than 10^6 K and X-ray emission.

References

Bujarrabal, V., Castro-Carrizo, A., Alcolea, J., & Sánchez Contreras, C. 2001, A&A, 377, 868

Chu, Y.-H., Guerrero, M. A., Gruendl, R. A., Williams, R. M., & Kaler, J. B. 2001, ApJ, 553, L69

Dunne, B. C., Chu, Y.-H., Chen, C.-H. R., Lowry, J. D., Townsley, L., Gruendl, R. A., Guerrero, M. A., & Rosado, M. 2003, ApJ, 590, 306

García-Segura, G., Langer, N., & Mac Low, M.-M. 1996, A&A, 316, 133

Gieseking, F., Becker, I., & Solf, J. 1985, ApJ, 295, L17

Gruendl, R. A., Chu, Y.-H., O'Dwyer, I. J., & Guerrero, M. A. 2001, AJ, 122, 308

Gruendl, R. A., Chu, Y.-H., & Guerrero, M. A. 2004, ApJ, 617, L127

Guerrero, M. A., Chu, Y.-H., & Gruendl, R. A. 2000, ApJS, 129, 295

Guerrero, M. A., Chu, Y.-H., & Gruendl, R. A. 2002, A&A, 387, L1

Guerrero, M. A., Chu, Y.-H., Gruendl, R. A., & Meixner, M. 2005, submitted to A&A

Guerrero, M. A., Chu, Y.-H., Gruendl, R. A., Williams, R. M., & Kaler, J. B. 2001, ApJ, 553, L55

Kastner, J. H., Soker, N., Vrtilek, S. D., & Dgani, R. 2000, ApJ, 545, L57

Kastner, J. H., Balick, B., Blackman, E. G., Frank, A., Soker, N., Vrtilek, S. D., & Li, J. 2003, ApJ, 591, L37

Kwok, S., Purton, C. R., & Fitzgerald, P. M. 1978, ApJ, 219, L125

Riera, A. 2004, Asymmetric Planetary Nebulae III, ASP Conference Series, eds. M. Meixner, J. H. Kastner, B. Balick, & N. Soker, 487

Sahai, R., Kastner, J. H., Frank, A., Morris, M., & Blackman, E. G. 2003, ApJ, 599, L87

Sahai, R., & Trauger, J.T. 1998, AJ, 116, 1357

Spitzer, L., 1962, Physics of Fully Ionized Gases, (New York: Interscience)

Weaver, R., McCray, R., Castor, J., Shapiro, P., & Moore, R. 1977, ApJ, 218, 377

Zhekov, S. A., & Perinotto, M. 1998, A&A, 334, 239

Zweigle, J., Grewing, M., Barnstedt, J., Goelz, M., Gringel, W., Haas, C., Hopfensitz, W., Kappelmann, N., Kraemer, G., Appenzeller, I., Krautter, J., & Mandel, H. 1997, A&A, 321, 891

THE HIDDEN LIFE OF MASSIVE STARS

A. Lenorzer
Anton Pannekoek Institute
Kruislaan 403, 1098 SJ Amsterdam, The Netherlands
&
SRON
Sorbonnelaan 2, 3584 CA Utrecht, The Netherlands

A. Bik
Anton Pannekoek Institute
Kruislaan 403, 1098 SJ Amsterdam, The Netherlands
&
ESO
Karl-Schwarzschild-Str. 2 D-85748 Garching, Germany

M.R. Mokiem, A. de Koter, and L. Kaper
Anton Pannekoek Institute
Kruislaan 403, 1098 SJ Amsterdam, The Netherlands

L.B.F.M. Waters
Anton Pannekoek Institute
Kruislaan 403, 1098 SJ Amsterdam, The Netherlands
&
Instituut K.U. Leuven
Celestijnenlaan 200B, 3001 Heverlee, Belgium

Abstract A considerable fraction of the massive stars in our galaxy is obscured by dust. Dust extinction prevents us from using powerful criteria, developed at ultraviolet and optical wavelength ranges, to infer physical properties of hot stars. At wavelengths longwards of about 5 microns, emission from warm dust typically dominates the spectra. It is, therefore, crucial to develop diagnostic tools in the near-infrared window to understand the nature and evolution of hot stars surrounded by dust. Such embedding in dust occurs, for instance, during the early stages of their lives in giant molecular clouds. To characterise the physical parameters of these hidden massive stars, and to find clues about their formation process, we developed spectroscopic diagnostics based on the near-infrared

J. C. del Toro Iniesta (eds.), The Many Scales in the Universe, 175–185.

wavelength range. Using a grid of 30 model atmospheres computed with CMFGEN, we find that the helium lines present in the J-, H- and K-band can be used for spectral typing of hot stars. The Brα line located in the L-band appears to be a good tracer of the wind density. Concerning stars that only show emission lines in the near-infrared part of their spectra, we propose a hydrogen line flux diagram to probe whether the circumstellar material is distributed in a disk or in a wind. The application of some of these diagnostics to NGC2024/IRS2 (one of the targets of our UC H II survey campaign) combined with a careful investigation of its spectral energy distribution shows that this young massive star is very likely surrounded by a gaseous disk. Whether this disk is the signature of a remnant accretion disk is, however, still under debate.

1. Introduction

Massive stars spend a significant fraction of their life embedded in their natal molecular cloud, hidden behind large amounts of visual extinction. It is estimated that at the time they become optically visible, already some 10 to 20 percent of their main sequence lifetime has passed. The near-infrared spectral window, where the extinction is strongly reduced, allows one to probe the properties of these embedded massive stars. Recent near-infrared observations of star-forming regions such as the survey presented in Sect. 2 have revealed the presence of near-infrared sources showing spectroscopic features originating from photospheric and/or circumstellar matter. The detection of the stellar photosphere allows, at least in principle, a direct determination of basic properties of the newly formed stars, such as their luminosity class (log g), effective temperature and projected rotational velocity. The near-infrared spectral lines potentially also contain information on the close circumstellar medium of massive young stellar objects, such as the geometry (i.e. disk and/or wind) and kinematic properties (i.e. rate and velocity of inflow and/or outflow). Whether this can really be achieved depends on the diagnostic potential of the spectral lines present in the near-infrared window, and requires careful calibration of such diagnostics using massive stars that can also be observed in optical light. We present the results of our investigations of the potential of near-infrared spectroscopy for the determination of the stellar (in Sect. 3) and circumstellar parameters (in Sect. 4) of massive stars and show their application to a young embedded massive star, NGC2024/IRS2, in Sect. 5.

2. Search for young massive stars

Ultra-compact H II regions (UC H IIs) represent the earliest stage indicating the presence of a newly born massive star inside its natal molecular cloud. Searching for just-born massive stars, Kaper et al. (2004) carried out a near-infrared survey of a sample of 44 IRAS point sources with colours characteristic of UC H IIs (see Wood & Churchwell 1989) using SOFI mounted at the

ESO *New Technology Telescope*. With narrow-band images, obtained in the J and K band, the deeply embedded, candidate ionising central star(s) were selected as the brightest and reddest objects. Follow-up K-band spectra were taken for about 60 candidates with the VLT/ISAAC (Bik 2004). Of these spectra, approximately 2/3 show typical atmospheric features similar to the ones observed for field massive stars (see Hanson et al. 1996); the other spectra only contain emission lines indicating the presence of dense circumstellar matter.

3. Stellar parameters from near-infrared spectra

To characterise the population of embedded massive stars showing photospheric signatures, we investigate to what extent the near-infrared lines can be used to determine the spectral type and mass loss of O-type stars using a grid of unified stellar photosphere and wind models for O-type stars of luminosity class V, III and Ia. This grid was constructed using the CMFGEN program of Hillier & Miller (1998), to which we refer for a full description. The grid consists of 30 models ranging in effective temperature, $T_{\rm eff}$, from $\sim$ 24 000 K up to $\sim$ 49 000 K, with 10 models for each luminosity class (see Lenorzer et al. 2004a). For the basic stellar parameters we employed the calibration from Vacca et al. (1996). The mass-loss rates incorporated in the CMFGEN models are from the theoretical predictions by Vink et al. (2000, 2001).

Spectral Type

The spectral sub-type of O-type stars can uniquely be defined using the ratio of the equivalent widths (EW) of the He I λ 4471 Å and He II λ 4542 Å lines (see Mathys 1988). When correlating optical and near-infrared line behaviour, we first notice that O-type stars are no longer *defined* by the presence of He II lines if one concentrates on only the near-infrared window. Indeed, at these wavelengths, most He II lines disappear around spectral type O8.5. Consequently, near-infrared He I/He II line ratios can only be measured for earlier spectral types. This is to be expected as the near-infrared continuum is formed further out in the atmosphere of O-type stars, where the temperature is lower.

We present predicted He I/He II EW ratios for lines in the J, H, and K band in Fig. 1. We do not find suitable helium lines in the L-band. We treat all luminosity classes on the same footing and plot values applicable for both medium- and low-resolution spectra. With medium spectral resolution (circles in Fig. 1) we mean $R \sim 6\,500$ and 4 500 for the J and K band, respectively. With low resolution (triangles) we imply $R \sim 1\,000$. In the J-band (top panel) we find that the ratio He I λ1.2788 μm/ He IIλ1.1676 μm may serve to determine the spectral type. This only works at medium resolution, because at low resolution the He I 1.2788 μm line starts to blend with C III 1.2794 μm, rendering this diagnostic unusable.

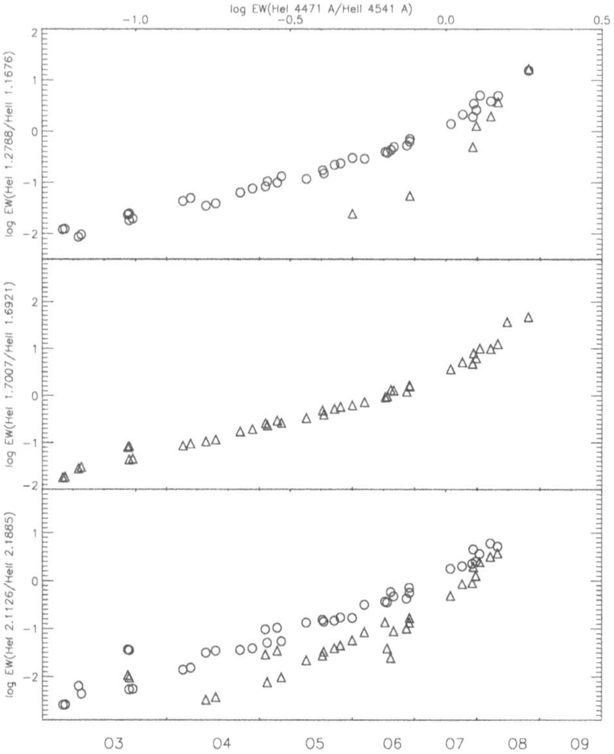

Figure 1. The correlation between optical and near-infrared He I/He II equivalent width (EW) ratios which may be used to calibrate spectral types. Circles denote EW predictions for a medium spectral resolution ($R \sim 4\,500 - 6\,000$); triangles indicate predictions for low resolution ($R \sim 1\,000$).

The predicted near-infrared EW ratios presented in Fig. 1 correlate well with the optical ratio and show a steeper dependence on spectral type compared to He I λ4471 Å over He II λ4541 Å. The values of the ratios are, however, to be taken with care as our models do not perfectly reproduce the observed line strengths (see Lenorzer et al. 2004a). Still they give a reasonably good idea of the observational requirements needed to derive quantitative information on the spectral type of hot stars from near-infrared spectroscopy alone. We conclude that a derivation of the spectral sub-type of O stars from near-infrared helium line ratios is in principle feasible for good quality spectra in the J, H, and K band and for stars that have spectral types in the range O4 to O8.

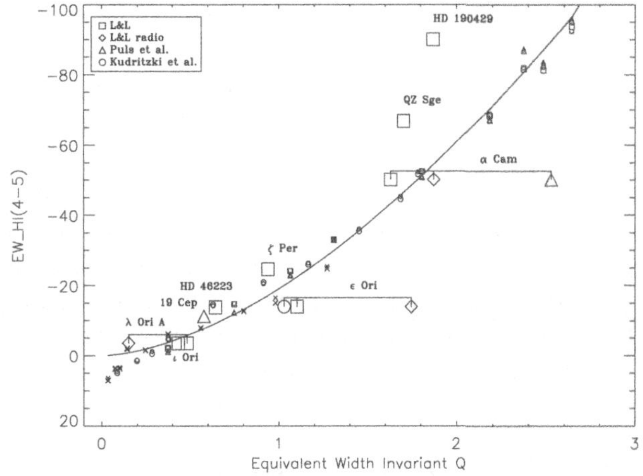

Figure 2. The predicted equivalent widths for dwarfs (denoted by crosses), giants (circles) and supergiants (squares) and lower surface gravity supergiants (triangles) are overplotted with observations from Lenorzer et al. (2002a), for which Q was derived using Lamers et al. (1999), Puls et al. (1996), and Kudritzki et al. (1999). Q is given in units of 10^{-20} $\mathrm{M}_{\odot}\mathrm{yr}^{-1}$ per $\mathrm{R}_{\odot}{}^{3/2}$ per K^2 per km s^{-1}. A best fit (solid line) to all negative EW values yields $\log \mathrm{EW}\,(\mathrm{Br}\alpha) = (32.4 \pm 0.5) + (1.55 \pm 0.03) \log Q$.

Wind density

Mass-loss rates of O-type stars can be determined from ultraviolet resonance and subordinate lines, Hα, and radio flux measurements (see Kudritzki & Puls 2000). A relatively simple method to derive $\dot{M}$ is to use the net EW of Hα (see e.g. Klein & Castor 1978, Leitherer 1988) and correlating it with the EW invariant $Q = \dot{M}/(R^{3/2}\mathrm{T}_{\mathrm{eff}}^{2} v_{\infty})$ (see Schmutz, Hamann & Wessolowski 1989, Puls et al. 1996 and de Koter, Heap & Hubeny 1998). In principle, the same strategy can be applied using the Brα line. Relations between the EW of Brα and the mass-loss rate have been proposed based on model predictions (Schaerer et al. 1996) and on observations (Lenorzer et al. 2002a). This line is intrinsically stronger than Hα, which means that the photospheric absorption has a smaller impact on its EW. We have opted not to correct for a photospheric contribution, as a "true" photospheric component of this line can only be defined if a core-halo approximation is adopted (i.e. a separate treatment of the stellar photosphere and wind), which is not physically realistic in the near-infrared regime since most near-infrared lines are formed mainly in the transition region between photosphere and wind.

The correlation between predicted Brα EW and Q is given in Fig. 2. The measurements are from 4.0 to 4.1 μm and include a number of blends from weak He I and He II lines. This causes most of the modest scatter in the model results; overall the correlation is very good. The divergence from the fit of models with mass-loss rates below about 10^{-7} $M_{\odot}yr^{-1}$ ($\sim Q/10^{-20} \leq 0.2$) marks the transition to profiles dominated by photospheric absorption. In the wind dominated regime the fit function recovers the Q value to within 0.05 to 0.15 dex for $0.5 < Q/10^{-20} < 1.0$, and to within 0.04 dex for larger values. The observed EWs of nine giants and supergiants derived with different methods show that the typical errors on individual measurements are about 0.3 dex in the Q parameter. In seven out of nine cases the determinations are in good agreement with the trend. For the two Iaf supergiants QZ Sge and HD190429, however, the observed EWs are clearly above the trend. These two stars show He II emission lines that are much stronger than the ones produced in our supergiant models, giving rise to the observed difference.

4. Geometry of the circumstellar material

Stellar photospheric parameters cannot be inferred for the embedded massive stars for which the spectra only show emission lines. These spectra originate in dense circumstellar material and do not contain direct information about the underlying star. As we are particularly interested in potential signatures of the formation process of massive stars such as a remnant of the accretion disk or abnormal wind properties, we devised a diagnostic tool that allow to probe the geometry of these emission line objects. Fig. 3 presents a diagram in which the line flux ratio of Hu(14-6)/Brα versus Hu(14-6)/Pfγ is plotted for different types of emission line objects. The aim of this diagram is to provide a simple means to investigate the nature of circumstellar gas in highly obscured sources. The diagram shows a clear trend. Both line flux ratios typically increase from LBV (hot massive post-main-sequence stars characterised by a dense ionised stellar wind) to B[e] (hot stars surrounded by large amounts of ionised and neutral/molecular material likely distributed in an extended or flared disk) and Be stars (hot stars surrounded by a gaseous disk). This trend can be understood in terms of the span in mass absorption coefficient between Hu(14-6), Pfγ and Brα (see Lenorzer et al. 2002b). In an optically thin medium one expects the line flux ratios to follow Menzel Case B recombination theory (indicated by the dashed area in Fig. 3, from Storey & Hummer 1995). However, in an optically thick medium the ratio becomes independent of mass absorption coefficient as the flux in any line is dominated by the size of the emitting surface. This limit is indicated by a filled star in Fig. 3 and falls close to the locus of the Be stars.

In first order, both Pfγ and Brα probe the emission measure of the gas. This explains why the stars are more-or-less concentrated around the diagonal. So, in principle one can do the analysis using only one of the two line

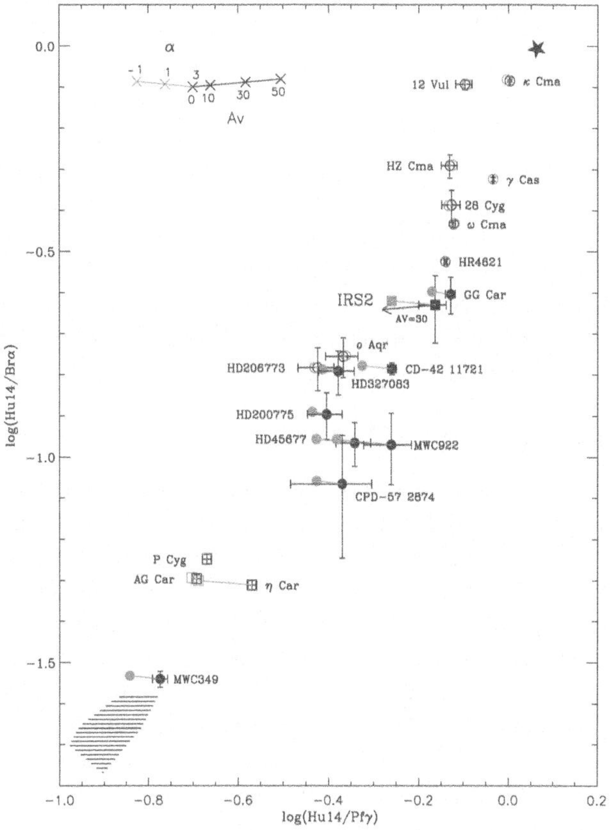

Figure 3. Hu(14-6)/Brα versus Hu(14-6)/Pfγ line ratio diagram for the hot stars observed with the Infrared Space Observatory (Lenorzer et al. 2002a). The different classes of objects, LBVs (squares), B[e] stars (filled circles) and Be stars (open circles), are well separated. The black symbols indicate the position of the line flux ratio, the grey symbols refer to the EW ratio. The thick asterisk indicates the position of optically thick black body emission; the striped region shows the range of ratios for Menzel case B recombination, including collisional de-excitation, for temperatures higher than 10^4K. Extinction and continuum slope will affect the position of the flux ratio and EW ratio, respectively, as indicated in the top left corner.

ratios. However, both lines have a specific advantage. The pro of Brα is that it is the strongest line, while the advantage of Pfγ is that it suffers less from possible contamination by nebular emission. Plotting the line fluxes relative to Hu(14-6), the strongest Humphreys series line in the L'-band, provides a better contrast between contributions from optically thin and thick material than does Brα relative to Pfγ.

Embedded massive stars suffer from large extinction that will affect their position in the line flux diagram. The line EW ratios, however are insensitive to extinction. The values of the EW ratios are also plotted in Fig. 3, they were shifted in both vertical and horizontal directions so that the EW ratios for P Cygni, which has negligible extinction, falls on top of its line flux ratio. The presence of an additional continuum source between 3.7 and 4.1 micron, e.g. due to dust, will affect the EW but not the line flux. Both the effect of extinction on line flux ratio and of continuum slope on EW are indicated at the top left corner of the diagram. If the spectra of embedded massive stars are absolutely calibrated, it is possible to plot both ratios, yielding an indication of the geometry of the ionised matter, as well as an estimate of the amount of extinction and dust continuum contribution.

5. Case study: NGC2024/IRS2

NGC2024/IRS2 is the near-infrared counterpart of the ultra-compact radio source G206.543-16.347 (Kurtz et al. 1994). Located in a star forming region (Orion B) and associated with one of the brightest IRAS sources following the UC H II criteria (despite an offset of 72.8$''$), it is a good candidate young massive star, possibly still showing signatures of its formation process. As such, this object was part of the near-infrared survey discussed in Sect. 2.

We obtained a K-band spectrum of IRS2 as part of the spectroscopic follow-up campaign conducted with VLT/ISAAC. This spectrum only shows emission lines originating in dense circumstellar material. To get a hint on the geometry of this material, we obtained an L'-band spectra with UKIRT/CGS4 and plotted the hydrogen line ratio in the diagram presented in Sect. 4. In Fig. 3, IRS2 falls in the upper right part of the diagram, close to the B[e] star GG Car. This position indicates the presence of optically thick gas and favours a disk-like geometry. The near-infrared spectra also contain lines of Mg II, Fe II, and Mg I indicating the presence of warm and dense material shielded from the ionising radiation of the star. These lines have full widths at half maximum comparable to that of the Pfγ and Humphreys series lines, i.e. about 400 km s^{-1}. Moreover the Brackett lines are narrower with full width at half maximum of about 200 km s^{-1}. This situation typically arises in a disk in which velocities decrease outwards and that can provide shielding close to the star. The observation of CO band-head emission at a temperature of about 4 000 K by Chandler et al. (1993,1995) is in agreement with this explanation. We therefore conclude that this star is surrounded by a gaseous disk that is dense enough to provide shielding and to reduce the temperature to 4 000 K in the mid-plane close to the star.

To investigate the evolutionary status of this disk we look for the signature of colder circumstellar material in the spectral energy distribution plotted in

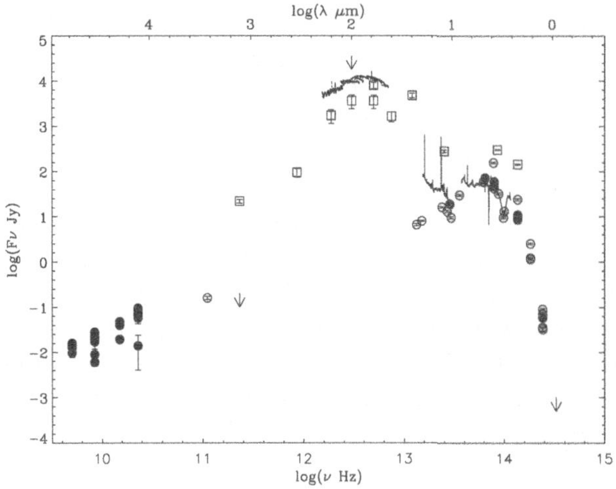

Figure 4. The spectral energy distribution of IRS2, compiled from published photometric data and SWS/LWS spectra taken with the Infrared Space Observatory (ISO). The lines represent high resolution ISO spectra. Open squares denote measurements made with apertures larger than 40″; open circles indicate aperture sizes of 5–20″; closed circles represent beam sizes less than 5″. Most often, the error bars on the data are smaller than the symbol size. The subdivision in aperture size clearly shows that the emission bump peaking at about 100 μm is not due to the nearby circumstellar medium of IRS2, i.e. it is not associated with the system itself, but must come from warm material in the vicinity of the star.

Fig. 4. The SED is built using data compiled from the literature (see Lenorzer et al. 2004b). The different symbols correspond to measurements that have been obtained with widely different apertures. Based on the comparison between the measurements taken at similar wavelength, in particular around 20 microns, we conclude that the bulk of the mid- and far-infrared emission observed in large apertures and peaking at $\sim$ 100 μm (T $\sim$ 30K) is *not* associated with the nearby circumstellar medium of IRS2, but arises from an extended source. No small aperture measurements are available at infrared wavelengths. However the low flux observed at 22 μm and 2.7 mm at resolution of 12 and 8″, respectively, puts strong limits on the amount of cold material in the vicinity of the star. Recent radio observations by L.F. Rodriguez (priv. com.) resolved the ultra-compact radio source. Its diameter is 0.8″ at 3.6 cm, corresponding to 145 AU at 360 pc (Brown et al. 1994). The amount of dust required to cause confinement of the ionised region would give rise to strong infrared emission, incompatible with the observations mentioned above. The limited size of the UCH II region therefore cannot be attributed to its youth, i.e. it is not an expanding bubble. More likely, the flow is dense enough to induce recombination. The slope and short timescale variability observed at radio wavelength also favours this hypothesis.

We conclude that NGC2024/IRS2 is surrounded by a very dense gaseous disk, denser than usually observed around classical Be stars, likely in the process of being evaporated. The region around this source is cleared up. There is no evidence for the presence of dust in the vicinity of the source nor for interaction between the disk and the molecular cloud. This may represent the final stage of a remnant accretion disk; however, we cannot exclude the possibility that the disk has a "stellar" origin as a result of e.g. rapid stellar rotation, pulsation and/or a magnetic field such as in Be stars. Other objects of the survey presented in Sect. 2 show similar results (Bik 2004). If these disks are remnant accretion disks, it would support the hypothesis that massive stars can form through accretion.

References

Bik, A. 2004, PhD Thesis: "The stellar content of high-mass star-forming regions"

Brown, A.G.A., de Geus, E.J., & de Zeeuw, P.T. 1994, A&A 289, 101

Chandler, C.J., Carlstrom, J.E., Scoville, N.Z., Dent, W.R.F., & Geballe, T.R. 1993, ApJ 412, L71

Chandler, C.J., Carlstrom, J.E., Scoville, N.Z. 1995, ApJ 446, 793

de Koter, A., Heap, S.R. & Hubeny, I. 1998, ApJ 509, 879

Hanson, M.M., Conti, P.S. & Rieke, M.J. 1996, ApJS 107, 281

Hillier, D.J. & Miller, D.L. 1998, ApJ 496, 407

Kaper, L., Bik, A., Comeron, F., Hanson, M.M. 2004, submitted to A&A

Klein, R.I. & Castor, J.I. 1978, ApJ 220, 902

Kudritzki, R.-P., Puls, J., Lennon, D.J. 1999, A&A 350, 970

Kudritzki, R.-P., Puls, J. 2000, ARA&A 38, 613

Kurtz, S., Churchwell, E., & Wood, D.O.S. 1994, ApJS 91, 659

Lamers, H.J.G.L.M., Haser, S., de Koter, A. & Leitherer, C. 1999, ApJ 516, 872

Lenorzer, A., Vandenbussche, B., Morris, P. et al. 2002, A&A 384, 473

Lenorzer, A., de Koter, A., & Waters, L.B.F.M. 2002a, A&A 386, L5

Lenorzer, A., Mokiem, M. R., de Koter, A., Puls, J.2004a, A&A 422, 275

Lenorzer, A., Bik, A., de Koter, A., Kurtz, S. E., Waters, L. B. F. M., Kaper, L., Jones, C. E., Geballe, T. R. 2004b, A&A 414, 245

Mathys, G. 1988, AAS 76, 427

Puls, J., Kudritzki, R.-P., Herrero, A. et al. 1996, A&A 305, 171

Schaerer, D., de Koter, A., Schmutz, W. & Maeder, A. 1996, A&A 312, 475

Schmutz, W., Hamann, W.-R. & Wessolowski, U. 1989, A&A 210, 236

Storey, P.J., Hummer, D.G. 1995, MNRAS 272, 41

Vacca, W.D., Garmany, C.D. & Shull, J.M. 1996, ApJ 460, 914

Vink, J.S., de Koter, A. & Lamers, H.J.G.L.M. 2000, A&A 362, 295

Vink, J.S., de Koter, A. & Lamers, H.J.G.L.M. 2001, A&A 369, 574

Wood, D.O.S. & Churchwell, E. 1989a, ApJ 340, 265

V. Trimble: This perhaps a question to you and all 3 previous speakers. Looking at the location in your 2D images of your ensemble of protostars, and the remaining cloud around them. Can you say whether they are likely to be the product of the kind of intermittent turbulence Dr. Nordlung talked about?
A. Lenorzer: We haven't done an in depth investigation of the morphology of these star forming regions. Some of them appear to show string like structures of H2 blobs. However, we interpret these structure as probing the edge of the molecular cloud and do not find indications that these shapes are remnants of possible filamentary structure predicted by MHD calculations.
K. Stepien: Why the presence of a disk-like structure should favor the formation of massive stars by an accretion process? Be stars possess disk-like structures which are not accretion disks.
A. Lenorzer: In principle you are right and the presence of a gaseous disk may not be a definitive proof that they are formed through accretion. Note however that the properties of the disks of stars like NGC2024/IRS2 and Be stars different: first the former are located in star-forming regions and are therefore likely to be younger than Be stars; second their disks seem to be much denser than Be stars disks.

WHAT CAN WE LEARN ABOUT THE SUN FROM OBSERVATIONS IN THE NEAR ULTRAVIOLET?

Ground based polarimetry and the role of SUNRISE

Achim Gandorfer
Max-Planck-Institut für Sonnensystemforschung
Max-Planck-Strasse 2, D-37191 Katlenburg-Lindau, Germany
gandorfer@mps.mpg.de

Abstract Solar magnetic fields can be observed using different techniques. While the Zeeman effect has proven to be an excellent diagnostic tool for strong isolated magnetic fields in the photosphere, a rigorous exploitation of the diagnostic potential of scattering polarization and its modification by the Hanle effect has started only recently. Scattering polarization is more pronounced at shorter wavelengths, due to a number of physical arguments. The near ultraviolet part of the solar spectrum therefore ideally complements the visible and near infrared portions of solar radiation, which are traditionally used in Zeeman imaging. In this talk I will review our recent achievements in near UV polarimetry and give a short outlook on the future polarimetric observations of chromospheric magnetism. In parallel, I will discuss the UV capabilities of the SUNRISE balloon-borne telescope, which will in an ideal way complement our ground based observations.

Keywords: Solar magnetic fields, Polarimetry, Scattering, Zeeman effect, Hanle effect

1. Introduction: magnetometry of the solar photosphere

The key to our understanding of the Sun's magnetic activity is the photospheric magnetic field. The photosphere is the main interaction region, where all energy densities are comparable, which leads to a rich variety of magnetic structures, from sunspots and active regions, down to the smallest observable magnetic elements. Here, in the photosphere, the Sun's magnetic field shows its highest complexity, and the question, why we try to understand the field where it is mostly complex, is indeed justified. The answer is found on the Sun itself: the photosphere, where the field is most complicated, provides us with a maximum of information: The Fraunhofer spectrum contains a multitude of spectral lines, and different magnetic effects leave their fingerprints, which we

J. C. del Toro Iniesta (eds.), The Many Scales in the Universe, 187–196.

can observe. Inversion codes connect our observables to physical quantities, that we can compare with the results of powerful numerical three-dimensional magnetohydrodynamical simulations.

In solar magnetometry we can use three different techniques: While Zeeman and Hanle effect based diagnostics both directly address the magnetic field via spectropolarimetry, proxy magnetometry is an indirect technique, where one tries to detect the signatures of magnetic fields without measuring the field itself.

Proxy magnetometry

Proxy magnetometry means the detection of signatures of magnetic fields without measuring the field itself. In the outer solar atmosphere, the best known proxies of magnetic fields are coronal loops. The loops that we see are not magnetic field lines, we only see matter. Our physical intuition tells us that the material is outlining magnetic fields. Realistic physical models can then be established to put our intuitive model on a firm basis. Not only in the outer layers of the Sun's atmosphere, but also in the photosphere we can use proxies: Imaging in molecular bands has proven to be a powerful technique to detect otherwise hidden magnetic elements in the quiet solar granulation. Mostly the so-called G-band (a molecular absorption band of the CH molecule around 430 nm) is used (for further information on G-band analysis we here refer to a recent paper by Shelyag et al. (2004) and references therein). Since the filter passbands can be chosen fairly wide (typically 1 nm), short time exposures near the diffraction limit of our current high resolution telescopes can be obtained.

Zeeman diagnostics

Zeeman diagnostics has been the standard magnetometric technique in solar research for decades. The Zeeman effect consists in the magnetically induced splitting of spectral lines. Magnetic field strength and orientation lead to a characteristic polarization signature over a spectral line. By analysing this signature one thus can deduce the magnetic field strength as well as its orientation. The Zeeman effect is an ideal remote sensing technique for isolated strong field components, since it works best when the Zeeman splitting is comparable to the Doppler width of the spectral line: This is typically the case for kG fields in the photosphere. Zeeman diagnostics is a high resolution technique, both in the spectral and the spatial domain, since the contributions from mixed polarity fields within the resolution element tend to cancel each other. The Zeeman effect is therefore not suitable for turbulent fields. Since the sensitivity of the Zeeman effect scales with wavelength, Zeeman effect based diagnostics is best performed in the near infrared.

Hanle diagnostics

The Hanle effect describes the polarization effects during scattering in the presence of a magnetic field. In a very simplistic picture (that shall help us to imagine the main manifestations of the Hanle effect) it can be understood as a precession of the scattering particle (atom or molecule) around a magnetic field line during the scattering process. For an observer of the scattered radiation, the Hanle effect manifests itself in two ways, in a depolarization of the scattered radiation as compared to the nonmagnetic scattering, and in a rotation of the polarization plane. While the rotation of the polarization plane depends on the field orientation, the depolarization is (to first order) independent of it. This means that for a turbulent field there will always be an observable depolarization effect. The other important aspect of the Hanle effect is the sensitivity range: The Hanle effect has its highest sensitivity when the corresponding Zeeman splitting is comparable to the *natural* line width. For typical lines we thus have sensitivities on the order of several Gauss, perfectly complementary to Zeeman diagnostics. For a detailed and up-to-date introduction to the Hanle effect and its application in solar physics we refer to Stenflo (1994) and Trujillo Bueno (2003) and references therein.

2. Scattering polarization and the hanle effect in the near UV

Before we can use the Hanle effect as a standard diagnostic tool for solar magnetic field research, we have to observe and understand the non magnetic line scattering polarization. Scattering polarization can best be observed at the limb of the Sun. The temperature and density stratification in the solar photosphere leads to an illumination anisotropy, which expresses itself mainly in the well known limb darkening. If we observe away from disk center, our line of sight and the illumination cone (which is symmetric around the radius vector of the Sun) define a plane and break the rotational symmetry. Observed scattered radiation will therefore be linearly polarized with the electric field in general being perpendicular to the scattering plane, thus parallel to the nearest solar limb. The polarization amplitudes are highest for large scattering angles, thus near the limb of the Sun.

The polarization signatures are generally very weak. They can be found in the continuum radiation as well as in the Fraunhofer lines. The linear polarization observed at the limb shows a strong wavelength dependence, since different physical mechanisms contribute to the spectral structuring: Atomic physics enters via the "polarizability" of the line. The continuum radiation is mainly dominated by Rayleigh scattering at hydrogen, showing the typical λ^{-4} increase when going to shorter wavelengths. Also the solar atmosphere comes into play, since the radiation anisotropy as the primary source of the polarization is wavelength dependent. In addition radiation transfer effects influence the spectral shape of the polarization signatures. All this leads to a whole spectrum of the degree of linear polarization, which is therefore nowadays called the "second solar spectrum".

A rigorous exploration of the second solar spectrum has been possible only after the advent of highly sensitive spectropolarimeters. The key instrument in scattering polarimetry has been the Zurich Imaging Polarimeter ZIMPOL (Povel 1995) that opened the new window for solar spectroscopy (Stenflo and Keller 1996). A high resolution atlas of the second solar spectrum has been published by Gandorfer (2000, 2002), based on observations with the ZIMPOL II instrument at the IRSOL facility in Locarno.

The near UV part of the solar spectrum is of particular interest for scattering polarimetry for different reasons: Firstly, the number density of lines dramatically increases when going to shorter wavelengths as a consequence of the higher density of states; secondly, the amplitude of the scattering signals in the lines increases, since the radiation anisotropy (which is the primary source of the polarization) increases with decreasing wavelength, and finally, the polarization of the continuous spectrum steeply increases when going to the near UV (Stenflo 2004). The structural richness of the scattering polarization in the UV can be seen exemplarily in Fig. 1.

Although thorough investigations of the Hanle effect have been carried out earlier in distinct spectral lines in the near UV (Bianda et al. 1998a,b) with the IRSOL spectropolarimeter, a systematic recording of the second solar spectrum in the wavelength range below 460 nm down to the atmospheric cut-off around 310 nm has only been feasible with the UV sensitive version of the ZIMPOL II polarimeter (Gandorfer et al. 2004). While the wavelength range from 3910 Å to 4630 Å was measured at the IRSOL facility (Gandorfer 2002), the part below 390 nm could only be registered at the 1.5 m McMath-Pierce telescope of NSO (Kitt Peak) (Gandorfer 2004).

After the rigorous systematic exploration of the second solar spectrum we now have to understand the various spectral signatures. Only if we understand the formation of the non-magnetic line polarization we will be able to use the second solar spectrum as a standard tool in solar magnetometry.

A lot of progress has been made in theoretically explaining the observed features and to further constrain the magnitude of the turbulent field (c.f. Landi degl'Innocenti 1998; Faurobert et al. 2001; Manso Sainz et al. 2004; Trujillo Bueno et al. 2004). Differential Hanle effect techniques help to minimize the influence of a specific model dependence on the magnetic field determination (c.f. Stenflo et al. 1998).

Most scattering signatures are too weak to allow for spatially resolved work (in most cases the polarization signals have been averaged along the spectrograph slit, which is then placed parallel to the nearest solar limb, to enhance

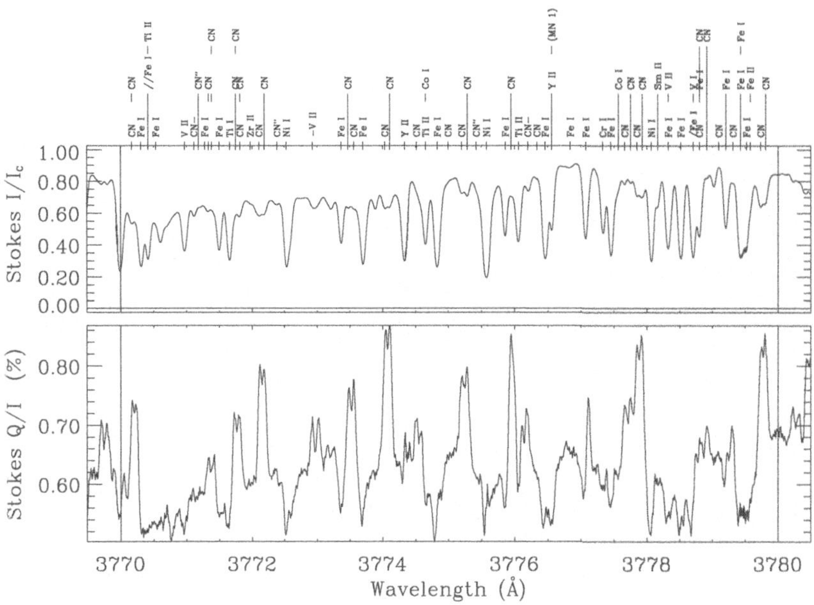

Figure 1. Detail of the Fraunhofer spectrum (top panel) and the second solar spectrum (lower panel) around 377 nm. In the fractional polarization many signals of the CN molecule are visible, providing complementary information in addition to normal spectroscopy of the solar atmosphere.

the signal-to-noise ratio while keeping control over the limb distance). Only in a very limited number of spectral lines with high polarization amplitudes investigations with moderate resolution have been possible. Since most of these strongly polarizing lines are found in the near UV part of the solar spectrum, the UV is the region of choice for Hanle diagnostics.

A spectral line that has proven to be particularly suited for this purpose is the Ca I line around 4227 Å(Bianda et al. 1998a). ZIMPOL II now allows investigations with much higher spatial resolution while even increasing the polarimetric sensitivity (Bianda et al. 2003). An example of such an observation is shown in Fig. 2. The measurement of the complete Stokes vector in the spectral region around the strong Ca I line around 4227 Å has been recorded with the ZIMPOL II polarimeter at the vertical spectrograph of the 1.5 m McMath-Pierce telescope of NSO (Kitt Peak) in March 2002. Observations like this one unveil complicated polarization signals of both, Hanle and Zeeman effect, and are therefore of particular interest for the understanding of chromospheric magnetic fields.

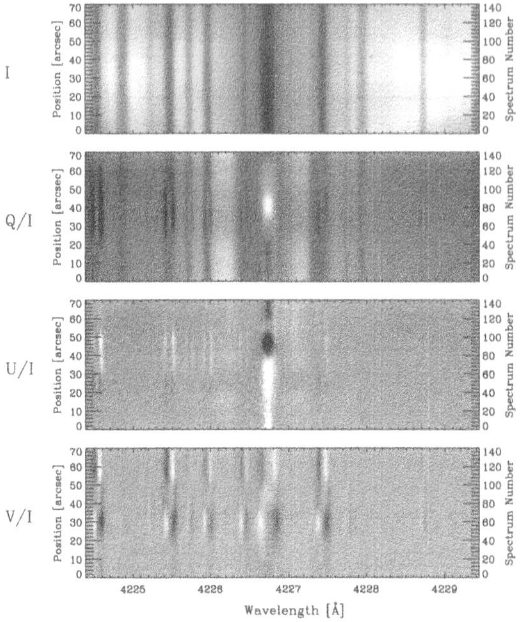

Figure 2. Scattering polarization signatures in the Ca I 4227 Å line, unveiling polarization effects of both, Hanle and Zeeman effect, being of particular interest for the understanding of chromospheric magnetic fields. The spectrograph entrance slit of the 1.5 m McMath-Pierce solar telescope has been set parallel to the solar limb at a distance of 10 arcsec inside the visible solar limb and covers 70 arcsec. The spectral field of view is centered on a line of neutral calcium. White and black correspond to $\pm 1 \times 10^{-2}$ in Stokes Q/I and V/I, and to $\pm 4 \times 10^{-3}$ in Stokes U/I.

The spatial topology of scattering polarization can be investigated by imaging polarimetry with narrowband filters. Those observations can in principle be done with very high spatial resolution.

As an example we show a region of moderate magnetic activity near the solar limb. The scene has been imaged with ZIMPOL II at the New Swedish Solar Telescope through a 1 Å bandpass interference filter centered at the core of the Ca II K line around 3933 Å. In contrast to the Ca II H line around 3968 Å the K line exhibits strong resonance polarization (c.f. the atlas of Gandorfer 2002). Both lines, however, are sensitive to Zeeman effect (Martínez Pillet et al. 1990). Since the transmission band is fairly wide, Zeeman polarization signals are expected to be below the detection limit, due to cancellation effects. The linear polarization caused by scattering, as well as its modification by local weak magnetic fields give, however, detectable linear polarization (see Fig. 3) not only in a narrow limb zone but also above the visible solar limb in spicules, as well as in regions corresponding to network boundaries. While circular polarization due to the Zeeman effect is not detected due to spectral smearing, signatures of Hanle rotation are prominent in the limb zone as well as in the observed spicules. Although the present example is of very preliminary character, observations of this type harbour great potential for diagnostics of chromospheric magnetic fields.

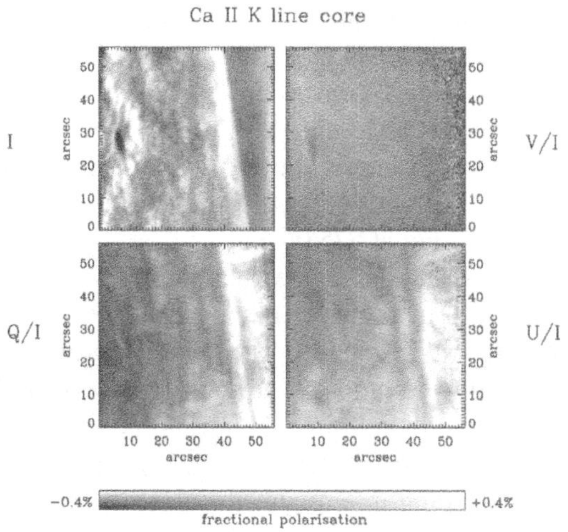

Figure 3. Images of a moderately active region at the solar limb. The four Stokes images were recorded with ZIMPOL II at NSST on La Palma in October 2003, using a 1 Å filter centered on the Ca II K line. The contrast in the intensity image has been enhanced by subtraction of the general center-to-limb variation. The gray scale of the fractional polarization images is calibrated at the bottom of the picture.

3. Solar magnetometry at high spatial resolution: The role of SUNRISE

SUNRISE is a balloon-borne solar telescope with an aperture of 1 m, working in the UV/VIS optical domain (Solanki et al. 2002). The main scientific goal of SUNRISE is to understand the structure and dynamics of photospheric and chromospheric magnetic field. SUNRISE will provide uninterrupted time series of diffraction-limited images and spectra of the photosphere and chromosphere with an unprecedented resolution down to 35 km at wavelengths around 220 nm. Focal-plane instruments are a spectrograph/polarimeter, a Fabry-Perot filter magnetograph, and a filter imager. The first stratospheric long-duration balloon flight of SUNRISE over Antarctica is planned in winter 2008. SUNRISE is a joint project of the Max-Planck-Institut für Sonnensystemforschung (MPS), Katlenburg-Lindau, with the Kiepenheuer-Institut für Sonnenphysik (KIS), Freiburg, the High-Altitude Observatory (HAO), Boulder, the Lockheed-Martin Solar and Astrophysics Lab. (LMSAL), Palo Alto, and the spanish IMaX consortium.

Observational philosophy. The SUNRISE postfocus instrumentation consists of 5 units, three out of which are science instruments, the other two are system units for image stabilization and light distribution.

While a detailed technical description of the individual science instruments is beyond the scope of this paper, we will list their specific roles here. For further information on the SUNRISE instruments we here refer to Gandorfer et al. (2004), and Martínez Pillet et al. (2004).

The *Imaging Magnetograph eXperiment* (IMaX) is an imaging vector magnetograph based upon a tunable narrow-band filter. The instrument, which is currently developed by the spanish IMaX consortium, will provide fast-cadence two-dimensional maps of the complete magnetic vector, the line-of-sight velocity, and continuum frames with unprecedented spatial resolution. IMaX images will be taken in two to four narrow wavelength bands in either wing of the photospheric spectral line of Fe I at 525.06 nm.

One of the main science goals of SUNRISE is the quantitative and accurate measurement of the strength and orientation of the magnetic field with appropriate spatial, spectral, and temporal resolution. The *Sunrise Polarimetric Spectrograph* (SUPOS) allows high-resolution vector-polarimetry, simultaneously providing photospheric magnetic field measurements (polarimetric branch) and diagnostic spectroscopy (diagnostic branch). SUPOS is based on an all-mirror scanning Echelle spectrograph in a modified Littrow configuration, similar to the POLIS instrument (Schmidt et al. 2001), now installed at the German Vacuum Tower Telescope. The polarimetric branch of SUPOS is dedicated to the determination of the magnetic field vector in the solar photo-

Figure 4. Sketch of the telescope in the gondola in landing position at zero elevation. The instrument platform is attached to the telescope central frame on top of the telescope. The instrument platform houses all three science instruments (SUPOS, SUFI, and IMaX) as well as the Image Stabilization and Light Distribution system and a Correlating Wavefront Sensor.

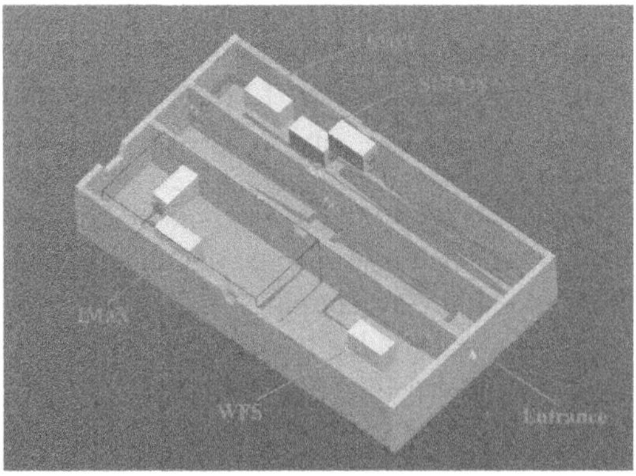

Figure 5. Sketch of the arrangement of the postfocus instrumentation on the instrument platform. The envelope of the optical light paths is visible. Individual optical components are omitted in this picture. The CCD cameras of the three science instruments and the wavefront-sensor camera are shown. Light from the main telescope enters from the right, as indicated.

sphere by measuring the full Stokes vector in the Fe I lines around 630.2 nm. In the so-called diagnostic branch spectra of Mg II k will be recorded. The Mg II k line at 279.6 nm is an excellent diagnostics tool for the temperature structure in the chromosphere.

The *Sunrise Filter Imager* (SUFI) samples the photosphere and chromosphere in four wavelength bands. The channel at 225 nm allows studies of the upper photosphere and lower chromosphere at a spatial resolution of 0.05 arcsec (35 km on the Sun). The OH-band at 313 nm and the CN-band at 388 nm provide high contrast, and thus sensitivity to thermal inhomogeneities in the photosphere (Steiner et al. 2001). They are therefore also ideal proxies of magnetic elements with even higher contrast as compared to G-band imaging (Zakharov et al., in preparation).

4. Summary

We are currently entering a golden age in solar magnetometry. Highly sensitive spectropolarimeters in combination with large ground based solar telescopes enable us to use Hanle effect diagnostics of weak and turbulent magnetic fields on the solar surface.

However, due to the intrinsic weakness of the scattering signals, Hanle diagnostics will always be a low spatial resolution technique.

Magnetometry at highest spatial resolution will in future be possible from space. On longer timescales ESA's Solar Orbiter mission will be a milestone on our way to understand solar magnetism. Within the following years, SUNRISE will push the frontier in solar magnetometry by allowing for proxy magnetometry and Zeeman diagnostics with unprecedented spatial resolution.

In my opinion we need a combined approach in solar magnetometry: Zeeman and Hanle diagnostics are ideally complementary techniques for quantitative magnetic field research. While high resolution observations, using both, proxy and Zeeman techniques, are best performed above the atmosphere, Hanle diagnostics can perfectly be done from ground. It will ideally complement our attempts to finally understand the Sun's magnetic field.

References

Bianda, M., Solanki, S. K., & Stenflo, J. O. 1998a, A&A, 331, 760

Bianda, M., Stenflo, J. O., & Solanki, S. K. 1998b, A&A, 337, 565

Bianda, M., Stenflo, J. O., Gandorfer, A., & Gisler, D. 2003, in *Current Theoretical Models & Future High Resolution Solar Observations: Preparing for ATST*, ASP Conference Series, 286, 61

Faurobert, M. & Arnaud, J. & Vigneau, J. & Frisch, H. 2001, A&A 378, 627

Gandorfer, A. 2000, *The Second Solar Spectrum*, Vol. 1. Zürich: vdf. ISBN 3 7281 2764 7

Gandorfer, A. 2002, *The Second Solar Spectrum*, Vol. 2. Zürich: vdf. ISBN 3 7281 2855 4

Gandorfer, A. 2004, *The Second Solar Spectrum*, Vol. 3. in preparation.

Gandorfer, A. M., Povel, H. P., Steiner, P., Aebersold, F., Egger, U., Feller, A., Gisler, D., Hagenbuch, S., Stenflo, J. O. 2004, A&A 422, 703

Gandorfer, A. M., Solanki, S. K., Schüssler, M., Curdt, W., Lites, B. W., Martínez-Pillet, V., Schmidt, W., & Title, A. M. 2004, SPIE, 5489, 732

Landi–Degl'Innocenti, E. 1998, Nature, 392, 256

Manso–Sainz, R. & Landi–Degl'Innocenti, E. & Trujillo–Bueno, J. 2004, ApJL, 614, L89

Martínez–Pillet, V., Garcia–Lopez, R. J., del Toro–Iniesta, J. C., Rebolo, R., Vazquez, M., Beckman, J. E., Char, S. 1990, ApJL, 361, L81

Martínez–Pillet, Bonet, J. A., V., Collados, M. et al. 2004, SPIE 5487, in press.

Povel, H. P. 1995, Optical Engineering 34, 1870

Shelyag, S., Schüssler, M., Solanki, S. K., Berdyugina, S. V., & Vögler, A. 2004, A&A, 427, 335

Schmidt W., Kentischer T., Bruls J., Lites B. 2001, in: *Advanced Solar Polarimetry — Theory, Observation & Instrumentation*, Ed. M. Sigwarth, ASP Conference Series, 236, 49

Solanki, S. K., Schüssler, M., Curdt, W., Lites, B. W., Martínez–Pillet, V., Schmidt, W., Title, A. M., & the Sunrise Team 2002, ESA SP-505, 27

Steiner, O., Bruls, J., & Hauschildt, P.H. 2001, in *Advanced Solar Polarimetry — Theory, Observation & Instrumentation*, Ed. M. Sigwarth, ASP Conference Series, 236, 453

Stenflo, J. O. 1994, *Solar magnetic fields: polarized radiation diagnostics*, Kluwer Academic Publishers.

Stenflo, J. O. 2004, A&A, in press.

Stenflo, J.O., Keller, C.U., 1996, Nature, 382, 588.

Stenflo, J. O., Keller, C. U., & Gandorfer, A. 1998, A&A 329, 319

Trujillo–Bueno, J. 2003, in: *Solar Polarization 3*, Eds. J. Trujillo–Bueno, & J. Sánchez–Almeida, ASP Conference Series, 307, 424

Trujillo–Bueno, J., Shchukina, N., Asensio–Ramos, A. 2004, Nature, 430, 326

OUR MAGNETIC SUN

E.R. Priest
Mathematics Institute, St Andrews University
St Andrews KY16 9SS, UK
eric@mcs.st-and.ac.uk

Abstract Solar Physics is at present in a highly dynamic state with many new observations from a series of major space satellites that are transforming our basic understanding. This review summarises some of the key recent discoveries about our Sun on fundamental topics which have many implications for fundamental plasma processes elsewhere in the universe. In particular, the solar corona is dominated by the subtle nonlinear behaviour of the Sun's magnetic field and its interaction in complex ways with plasma.

Keywords: Sun, MHD

1. Introduction

It is a great pleasure for me to attend JENAM. As a solar physicist, I have always enjoyed collaborating with many friends in other European countries and encouraging closer links between solar physics and astronomy. I work at St Andrews University, founded nearly 600 years ago in 1411, when the Moors were at their height in Granada.

For centuries humans have worshipped the Sun, and indeed many flock to the Costa del Sol to continue that tradition, but why should we study the Sun today? Firstly, because it is of very great interest in its own right. Secondly, because it has profound influences on the Earth and its climate. Thirdly, because it is of key importance for astronomy as a whole, since we can study fundamental cosmic processes at work on the Sun. For me, however, I study the Sun simply because I am captivated and fascinated by it.

Many of the basic properties of the Sun are still a mystery. For example, we do not know: how the Sun's magnetic field is generated; how the solar wind is accelerated; how the corona is heated; how ejections of mass occur or how particles are accelerated in solar flares. Nevertheless, great progress on each of these questions has been made over the past five years and they have all been greatly refined and unpacked. So, the Sun is at present one of the liveliest branches of astronomy and many students are being attracted to study it.

J. C. del Toro Iniesta (eds.), The Many Scales in the Universe, 197–210.

Figure 1. A picture of James Gregory in his laboratory.

Traditionally, there has always been a close link between mathematics and astronomy. This was certainly true of Andalucian astronomy here in Granada in the fifteenth century, and it was also true in St Andrews in the seventeenth century. I happen to hold the Gregory chair of mathematics, named after James Gregory (1638-1675), who invented the gregorian telescope and was also one of the co-founders of calculus (with Newton and Leibniz). He was elected an FRS (an academician) in 1668 at the age of 30, and the same year was appointed the first regius professor of mathematics in St Andrews. He was given what is now known as Upper Parliament Hall (Fig. 1) as his laboratory, and you can still see today the meridian line on the floor along which he lined up his telescope. James Gregory died in 1675 at the age of only 37, but before that he had been the first to discover many basic elements of calculus that we take for granted today, such as: the general binomial theorem, Taylor expansions (40 years before Taylor), the ratio test for convergence of a series, the series expansions for sin x and tan x, the integrals of sec x and log x, the fact that differentiation is the inverse of integration, and how to use a change of variable in integration.

In my view the role of theory should not be to reproduce images or to explain every observation, but rather to understand the basic processes, and to do so in a step-by-step manner, starting with a simple model and gradually making it more and more sophisticated and hopefully more realistic. In addition, the different types of theory complement one another, namely, analytical theory, computational experiments and data analysis.

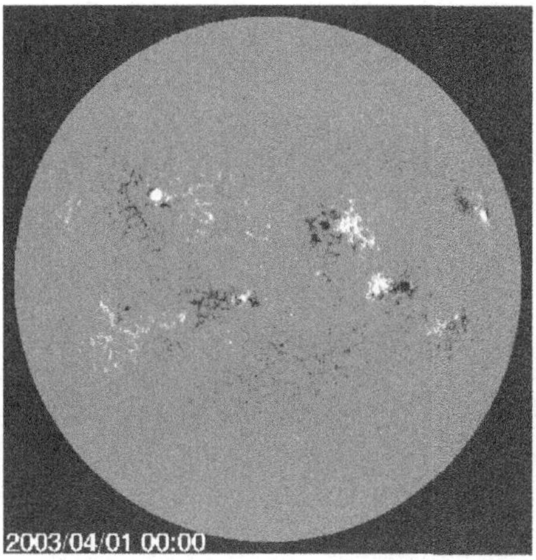

Figure 2. A magnetogram of the photospheric magnetic field.

My title today is "Our Magnetic Sun" in the sense both that magnetic fields are crucial for the Sun and also that the Sun is a highly attractive object of study. After introducing you to the structure of the Sun (Sect. 1) and sunspots and the corona, I shall describe briefly MHD and reconnection (Sect. 3) and will go on to discuss three topics on which there has been progress recently, namely, the solar interior, erutpive solar flares and coronal heating (Sect. 4). My overall theme is that the Sun is full of surprises and a subtheme is that many of these surprises are caused by the magnetic field. The Sun is of course a plasma and so is coupled in an intimate, subtle way to the magnetic field, which exerts a force on a plasma and can store substantial amounts of energy.

2. Overall structure of the Sun

The interior of the Sun consists of a core (extending out to about $0.25R_0$, where the radius (R_0) of the Sun is about 700 Mm), a radiative zone, and a convective zone (extending between $0.7R_0$ and the surface). The atmosphere consists of the photosphere (the surface layer at about $6000K$), the rarer chromosphere, and the corona (which reaches to the Earth and beyond).

The photosphere is covered with turbulent convection cells, namely, granulation (with typical diameters of 1 Mm) and supergranulation (with diameters of 15 Mm). It rotates differentially, with the equator rotating more rapidly than the polar regions. A map of the photospheric magnetic field (Fig. 2) shows the line-of-sight magnetic field, with white being directed towards you and black away from you. The active regions around sunspots show up as large-scale bipolar regions.

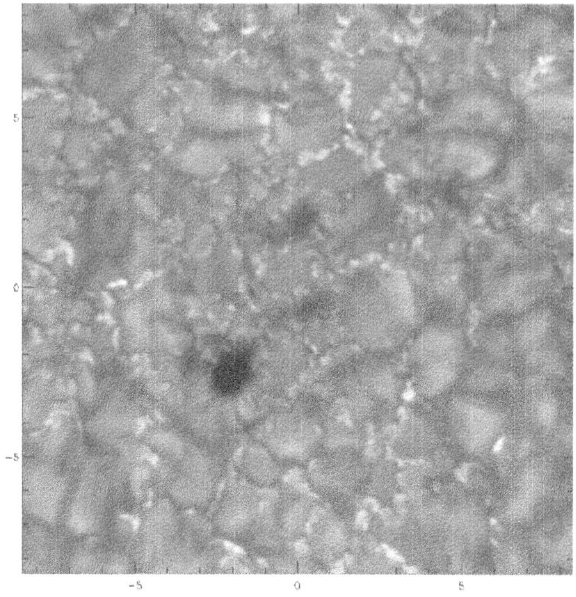

Figure 3. A close-up of a few granule cells in the photosphere (Swedish Solar Telescope, La Palma, G. Scharmer).

One surprise is that outside active regions intense magnetic fields cover the whole Sun, concentrated at the boundaries of supergranules, whence they are carried by the supergranule flow. Another is that active regions form a global pattern, with black polarity to the left in the northern hemisphere in Fig. 2 and to the right in the southern. This pattern reverses with the start of a new sunspot cycle. It occurs because differential rotation in the interior shears up poloidal flux and creates toroidal flux of one polarity in the north and the opposite polarity in the south. It is when such toroidal flux rises by magnetic buoyancy that it creates a pair of sunspots where it breaks through the surface.

Amazing images in white light at 0.1 arcsec from the Swedish Solar Telescope reveal incredible detail in and between the granulation (Scharmer et al. 2002, van der Voort et al. 2004). Tiny bright points are probable locations of intense magnetic flux tubes at the edges of supergranules, where it was thought until this year that 90 % of the quiet-Sun flux is located. However, close-ups of a few granules (Fig. 3) reveal for the first time bright points, "flowers" and ribbons in the intergranular lanes around granules and suggest the presence of many more intense magnetic tubes throughout the centres of supergranules at the granule boundaries. Indeed, Trujillo-Bueno et al. (2004) have suggested five or six times as much magnetic flux there than we thought previously.

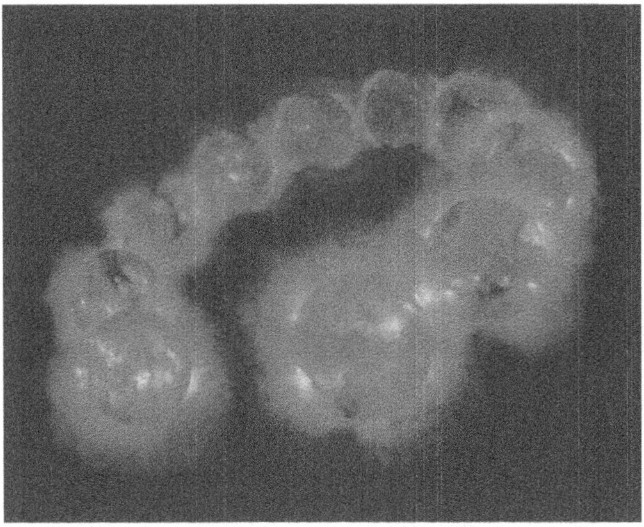

Figure 4. The corona over a solar cycle from Yohkoh (S Tsuneta).

Sunspots are dark because they are cool, which in turn is because the magnetic field stops granulation. They vary with the 11-year cycle. The dark umbra of a spot is the central part where the magnetic field is close to vertical. The striated penumbra surrounds the umbra and possesses a spreading magnetic field that is far from vertical. However, the penumbra is certainly not simple: recent observations from the Swedish Solar Telescope (Scharmer et al. 2002) have revealed bright flows moving both inward and outward, together with strange dark cores. Also, the bright penumbral filaments are thought to be at intermediate angles to the vertical and to represent magnetic field lines that go far from the sunspot, whereas the dark filaments are lower-lying and so return to the solar surface close to the spot, probably held down by granular pumping (Thomas et al. 2002, Weiss et al. 2004).

3. The corona

The corona can be viewed during a solar eclipse, and an early surprise was the discovery by Edlen (1940) that the coronal temperature is a million degrees or so. In the corona, the magnetic field dominates the plasma, both heating it somehow and creatings its beautiful structure. The corona can also be observed direct with an X-ray or euv telescope (Fig. 4), and indeed Yohkoh has revealed it to be a magnetic world with an amazingly rich variety of MHD phenomena.

Earlier rocket images and images from Skylab showed that the corona consists of dots called x-ray bright points, together with coronal holes (dark regions from which the fast solar wind escapes), coronal loops and active re-

Figure 5. Coronal loops imaged by TRACE.

gions which are rather fuzzy. However, the TRACE mission has shown active regions in incredible detail (Fig. 5) and that the corona is made up of intricate loops of plasma aligned along the magnetic field.

4. MHD and reconnection

Magnetohydrodynamics models the interaction between a magnetic field and a plasma treated as a continuous medium. It comprises a set of partial differential equations for the behaviour of the plasma velocity ($\mathbf{v}$), magnetic field ($\mathbf{B}$), pressure (p), temperature (T) and density (ρ) as functions of (x, y, z, t). The equations are nonlinear and to me as beautiful as the Sun itself.

For example, the induction equation

$$\frac{\partial \mathbf{B}}{\partial t} = \nabla \times (\mathbf{v} \times \mathbf{B}) + \eta \nabla^2 \mathbf{B}, \tag{1}$$

describes how the magnetic field changes in time due to two terms on the right-hand side, which represent the transport of magnetic field with the plasma and the diffusion of magnetic field through the plasma. The electric current is then determined as a secondary variable (once the magnetic field has been found) from

$$\mathbf{j} = \nabla \times \mathbf{B}/\mu.$$

In most of the universe the transport term in (1) is very much larger than the diffusion term and so the magnetic field is frozen to the plasma and hangs on to

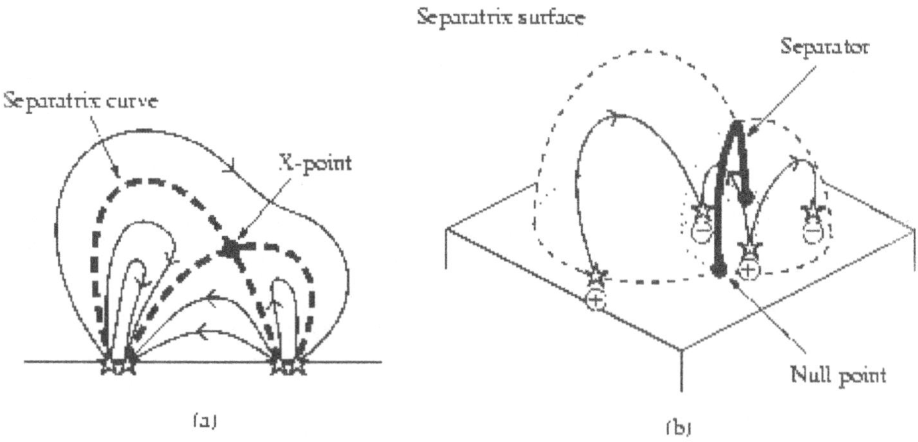

Figure 6. Complex topology in (a) 2D and (b) 3D.

its energy. The exception is in singularities where the magnetic gradient (and therefore the electric current) is extremely large.

Such singularities can form at null points of the magnetic field (where **B** vanishes and near which the magnetic field lines have an X-type topology). In the singularities, magnetic field lines break and reconnect and the inflowing magnetic energy is converted into heat, kinetic energy and fast particle energy. The singularities take the form of sheets containing extremely large currents and therefore being the sights of large ohmic heating. Such reconnection is thought to be at the core of solar flares and of coronal heating events.

Recently, attention has turned to trying to develop theories for reconnection in three dimension (e.g., Priest & Forbes 2000). However, so far the theory is in a rudimentary state since many features of 3D reconnection are completely different from 2D. For example, null points have a different structure and reconnection can occur either at nulls or in the absence of nulls. Also, the topology of the field is more complex. In 2D the field due to four sources of alternating sign in a line possesses an X-type null point (Fig. 6a), from which emanate four separatrix curves: these special field lines divide the plane up into topologically distinct regions, in the sense that in each region all the field lines start from the same positive source and end at the same negative source. In 3D, in contrast, the field due to four sources (such as sunspots) on a plane possesses two separatrix surfaces (or *separatrices*) in the form of domes (Fig. 6b) which separate the volume into topologically distinct regions. These surfaces can intersect in a special magnetic field line called a *separator* which joins one null point to another.

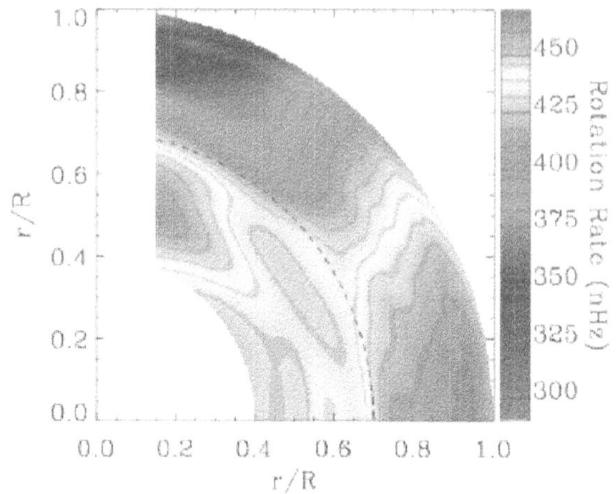

Figure 7. The Sun's internal rotation deduced from the MDI instrument on SoHO (A. Kosovichev).

5. The solar and heliospheric observatory (SOHO)

SOHO was launched in 1995 and is orbiting the Sun at the L1 point in phase with the Earth. A joint ESA-NASA mission, it is observing the Sun continuously for the first time and has transformed our understanding of the Sun. It has produced many surprises and I only have time to describe three of them today.

The Solar Interior

The first question is: what is the structure of the solar interior? Just as a Spanish guitar string can oscillate beautifully in different normal modes, so can a three-dimensional object such as the Sun. Indeed, several million different normal modes of vibration of the Sun have now been discovered, and with the MDI instrument on SoHO the line-of-sight velocity at a million points on the solar surface per minute is being measured. By summing such observations for several months and using techniques of solar seismology, the temperature inside the Sun as a function of radius has been measured and found to agree with the standard model to within 1%.

More interestingly, the MDI observations have also been used to deduce the Sun's internal rotation. At the solar surface we know the equator rotates faster than the polar regions, but how is this differential rotation continued into the solar interior? The expectation was that the rotation would be constant on cylinders around the Sun's axis of rotation and that the magnetic field would be generated by dynamo action throughout the convection zone. The surprise from MDI (Fig. 7) is that the rotation tends to be constant along cones in the convection zone and is then fairly uniform in the radiative zone, which implies the existence of a strong shear layer at the base of the convection zone (known as the tacocline). In addition, it is now thought that the magnetic field of sunspots and active regions is generated in and near the tacocline, so that theorists are now attempting to build tacocline dynamo models.

In another important development, methods of local helioseismology are being used to probe the structure below sunspots and it has been discovered that there is a region of reduced wave speed by 10 % below a sunspot for a distance of about a sunspot radius, presumably because of the reduced temperature there, and a second region of enhanced wavespeed for a distance of about 3 sunspot radii, presumably because the temperature is no longer cooler but the magnetic field is enhanced.

Solar Flares and Coronal Mass Ejections

A second important question is: how do eruptive solar flares and coronal mass ejections occur? The LASCO instrument on SoHO is a coronagraph which has discovered huge ejections of mass called *Coronal Mass Ejections (CME's)*, which can sometimes reach the Earth and disrupt communications and space satellites.

On October 28 last year an incredibly large and complicated group of sunspots was crossing the solar disc and spawned the 3rd largest solar flare ever recorded. It produced a halo CME, namely, one that produces a halo round the Sun (Fig. 7) since it is either coming right towards the Earth or is moving away from it. It was travelling at 2000 $km\ s^{-1}$, five times faster than normal. High-energy particles taking only an hour to reach SoHO (by comparison with the CME itself, which takes a couple of days) produced "snow" as they bombarded the CCD detector plates (see Fig. 8), and when the CME did reach Earth it produced beautiful aurora that we viewed eagerly in St Andrews for a couple of nights. One week later, when the sunspot group had reached the limb of the Sun, the fireworks continued as it gave birth to the largest solar flare ever recorded.

The overall picture of what happens in an eruptive flare is that a sheared and twisted magnetic tube with an overlying arcade either loses equilibrium (Priest & Forbes 1990, Forbes & Priest 1995) or goes kink unstable or breaks out (Antiochos et al. 1999, Maclean et al. 2004). As the tube erupts, it drives reconnection in the arcade under the erupting tube. The reconnection heats a loop to high temperatures, which then cools down and drains as new loops are heated and the reconnection location rises. The result is the appearance of a rising arcade of hot loops with cool loops beneath them. A particularly fine example was caught by the TRACE satellite and the RHESSI flare satellite on April 21, 2002 (Fig. 9). The RHESSI contours of hard x-ray flux at 12-25 keV

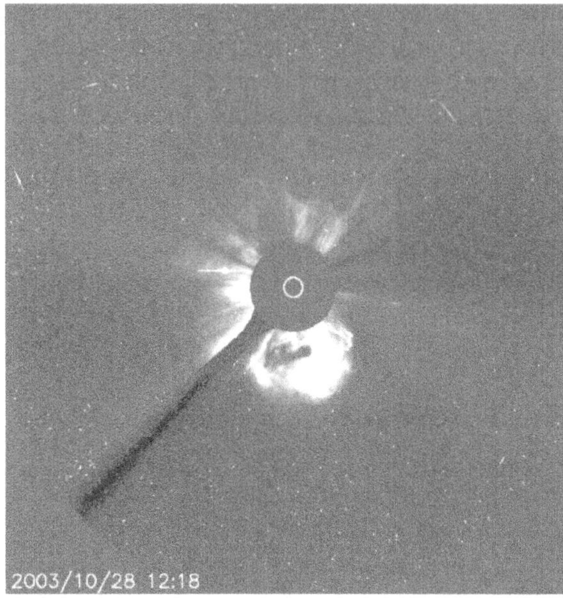

Figure 8. CME on October 28, 2003, viewed by LASCO.

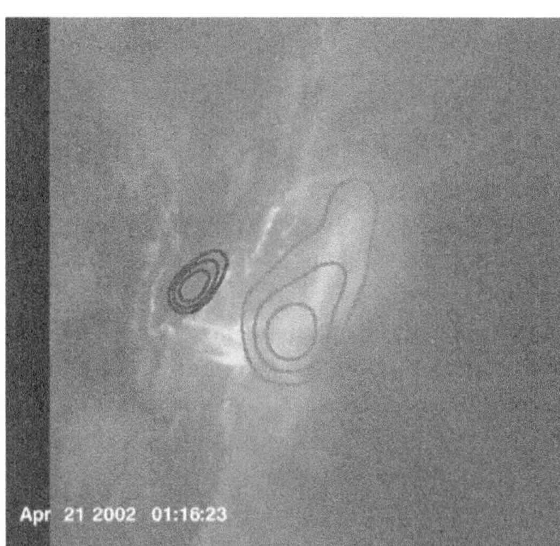

Figure 9. Overlay of RHESSI contours of hard x-ray flux and TRACE image in 195 A (P. Gallagher).

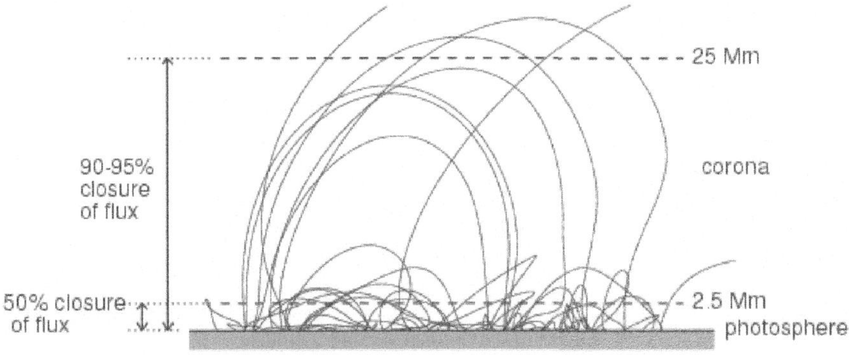

Figure 10. Magnetic field lines in the quiet Sun (Close et al. 2004).

show emission from the reconnecting current sheet above the 1.5 MK TRACE loops, while the 50-100 keV contours show emission from high-energy electrons accelerated in the reconnection and having travelled down to the feet of the loops. Particle acceleration is thought to be partly by DC acceleration in the current sheet and partly by Fermi and betatron acceleration in the field lines that are springing downwards after reconnection.

Coronal Heating

A third questions is: how is the solar corona heated to several million degrees by comparison with the photospheric temperature of only 6000 K? We know that the magnetic field is responsible and the mechanism is likely to be magnetic reconnection, but the exact process is still uncertain. A key discovery from SoHO is, however, the existence of the *magnetic carpet* (Schrijver et al. 1997), the fact that the photospheric sources of the coronal magnetic field are highly fragmentary and concentrated into intense flux tubes threading the solar surface. These sources are also highly dynamic, magnetic flux emerging continually in the quiet Sun and then undergoing processes of fragmentation, merging and cancellation, in such a way that the quiet Sun flux is reprocessed very quickly, in only 14 hours (Hagenaar 2001).

Recently, from observed quiet-Sun magnetograms from the MDI instrument on SoHO, Close et al. (2004) have constructed the coronal field lines and studied their statistical properties. For the region they considered, 50 % of the flux closed down within 2.5 Mm of the photosphere and 95 % within 25 Mm, the remaining 5 % extending to larger distances or being open (Fig. 10). They then tracked the motion of individual magnetic fragments in the magnetogram and recalculated the coronal field lines and their connectivity. In doing so, they

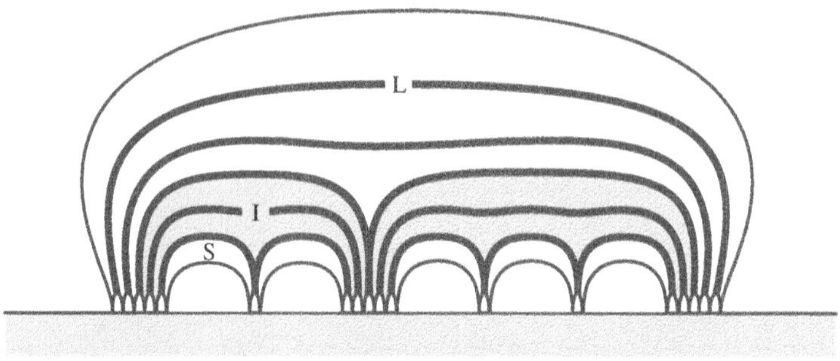

Figure 11. Coronal Tectonics model.

discovered the startling fact that the time for all the field lines in the quiet Sun to change their connections is only 1.5 hours. In other words, an incredible amount of reconnection is continually taking place - indeed, enough to provide the required heating of the corona.

Furthermore, a *Coronal Tectonics Model* has been proposed (Priest et al. 2002), which seeks to determine the effect of the magnetic carpet on coronal heating. Each observed coronal loop reaches down to the surface in many sources, so the flux from each of these tiny sources is separated by separatrix surfaces (*separatrices*). As the sources move around, they generate current sheets on the separatrices and separators, where reconnection and heating takes place (Fig. 11). In other words, the idea is that the corona is filled with myriads of separatrix and separator current sheets continually heating impulsively.

6. Conclusions

Solar physics is currently in a golden age of discovery, driven by a wealth of new observations from space and ground-based telescopes. In addition, the development of MHD theory is playing an important role. Magnetic reconnection, in particular, is likely to be at work in solar flares and coronal heating.

The current sense of vitality is certain to continue in future. The present missions are still working well, such as SoHO (launched in 1995), TRACE (in 1998) and RHESSI (in 2002). However, a fleet of new missions is planned to build on this success and answer new questions: these include STEREO (a NASA mission to be launched in 2006 consisting of two satellites, one moving ahead of the Earth and the other lagging behind, which will provide stereoscopic images of the corona), SolarB (a Japan-US-UK mission, to be launched in 2006, which will focus on understanding the subtle connections between the solar surface and the corona), Solar Dynamics Observatory (to be launched in 2008 as a successor to SoHO with a super-MDI and a super-TRACE), and Solar Orbiter (an ESA mission scheduled for launch in 2013, which will go three times closer to the Sun than Mercury). However, the future depends most on inspiring a new generation of active young researchers who can take advantage of an understand the new surprises expected from these future missions.

Acknowledgment

I am most grateful to ESA and PPARC for financial support.

References

Antiochos, S., DeVore, R. & Klimchuk, J. 1999, ApJ, 510, 485
Close, R., Parnell, C.E. & Priest, E.R. 2004, ApJ, in press.
Edlen, B. 1945, MNRAS, 105, 323
Forbes, T.G. & Priest, E.R. 1995, ApJ, 446, 377
Hagenaar, H.J. 2001, ApJ, 555, 448
Maclean, R., Beveridge, C., Longcope, D., Brown, D. & Priest, E.R. 2004, Proc. Roy. Soc., in press.
Priest, E.R. & Forbes, T.G. 1990, Solar Phys., 126, 319
Priest, E.R., Heyvaerts, J. & Title, A. 2002, ApJ, 576, 533
Priest, E.R. & Forbes, T.G. 2000, in *Magnetic Reconnection* Cambridge University Press, Cambridge, UK.
Rouppe vander Voort, L.H.M., Lofdahl, M.G., Kiselman, D. & Scharmer, G.B. 2004, A&A, 414, 717
Sanchez-Almeida, J. and Lites, B. 2000, ApJ, 532, 1215
Scharmer, G.B., Gudiksen, B.V., Kiselman, D., Lofdahl, M.G. & Rouppe vander Voort, L.H.M. 2002, Nature, 420, 151
Schrijver, C.J., Title, A.M., van Ballegooijen, A.A., Hagenaar, H.J. & Shine, R.A. 1997, ApJ, 597, L165
Schrijver, C.J. & Title, A.M. 2003, ApJ, 487, 424
Thomas, J.H., Weiss, N.O., Tobias, S.M. & Brummel, N.H. 2002, Nature, 420, 390
Trujillo-Bueno, J., Shchukina, N. & Asensio-Ramos, A. 2004, Nature, 430, 326
Weiss, N.O., Thomas, J.H., Tobias, S.M. & Brummel, N.H. 2004, ApJ, 600, 1073

Discussion

Virginia Trimble: When astronomers study magnetic fields they worry about having enough electric current to produce the field and how it closes - is this a problem on the Sun ?

Eric Priest: For processes where MHD is valid, which is often the case on the Sun, no, there is no problem, since, once you have calculated the magnetic field, the electric current follows from Ampere's law and automatically its divergence vanishes, so the electric current closes. In other words, in MHD the basic physics lies in the equations describing the magnetic field and plasma velocity, and the electric current and electric field are secondary variables which follow automatically from Ampere's and Ohm's laws.

Thierry Montmerle: How do you determine the magnetic diffusivity (η) on the Sun ?

Eric Priest: This is a very important issue. Usually, people just assume a classical value, but a better approach in the corona would be to try and link properly the macroscopics and microscopics of what is going on. In the corona a reconnecting diffusion region is certainly collisionless and so in a full analysis MHD would determine the global environment, while collisionless plasma physics would determine the detailed microscopic behaviour within that global MHD mould. However, the collisionless dissipation would in turn react back on the MHD in subtle ways, so that the macroscopics and microscopics are nonlinearly coupled. So far, collisionless dissipation has been studied only in relatively simple environments. However, it is not even clear that collisionless dissipation can be described in terms of a collisionless or anomalous diffusivity. Certainly, there are likely to be a number of different types of dissipation, depending on the parameter regimes. A start has been made, moreover, by several groups, as reported at the Isaac Newton Institute Progamme on Magnetic Reconnection Theory, held this summer in Cambridge and organised by Joachim Birn, Terry Forbes and myself. It included MHD and collisionless theory in the Sun and the Magnetosphere, and the main findings will be reported in a new book (Birn & Priest 2005).

Valentin Martinez-Pillet: What do you feel will be the consequences of the new "hidden flux" that has been discovered in the centres of supergranule cells ?

Eric Priest: The suggestion by Schrijver et al. (2004), Trujillo-Bueno et al. (2004) and Sanchez-Almeida & Lites (2000) that in the quiet Sun there may be five times more magnetic flux than we thought, due to the presence of excess flux between granules in supergranule cell centres is likely to have several consequences for the overlying atmosphere. The first is that the coronal field will be even more complex than Priest et al. (2002) considered in their Coronal Tectonics Model, and so there will be many more reconnecting current sheets low down in the atmosphere than in their estimate: indeed, they pointed out that future higher-resolution observations would probably reveal an even more fragmented web of separatrices, so all this makes their model even more effective. The second consequence is that the resulting addition of low-lying loops and separatrices would contribute substantially to the heating of the chromosphere. The third is that the additional low-lying loops and separatrices would provide extra heating for the corona. For the high corona I doubt that the heating would be affected much, although the connectivity of the fields would be changed, in the sense that half of the high coronal flux would now be connected to the supergranule cell boundaries and the remainder to the cell centres (Schrijver et al. 2004).

MOIST CONVECTIVE STORMS IN THE ATMOSPHERES OF JUPITER AND SATURN

Atmospheric storms in Jupiter and Saturn

Ricardo Hueso and Agustín Sánchez-Lavega
Física Aplicada I, E.T.S. Ing., Universidad del País Vasco
Alda. Urquijo s/n, 48013, Bilbao, Spain
wubhualr@lg.ehu.es; wupsalaa@lg.ehu.es

Abstract Moist convective storms might be a key constituent of the global energy budget in the atmospheres of the Giant Planets. The storms extract their energy from the release of latent heat produced in the condensation of water which is only abundant hundreds of kilometers below the observable cloud deck. Because these atmospheres are made of hydrogen and helium, dry air is lighter than moist parcels, providing a strong stabilization against vertical motions in the atmosphere. However, very large-scale convective storms have been observed in the atmospheres of the giant planets. Among them, Jupiter is the most convectively active, showing frequent storms with sizes on the order of 3000 km that occasionally trigger planetary scale disturbances. Observations from Voyager, Galileo and Cassini spacecrafts confirm the overall convective activity of Jupiter through observations of lightning flashes below the upper ammonia cloud deck. The energy associated to these storms is large enough to constitute a relevant fraction of the total internal heat source of the planet. Although Saturn presents a more quiescent atmosphere where storms are rarely observed, about once every 30 years, a giant storm has been observed to develop with sizes of 20000 km also triggering a planetary scale disturbance. We will review the current observational background of these giant storms in both Jupiter and Saturn presenting also numerical results obtained by different teams in simulating this vigorous meteorology. In both planets water storms may develop upward velocities of 50-150 m/s. The interaction of the storms with the powerful winds are not clear. In Saturn the giant storm of 1990 could have played a key role in originating the recently discovered change of 200 m/s in the broad and intense equatorial jet.

Keywords: Atmospheres: Jupiter, Saturn, Atmosphere: Dynamics, Meteorology

1. Introduction

The giant planets are fluid objects. They have no solid surfaces and their atmospheres gradually become denser in the interior until the distinction

J. C. del Toro Iniesta (eds.), The Many Scales in the Universe, 211–220.

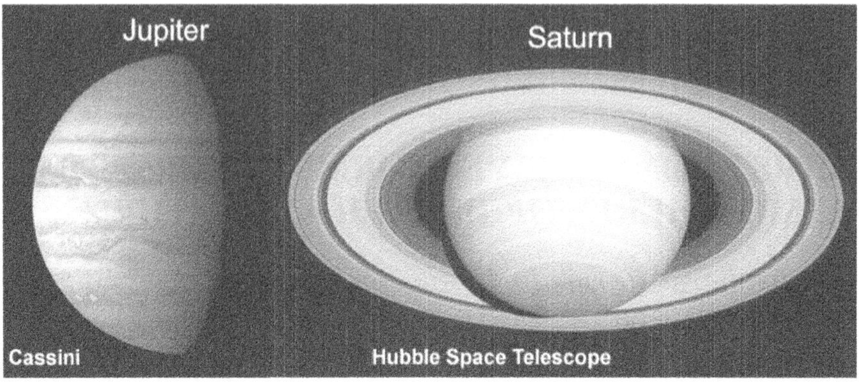

Figure 1. Visual aspect of the upper clouds of Jupiter and Saturn.

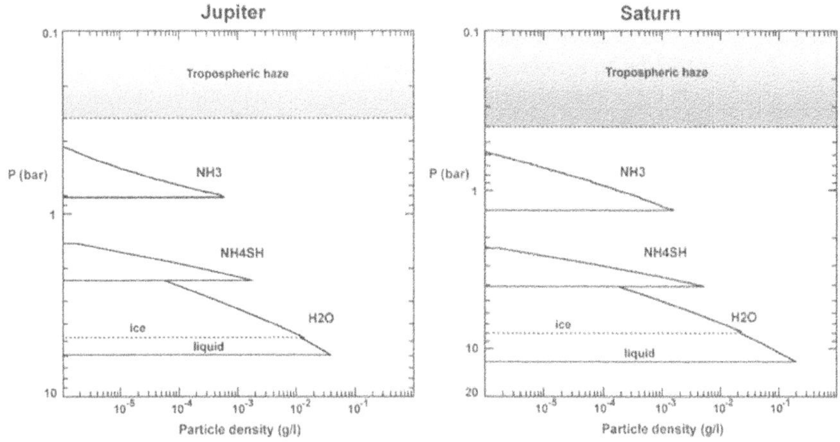

Figure 2. Average vertical cloud structure in Jupiter and Saturn. The upper hazes are derived from the observations and are probably of photochemical origin. The lower curves correspond to thermochemical calculations representative of the average cloud structure. Differences in cloud levels in both planets are mainly due to the colder temperatures of Saturn with respect to Jupiter with temperatures at the tropopause near the 100 mb level of 113 and 84 K respectively and, 170 and 135 K at P=1 bar near the top of the nearly adiabatic deeper atmosphere.

between gas and liquid becomes meaningless. In this chapter we will discuss the formation of convective storms in the upper atmospheres of Jupiter and Saturn. Water is in both cases the main candidate for powering the convective structures. Uranus and Neptune have also cloud structures of probable moist convective origin made of methane (Stoker and Toon, 1989).

In the tropospheres of Jupiter and Saturn the range of temperatures and chemical composition allow the formation of three distinct cloud layers. The main atmospheric features are observed in the upper cloud, close to the 1 bar level, giving the planets their caracteristic visual aspect (Fig. 2). The vertical cloud structure can be studied in terms of simple thermochemical models based on the Clausyus-Clapeyron equation (Sánchez-Lavega et al. 2004a). There are three main clouds expected to form. The upper cloud is made of ammonia condensed particles, below there are an intermediate ammonia hydrosulfide cloud and a lower and denser water cloud deck (Weidenschilling & Lewis, 1973). The particle density and the vertical location of the cloud depend on the condensable abundance and may vary from one location to other. The overall cloud structure is summarized in Fig. 2. The water cloud is specially interesting because water has a large latent heat. In both planets the atmospheres are heated not only by the solar insolation but also by the release of internal heat remanent from their planetary formation. For Jupiter and Saturn this heat is equivalent to 1.7 and 1.8 the heat absorbed from the Sun.

2. Observations of convective storms in Jupiter and Saturn

Jupiter

Ground-based observations have shown the regular development of mid-scale storms (1.000-5.000 km) at specific latitudes with the ocassional development of planetary scales disturbances triggered by the appearance of convective clouds (Sánchez-Lavega et al. 1996). The Voyagers and the Galileo and Cassini spacecrafts have obtained detailed observations of these convective features together with the detection of lightning strikes in long-term exposures of the night side of the planet (Little et al. 1999). The lightning has been shown to reside on deep levels of $\sim$ 5 bars close to the expected water cloud deck. In some cases day-side images of the lightning regions show convective clouds in the upper levels implying an activity which extends 150 km (Gierasch et al. 2000). The frequency of lightning strikes, the planetary area covered by the associated storms and the large energy released in the condensation of water and vertical ascension have led these authors to suggest that a significant proportion of Jupiter's inner heat could be transported to the upper troposphere by means of moist convective storms (Ingersoll et al. 2000). Therefore, the storms migh be an important constituent of the overall atmospheric dynamics of these planets.

Saturn

Mid-scale bright clouds of probable convective origin arise occasionally at the equator and at mid-latitudes and were best viewed during the Voyagers fly-bys in 1980-1981. Very rarely, about once every 30 years, a giant storm arises on the planet (Sánchez-Lavega, 1982; Sánchez-Lavega & Battaner, 1987). The last one, known as the 1990 Great White Spot (GWS) developed a 20,000 km size massive cloud system over a period of a month and evolved into a planetary scale disturbance (Sánchez-Lavega et al. 1991; Barnet et al. 1992). One of the mysteries of these storms is at which cloud layer they originate and how the seasonal cycle of 29.5 years of Saturn can influence their development. The other main mystery is if the large storms arising in Saturn's equator in the early 90s could have changed the local winds as to explain the wind decrease of 200 m/s in the equatorial current found between the Voyagers data obtained in 1981 (Sánchez-Lavega et al. 2000) and the 1994-2004 Hubble Space Telescope observations of Saturn (Sánchez-Lavega et al. 2003, 2004b).

3. Modelling moist convective storms

For any hydrogen based atmosphere, condensables are almost 7-8 times heavier than the hydrogen-helium atmospheric air with mean molecular weight ~ 2.2 g mol^{-1}) and the ascension of humid air is obstructed by the difference on molecular weight between dry and humid air. This factor alone is responsible for the main difficulties in originating moist convection in the giant planets. The development of moist convection becomes a competition between the heating produced by the condensation of volatiles in their vertical ascension and the heavier humid air.

Stoker (1986) developed the first quantitative model of moist convection for the Jupiter atmosphere. Her model was a 1D thermodynamical model of a raising parcel subject to entrainment with outside colder air. This basic model allowed to place upper limits for the vertical velocities and cloud tops to be expected in Jupiter storms. This model produced intense storms under favorable conditions (high relative humidity of water) able to fit the cloud tops observed by the Voyagers in the equatorial plumes. The model was also applied by Sánchez-Lavega & Battaner (1987) to study the onset of the 1990 GWS finding water storms were able to fit the cloud tops in the GWS only under favorable conditions.

Yair et al. (1992, 1995) performed 2D dynamical models of convective storms in Jupiter. Their results suggested very weak convection unable to be the cause of most of the meteorological phenomenology assumed to be of convective origin in Jupiter. This discrepancy can be explained by realising their initial conditions were typical of Earth cumulus development and may be also of average conditions in Jupiter which were shown to be very disfavorable for the development of storms in Jupiter, but probably not of the few spots where convective cumulus grow.

Hueso & Sánchez-Lavega (2000) developed a 3D model of moist convection for the giant planets by fully integrating the Navier-Stokes equations under the anelastic approximation. The microphysics were approached in the model by assuming condensation acts instantaneusly removing a fixed proportion of the condensed particles. The model has been applied to the water and ammonia clouds of Jupiter and Saturn. In the jovian atmosphere their results suggest that under favorable atmospheric conditions water-based moist convective storms develop and can become very energetic, being able to ascend 150 km from the 5 bar level to the observed cloud tops of 200-400 mbar at storm locations. Expected vertical velocities are on the order of 50-150 m/s. Ammonia on the other hand does not show the same activity since the clouds form in a more stable part of the atmosphere and its lower abundance and low latent heat do not allow for strong convective motions. The water based updrafts can ascend very fast because, contrary to the Earth case, they have a lot of vertical space to accelerate progresively. These simulations encompased an area one order of magnitude smaller than real observed storms. Hueso et al. (2002) presented a 2D model of the cloud tops in which they integrated the mass continuity equation with divergent sources and Coriolis forces to try to reproduce observations of large scale storms observed by the Voyagers. The results of these two works are partially summarized in Fig. 3.

Simulations of single cell storms in the atmosphere of Saturn were presented by these authors more recently (Hueso & Sánchez-Lavega, 2004). In Saturn's colder atmosphere ammonia clouds are at lower levels where the static stability of the atmosphere is not so high and ammonia storms are possible. Still much more intense convection can be attained if the updrafts originate at the water cloud deck buried down at 9 bars, 300 km below the observed cloud tops. As in Jupiter's case, the large vertical space allowed to the convective updrafts translates in powerful motions with updrafts on ther order of 150 m/s. These results are summarized in Fig. 4.

Because convection depends very strongly on the initial conditions both studies performed several simulations under different atmospheric conditions varying basic factors as the environment relative humidities, total weight of the condensates and total water and ammonia abundances. A brief summary of the sensitivity analysis of storms in both planets is presented on Table 1.

The main difficulty in originating water storms in both planets is the environment relative humidity which must be higher than 75% and 80% in Jupiter and Saturn respectively for water convection to originate. Under a lower humid environment, convection can not initiate because, in spite of the release of latent heat, saturated air is heavier than the environment colder air. Also, precipitation must act efficiently in both planets or the saturated updraft would

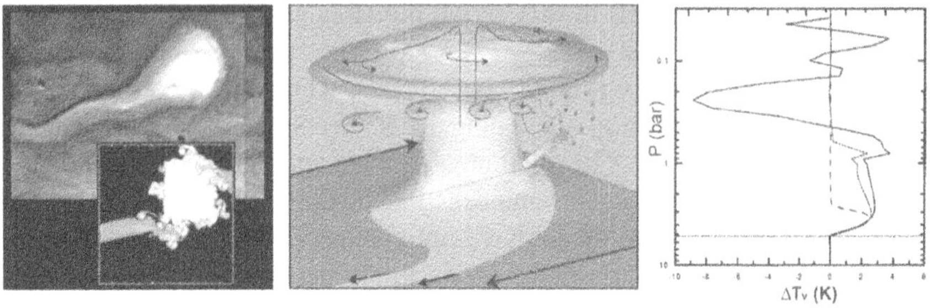

Figure 3. Jupiter storms, observations and models. The left image shows a mature stage of a South Equatorial Belt perturbation. The storm core is 5000 km wide. An inset shows results of simulating this particular storm by using a 2D model composed of 120 storm cells. The characteristics of the 2D storm cells are taken from more detailed 3D calculations. The center image shows the 3D structure of water storms on Jupiter. The right pannel is reserved to display the virtual temperature differences between the storm and its environment, being this thermal difference the atmospheric engine powering storm convection.

get negative buoyancy caused by the weight of the condensed particles. A precipation efficiency able to lose $\sim 25\%$ of the total condensates is required in both planets to develop moist convection. The total water abundance is also a key factor in determining convection strength but unfortunately this factor is not well constrained by the observations. Cloud tops at storm locations in Jupiter have been found to lie higher than 400 mbar in Jupiter and close to 200 mbar in Saturn's Great White Spots. This constrains weakly the deep water abundances. Jupiter storms can be explained with water abundance as low as 0.2 solar, while a higher deep water abundance would yield stronger convection more difficult to initiate. In Saturn the GWS cloud tops could be explained by water storms but also by ammonia clouds if ammonia were overabundant with 10 times solar abundance.

4. Storm locations and wind relation

Storms in Jupiter have only been found in regions of cyclonic shear close to a minimum in the wind speed. They are specially abundant in the South Equatorial Belt (SEB) at 16° S at the turbulent wake of the Great Red Spot and at some less frequent at the North Equatorial Belt (NEB) at 16° N. In Saturn most of the smaller scale storms in the last years have been observed at latitudes of 42° North and South with large scale storms arising in the equator in 1990 and 1994. Since storms are so energetic they may interact with the zonal wind system powering or weakening it. Saturn's equator has indeed experienced a dramatic wind change since the Voyager observations (Sánchez-

Table 1. Main characteristics of simulated storm convective plumes under different atmospheric properties. Abundances are measured relative to solar composition, h is the atmospheric relative humidity, f_c is the proportion of condensate particles carried upward by the storm with (1-f_c) the proportion of condensates that instantaneously rains out of the parcel, W_{max} is the maximum ascending velocity attained by the storm, P_{top} is the higher level where cloud material arrives and $Time$ is the average time to develop a mature stage convective cell. Note: (*) denote cases in which convection could be initiated only under large initial thermal perturbations and could not sustain long-term convection. The appearance of these cases close to stable convective storms means that convection, though difficult to initiate, must be very energetic once the required conditions are met.

Abundance	h (%)	f_c	W_{max} (m/s)	P_{top} (mb)	$Time$(hr)
Jupiter water storms					
2.0	99	0	210	100	1.1
1.0	99	0	145	140	1.4
1.0	99	0.25	115	200	1.5
1.0	99	0.50	75	250	2.0
1.0	75	0	70	330	(*)
1.0	50	0	30	500	(*)
0.2	99	0	50	250	2.2
Saturn water storms					
3.0	99	0	260	120	3.1
1.0	99	0	150	210	4.2
1.0	90	0	140	240	3.1
1.0	80	0	90	400	(*)
1.0	99	0.2	100	310	3.5
1.0	99	0.5	60	530	5.4
Saturn ammonia storms					
10.0	99	0	90	240	1.1
3.0	99	0	40	390	1.1
1.0	99	0	15	500	1.5

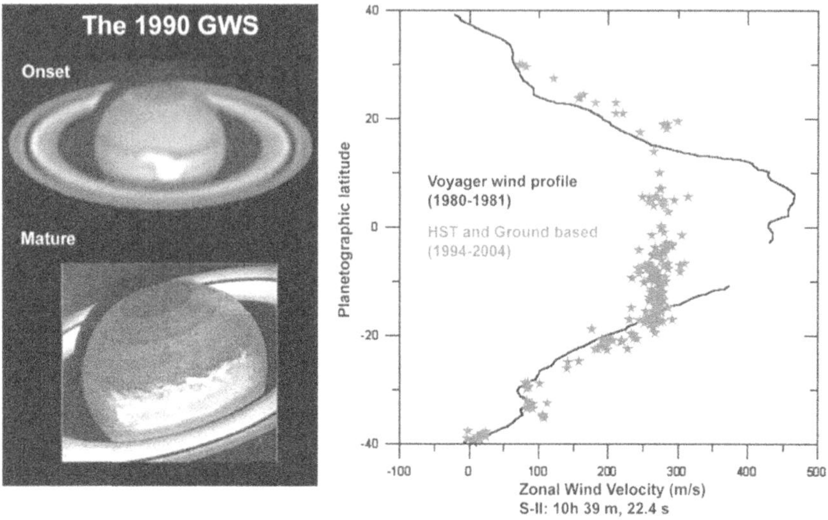

Figure 4. On the left: Images from the 1990 Great White Spot onset and mature stages. Images are from the Pic du Midi 1m telescope and the Hubble Space Telescope. On the right: Saturn's equatorial wind profile at two different periods from data obtained with the Voyagers in 1980-1981 and ground-based and Hubble Space Telescope observations on the 1994-2004 period.

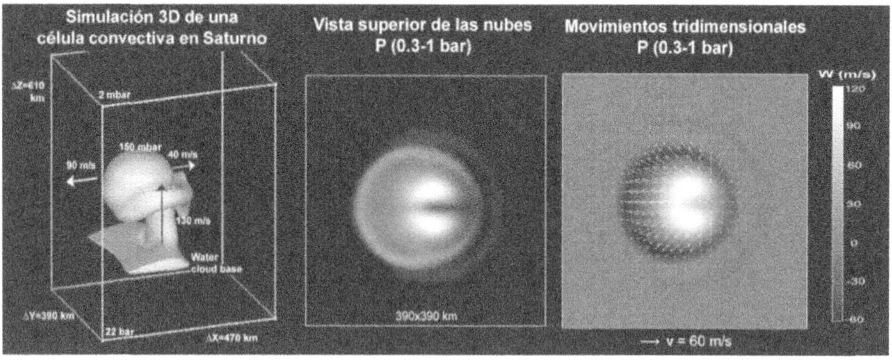

Figure 5. Detailed schema of an equatorial Saturn storm simulation. The left figure shows the 3D structure of a water-ammonia cloud storm cell. The center image is an XY map of cloud density in the 0.3-1 bar levels and shows how a 400 km storm would be seen in high resolution observations. The right image shows the vertical and horizontal motions expected if the upper flow is not turbulent. A net conversion of upward momentum to westward momentum is clearly seen in the simulation, a consequence of the three-dimensional Coriolis forces.

Lavega et al. 2003, 2004b). The change seems to have originated after the 1990 storm (Fig. 4).

Our simulations show that equatorial updrafts ascending from the water cloud base at 9 bars take enough time to be deflected to the west by means of Coriolis forces transforming up to a 10% of their kinetic energy to westward momentum (see Fig. 5). An order of magnitude evaluation of this effect, compatible with the size and intensity of the 1990 GWS, seems to indicate that the storm may have directly decreased the equatorial winds by 20-30 m/s with a larger contribution of kinetic energy to turbulent motions. In order to produce a larger change of the order of the observed decrease ($\sim$ 200 m/s) the GWS convective core should have stayed active for a whole year and most of the equatorial atmospheric material initially at 9 bars should have had to be transported to the upper troposphere. On the other hand, if the 1990 GWS was indeed a giant storm system at the upper ammonia cloud there would not have being any significant momentum transformation because of the lower vertical scales involved. The 1990 storm may however have altered deeply other atmospheric properties that eventually contributed to change the equatorial winds and overall dynamics. A work devoted to the exploration of the mechanisms responsible for the wind change will be presented elsewhere.

5. Conclusions

We have presented the basic phenomenology of convective storms forming in the atmospheres of Jupiter and Saturn. Detailed three-dimensional models have been used to explore the dynamics of such storms finding that only water can be responsible of Jupiter's convective storms while Saturn may have both water and ammonia storms. Saturn's ammonia storms may exhibit characteristics typical of water storms if ammonia is abundant enough in the atmosphere (10 times solar). In both, Jupiter and Saturn, Coriolis forces may transform upward momentum in the storms to zonal motions with an efficiency of the released energy of 10% in the equator. The relationship between the overall zonal winds and the moist convective activity remains obscure due to the complexity of the problem and the unknowns of some basic atmospheric properties like water abundance and overall planetary convective activity.

Acknowledgments

This work has been supported by Spanish MCYT research project PNAYA2003-03216, FEDER and Grupos UPV 13697/2001. RH thanks a postdoctoral fellowship from Gobierno Vasco.

References

Barnet, C.D., J.A. Westphal, R.F. Beebe & L.F. Huber, 1992, Icarus, 100, 499

Gierasch, P.J., A.P. Ingersoll, D. Banfield, S.P. Ewald, P. Helfenstein, A. Simon-Miller, A. Vasavada, H.H. Breneman, D.A. Senske, & the Galileo SSI Team, 2000, Nature, 403, 628

Hueso, R.,& A. Sánchez-Lavega, 2001, Icarus, 151, 257

Hueso, R.,& A. Sánchez-Lavega, 2002, JGR: Planets, 107, E10, DOI 10.1029/2001JE001839

Hueso, R.,& A. Sánchez-Lavega, 2004, Icarus, in press

Ingersoll, A.P., P.J. Gierasch, D. Banfield, A.R. Vasavada,& the Galileo SSI Team, 2000, Nature, 403, 630

Little, B., C.D. Anger, A.P. Ingersoll, A.R. Vasavada, D.A. Senske, H.H. Breneman, W.J. Borucki & the Galileo SSI Team, 1999, Icarus 142, 306.

Sánchez-Lavega, A., 1982, Icarus, 49, 1

Sánchez-Lavega, A.& E. Battaner, 1987, A&A, 185, 315

Sánchez-Lavega, A., F. Colas, J. Lecachjeux, P. Laques, D. Parker & I. Miyazaki, 1991, Nature, 343, 397

Sánchez-Lavega, A., J.M. Gómez, J. Lecacheux, F. Colas, I. Miyazaki, D. Parker& J. Guarro, 1996, Icarus, 121, 18

Sánchez-Lavega, A., J.F. Rojas& P.V. Sada, 2000, Icarus, 147, 405

Sánchez-Lavega, A., S. Pérez-Hoyos, J.F. Rojas, R. Hueso & R.G. French, 2003, Nature, 423, 623

Sánchez-Lavega, A., S. Pérez-Hoyos & R. Hueso, 2004a, American Journal of Physics,72, 767

Sánchez-Lavega, A., R. Hueso, S. Pérez-Hoyos, J.F. Rojas & R.G. French, 2004b, Icarus, 170, 519

Stoker, C.R., 1986, Icarus, 67, 106

Stoker, C.R.,& O.B. Toon, 1989, Geophysical Research Letters, 16, 929

Yair, Y., Z. Levin,& S. Tzivion, 1992, Icarus, 98, 72

Yair, Y., Z. Levin,& S. Tzivion, 1995, Icarus, 114, 278

Weidenschilling, S.J.& J.S. Lewis, 1973, Icarus, 20, 465

EXPLORING THE SOLAR SYSTEM BEYOND NEPTUNE

Jose L. Ortiz and Pablo Santos-Sanz
Instituto de Astrofísica de Andalucía, CSIC
Apdo. 3004, 18080 Granada, Spain
ortiz@iaa.es; psantos@iaa.es

Abstract Twelve years after the discovery of the first Trans-Neptunian Object (apart from Pluto), more than 900 bodies have been found in the belt beyond Neptune. This disk, usually referred to as the Kuiper or Edgeworth-Kuiper Belt is now thought to contain an even larger population of small bodies than the main asteroid belt. In this work, a summary of the current knowledge on the physics of these objects is presented, based on data obtained from various means such as CCD photometry, spectroscopy, millimetric observations, adaptive optics etc.

Keywords: Kuiper Belt, Trans-Neptunian Objects

1. Introduction

The aim of this paper is to carry out a review on the topic of the Trans-Neptunian belt for the non-expert. This belt is now thought to be the main source of the Short-Period Comets and comprises more minor bodies than the asteroid belt within Mars and Jupiter. The icy Trans-Neptunian Objects that populate the belt have been the subject of intense research during the last 12 years since the discovery of the first body (apart from Pluto) because as remnants of the formation of the solar system, they carry important information on the early solar system and its evolution. Trans-Neptunian objects have even reached the public. The media has widely covered the announcement of the discovery of Sedna, and the debate on whether Pluto should remain in the list of the major planets or not.

The first ideas on the existence of a belt of bodies beyond Neptune are traditionally ascribed to Gerard Kuiper (Kuiper 1951), although his ideas were about a disk beyond Pluto, not Neptune, and were based on a somewhat heuristic reasoning in which the surface density of the original solar nebula should not have an abrupt end beyond Pluto. If one plots the mass of the main bodies of the solar system (adding the lost volatiles) divided by the area of the annu-

J. C. del Toro Iniesta (eds.), The Many Scales in the Universe, 221–233.

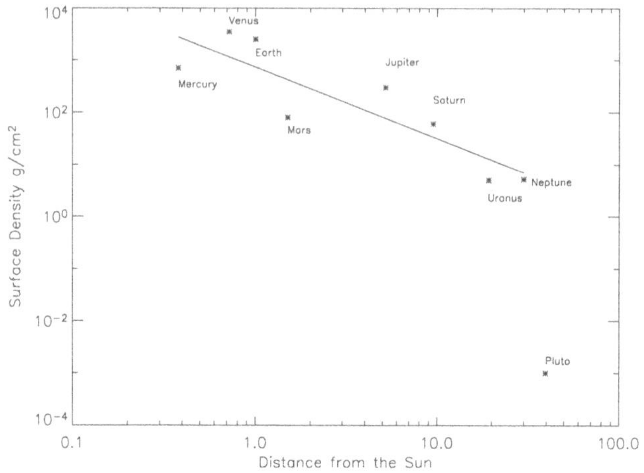

Figure 1. Surface density of the Solar Nebula as a function of radial distance to the Sun (in AU), with a fit superimposed.

lus they cover versus the radial distance, one gets a surface density that decays smoothly until Pluto is reached (Fig. 1). There, the surface density has an abrupt end, which Kuiper did not like. Therefore Kuiper thought that some icy planetesimals should be there to avoid the sharp end of the nebula. Some investigators claim that even before Kuiper, Kenneth Edgeworth mentioned the existence of a belt (Edgeworth 1943), but his work was not quantitative and appeared more a conjecture than a true theory. Others appear to have thought on the possibility of Trans-Plutonian bodies even before Kuiper or Egdeworth, but none of these ideas were close to the idea of the belt that we have today.

The first prediction of the correct belt based on some real physics (of the Short Period Comets) was made by Uruguayan dynamicist Julio Fernandez (Fernandez 1980). In 1980 Fernandez wrote a paper whose title was explicit and the content was indeed physically plausible. In his paper entitled "On the existence of a comet belt beyond Neptune" he presented calculations which showed that a belt beyond Neptune would be a source of incoming Short Period Comets (SPCs) by means of close approaches of the bodies within the belt, and he argued that this was much easier and efficient than having the comets evolve from the Oort cloud. In 1988 detailed numerical computations showed that the Oort cloud could not be the source of the SPCs basically because the SPCs inclinations are more or less confined to the ecliptic, whereas those comets from the Oort cloud would have random inclinations (Duncan et al. 1988). Finally, the first detection of a Trans-Neptunian Object (apart from

Pluto) was made in 1992, by David Jewitt and Jane Luu. The object received the designation $1992QB_1$.

The belt is usually called the Kuiper Belt, Edgeworth-Kuiper Belt or even Edgeworth-Kuiper-Whipple Belt. To skip the name problem, the disk is best referred to as the Trans-Neptunian Belt or Trans-Neptunian disk, although the other names are also used in the literature. The objects that populate the disk are therefore named in very different ways: Trans-Neptunian Objects (TNOs), Kuiper Belt Objetcs (KBOs), Edgeworth-Kuiper Belt Objects (EKOs) or even EKBOs. These objects are indeed very faint. A typical 50km body, assumed to have an albedo p around 0.04 and located at 40 Astronomical Units from the Sun and Earth would have a typical 25 magnitude. In this paper we review some of the knowledge that we have gained in the 12 years of observations since the discovery of $1992QB_1$.

2. Current knowledge

We now know several families of TNOs, grouped by special features in their orbital parameters. More than 900 TNOs have been discovered by different telescopic surveys and although a large fraction of them are lost due to lacking astrometry, a large number of orbits are known with the required precision. Each of the telescopic surveys has its own biases and therefore the orbits currently available are probably not bias free, but at least some orbital families can be distinguished so far. Fig. 2 shows inclinations and eccentricities of well known orbits versus their semimajor axes.

The so-called classical objects or cubewanos have low eccentricity orbits and are outside the mean motion resonances shown as vertical lines. Within the classicals, two different populations appear to exist: a "hot" and a "cold" population, with inclinations above or below 4.5 degrees from the ecliptic, according to the latest results although some give the limit as 7 degrees. The cold population appears to have redder colors than the hot population. It is also important to note that the largest bodies discovered so far belong to the hot population. A large number of TNOs are trapped in the 2:3 mean motion resonance with Neptune, like Pluto (which completes two revolutions around the Sun in the same amount of time that Neptune requires for exactly three revolutions). All the bodies trapped in this resonance were named after Pluto and that is why they are called "Plutinos". There are other resonances, and the objects trapped in the resonances are generally called "resonant objects".

In addition to these families, there is a scattered disk of objects with very large eccentricities. These SDOs are probably as numerous as the rest altogether, but their detection is difficult because they spend most of their time far from 40 AU, were they are detectable. The SDOs have had at least one close encounter with Neptune so that the objects were ejected outside the main belt.

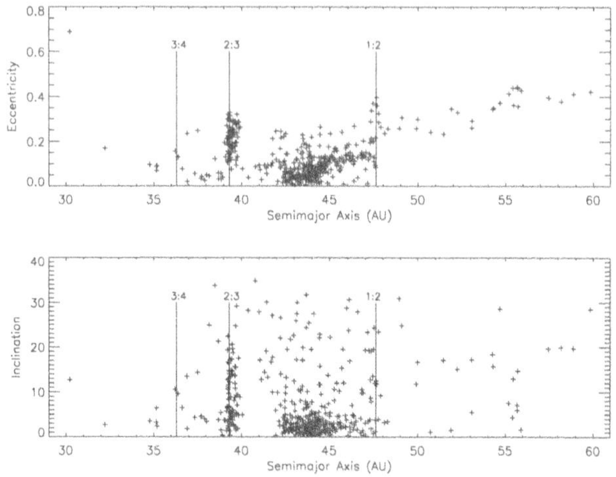

Figure 2. Eccentricities and inclinations of well-determined orbits as a function of the semi-major axis or the orbits. The main mean motion resonances are indicated with the vertical lines.

Similarly, the orbits of some objects can be perturbed to enter the major planets domain and become the so-called "Centaurs" which are bodies with perihelia and aphelia within those of Jupiter and Neptune's respectively. Strictly speaking the Centaurs are not TNOs, because they are not further away than Neptune, but they are very closely related to the SDOs. The Centaurs are perturbed by the large planets and finally will become Short-Period Comets. Some of the SPCs may end up even as extinct comets in the near-Earth region and become dangerous NEOs. Some calculations show that 10 to 20% of the NEOs can be of this type, although there is still debate on this.

There are a few objects that have large perihelion distances not controlled by Neptune and do not belong to any of these families. Sedna, the most widely known Trans-Neptunian object apart form Pluto, would appear to belong to the Scattered Disk family, because it has a very eccentric orbit, and in fact it is listed by the Minor Planet Center in their Centaur-SDO list, but there is no way to raise Sednas perihelion distance from the Kuiper Belt to its present location by any means, so Sedna is not a Trans-Neptunian body of any of these families. According to Fernandez (private communication), the 76 Astronomical Units of its perihelion distance cannot be reached even invoking the Kozai effect. So the formation of Sedna followed a different process to the rest of TNOs discovered so far. The objects in this group form the so called "Extended Scattered Disk" (Gladman et al. 2002).

According to Fernandez, Sedna could be a member of an inner core Oort cloud defined as a structure formed in our early solar system from the interaction of nearby stars while the Sun was being formed within a cluster. This is the preferred explanation by some dynamicists, although the existence of a massive enough planet further away has also been suggested. Morbidelli & Levison (2004) have recently proposed five different mechanisms but they claim that the close star passage is the most likely one. Indeed Sedna is a Trans-Neptunian Object in the sense that it is beyond Neptune, but Sedna is not a classical belt object or a Scattered Disk object.

It now appears that the large number of objects trapped as plutinos in the 2:3 mean motion resonance was the result of Neptune's outward migration. Several possibilities have been suggested and resonance sweeping was one of them (Malhotra 1995). Different numerical models explain the accumulation of plutinos in slightly different ways (Gomes 2003), but all of them require Neptune's outward migration. Most models require a 10AU outward migration. Now there is even evidence of the corresponding Jupiter's inward migration in the Hildas asteroid family (Franklin et al. 2004). Planet migration is also thought to occur in extrasolar environments, as many of the extrasolar planets discovered so far are very close to their stars, so planetary migration is no longer an odd issue.

Computer simulations of the evolution of the solar system can predict different amounts of TNOs in different families and therefore the relative numbers of bodies in the different families can give important clues on the formation and evolution of the solar system. However, the relative number of bodies in each population should be calculated from bias-free surveys and currently there are none. A possibility is to de-bias the current surveys, as it has been done for the NEO surveys, but this has not been done for TNOs yet.

The size distribution of the TNOs and the total mass is inferred from the surveys in different ways (Trujillo et al. 2001a,b, Larsen et al. 2001, Sheppard et al. 2000, Jewitt et al. 1998). The usual approach is to fit a straight line to the logarithm of the cumulative luminosity function that is derived from the surveys. The cumulative number of objects brighter than a given magnitude, per square degree near the ecliptic can be computed from the surveys. The slope of this fit, α, is directly related to the index of a power law differential size distribution by this simple expression $q = 5\alpha + 1$. The index q from the different surveys appears to be close to 4.2 because α, the slope of the plot is nearly 0.66 (Trujillo et al. 2001b). However, this steep value cannot hold for all possible sizes or the mass of the Kuiper Belt would diverge. Thus, it is usually assumed that the q index drops at the tens of km size. The total mass of the classical Kuiper Belt can be obtained integrating the size distribution within the assumed maximum and minimum sizes multiplied by the volume and density. The result is around 0.1 Earth masses. Since a similar number of SDOs appear to exist as the number of classicals, the total mass of the disk can be double than that. Therefore, the total mass of the disk was thought to be in the range of 0.1 to 0.2 Earth masses. Things have changed a bit very recently. There are now hints that the q index drops at the 30 km size range from pencil beam surveys carried out at Subaru, and there is now a work in press by Bernstein et al. (2004) based on a very deep HST survey which clearly demonstrates that the number of small KBOs is much lower than predicted from the $q = 4$ extrapolation. This is shown in their Fig. 3 as a deviation from the straight line. That work shows that a two-parameter power law is consistent with the data and it also shows that the classical belt has a different size distribution than the excited population. Other investigators have pointed out that the size distribution appears to be a function of the perihelion distance (Hughes 2003). The total mass can be obtained using Bernstein et al double power law with no assumptions on maximum and minimum diameters, as the integral no longer diverges. The total mass obtained by Berstein et al. (2004) is 0.01 Earth masses for the classical belt. However, at least two to three orders of magnitude more mass had to be present in the belt so that large bodies like Pluto and Quaoar could have grown through accretion. Besides, integrating the surface density of the solar nebula within 30 and 50AU one gets in the order of 10 Earth masses. So the question is, where did all that mass go?.

The mass depletion of the Trans-Neptunian Belt is now thought to have occurred by one of the following mechanisms: dynamical depletion or collisional grinding to dust. The dynamical depletion could have been caused in two different scenarios. One of the scenarios is the repeated passage of planetary embryos scattered by Neptune. These passages could excite the inclinations and eccentricities of the KBOs until they became Neptune crossers and therefore were ejected by Neptune out of the solar system or injected in the inner solar system. This would clear the region of bodies. A different dynamical depletion scenario would have been caused by the passage of a star through the Kuiper Belt which could have kicked the planetesimals directly to hyperbolic orbits or Jupiter crossing orbits without interacting with Neptune.

Collisional grinding of the TNOs to dust and subsequent ejection through radiation pressure and Poynting-Robertson drag would also severely deplete the mass of the initial Kuiper Belt, but the initial TNO size distribution would have to be very odd, with most of the mass in small bodies.

From the extrapolation of the current results of a wide area survey targeted at the brightest TNOs, Trujillo et al. 2003 conclude that there could be ten 1000-km bodies awaiting discovery and even a Pluto-sized body. The analysis by Bernstein et al. is not as optimistic and apparently they conclude that the only Pluto-sized object that can exist is Pluto. Some accretion models still predict two or more Pluto-sized objects so the issue is not settled yet until the wide area survey is not complete.

There is an edge to the classical belt at 48AU. This is not a bias effect caused by the fact that objects further away than 48AU would be faint. The edge is now a firm result from the surveys by different teams. A second Trans-Neptunian disk (even more massive than the current belt) could exist beyond 85 AU and remain undetected. The origin of this edge is still a mystery, although some mechanisms have been proposed. A close stellar passage (e.g. Ida et al. 2000), the existence of a massive planet further away (e.g. Brunini & Melita 2002) the formation of the whole system within 30AU and subsequent migration (Levison & Morbidelli 2003) have been proposed to explain the edge. The passing star would have left a clear signature in the inclination of the TNOs (Brown 2001), which is not observed and has other weaknesses. The planetary-sized body has also a number of weaknesses, including the fact that it should have been detected already. It is worth noting that the passing star and the nearby Earth-sized or Mars-sized planet scenarios have also been proposed to produce objects like 1999CR_{105} and Sedna and even to cause the mass depletion. Some investigators are not comfortable with the idea of a sharp end, so they are expecting that another belt will be discovered sooner or later perhaps in the 100AU range. Kuiper speculations are again applicable here, so we are left with a similar situation as we were in the fifties.

Concerning the colors of these objects, there is a wide diversity of colors measured in the BVRI bands by several independent groups (e.g. Peixinho et al. 2004, Doressoundiram 2003, Barucci et al. 2001, Boehnhardt et al. 2001, Jewitt & Luu 1998, Tegler & Romanishin 1998). Narrow filters cannot be used because of the faintness of the objects. There is no bimodal color distribution as initially reported by some authors when the data were still scarce, but as the number of measurements grew, the color gaps were filled and if Centaurs are removed, no bimodal distribution remains (Peixinho et al. 2003). An example of B-V versus V-R colors for the classical objects is shown here. There is a correlation of color and inclination in the classical belt (e.g. Trujillo & Brown 2002) in such a way that the redder the object, the lower the inclination. This was the first hint to the existence of two different populations within the classical objects. The hot population, whose inclinations are excited to 4.5 degrees or more, might have different colors because of different reasons, but the currently preferred explanation is the fact that the hot population might have formed closer to the Sun than the cold population. There is a variety of processes that can give rise to the color diversity. Different compositions at formation, collisional resurfacing, cometary-like activity, space weathering, meteoroidal impacts etc have been proposed as mechanisms capable of creating the diversity of colors seen. It is very likely that no single process is responsible for the coloring of all the TNOs, but a mix of some of them.

There are other correlations of colors with other orbital parameters (e.g. Peixinho et al. 2004, Doressoundiram 2003). It appears that a color corre-

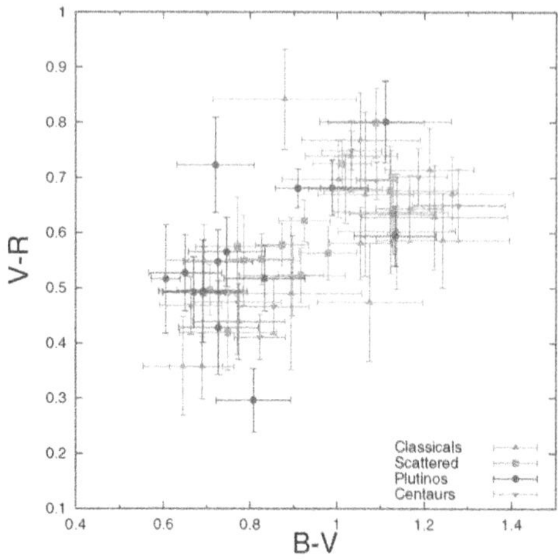

Figure 3. B-V versus V-R colors of the different families of objects.

lation with perihelion distances is present for the classicals, but just for the intrinsically bright ones. There are other correlations difficult to interpret and there are even lacks of correlations which can be used to rule out some theories. For example, there is no color correlation with aphelion distance within the SDOs whereas one would perhaps expect such a correlation because these bodies leave the solar wind termination shock at around 80AU and are more exposed to galactic cosmic rays. Therefore one would naively expect that the long aphelion bodies would be more extensively altered than the short aphelion ones, simply because the latter spend a shorter time beyond the bow shock limit. The SDOs with inclinations higher than 12 degrees appear to show very similar colors to the classicals which might indicate a common origin.

Other photometric results are related to the so called phase functions. The measurements of magnitude of some TNOs at different phases around the Sun show that there is a steep increase in brightness near opposition, when the phase angle is nearly zero. The increase is almost linear and is, on average around 0.15mag per degree (Sheppard & Jewitt 2002, Schaefer & Rabinowitz 2002). This is a much steeper behavior than that of Pluto. Low albedo asteroids also show this type of steep dependence so this might be an indication of low albedos for the TNOs, as initially suspected because the Short-Period comets have very low albedos.

The spectra of most TNOs are featureless in the visible. Some TNOs show broad and weak absorption features possibly due to aqueous alteration of minerals (Fornasier et al. 2004, de Bergh et al. 2004, Lazzarin et al. 2003). Some of them have extreme red slopes as already suggested from the photometry, especially the Centaurs. The different slopes are consistent with the variety of colors already seen by regular photometry programs. The VLT large programme on TNOs (Boehnhardt et al. 2003) has been responsible for most of the TNO spectra obtained so far and has shown the broad band absorption features possibly related to aqueous alteration of minerals. The programme has also shown clear heterogeneous compositions revealed by the rotation of the bodies.

The near-IR spectra are powerful tools to reveal absorption features due to different molecular species like water ice. Particularly, the water ice features at 1.5 and 2.0 microns are easy to detect on most solar system bodies and were expected to be detectable on TNOs, because the composition of the TNOs is thought to be water ice in a large percentage. However, the signal to noise is usually very low even with the largest telescopes and the largest TNOs. Therefore the idea of detecting water ice on TNOs was a challenging one. The first water ice detection was done with the Keck telescope (Brown et al. 1999). So far water ice has been detected in only a few TNOs and in a few Centaurs.

Models with different surface compositions can fit the spectroscopic data. In other words the models are not unique. Most of the surface compositions are consistent with mixtures of organics with different amounts of water ice and even pure Carbon. There are hints of aqueous alterations of some minerals in several TNOs and little more can be said. When geometric albedo measurements exist, the models can be further constrained (e.g., Cruikshank & Dalle Ore 2003), but different mixtures can still fit the data in some cases. The albedo measurements are therefore very important. There are different types of albedos. The albedo we usually referred to is the geometric albedo p, in the V band, which is related to the more widely known concept of albedo, A, through a simple equation $p = Aq$ in which q, the phase integral, is a well constrained value. To derive geometric albedos from the observations one needs data in the visible and in the thermal part of the spectrum. Through the use of simple thermal models, one can obtain the unknowns (A, p and r, where r is the radius of the body). The rest of the variables in the thermal models are parameters known entirely or at least to some extent and the observed magnitudes. The thermal observations for the TNOs should ideally be carried out near the 60 to 70 micron range, as the expected TNO temperatures are around 50K. This could be done with ISO for one object and will be feasible for at least some TNOs with the infrared telescope Spitzer. From the ground, the closest one can get is the millimetric range. The IRAM-30m has been used for a number of objects, although signal from only three objects has been clearly detected. For the rest we only have lower limits to the albedos. Another technique to measure albedos is by direct measurement of the size of the body using the HST, but only Quaoars albedo has been measured in this way and indeed only the largest objects have resolvable disks by HST. The object 2004DW could potentially be larger than Quaoar, but its albedo has not been measured yet.

Compiling all the results obtained so far (Table 1) it is clear that most of the objects have geometric albedos above the canonical 0.04 value, which is usually adopted for comets. A more appropriate value for TNOs appears to be 0.08. Therefore, TNOs are dark, but not as dark as comet nuclei. As of this writing, we have just learnt that some albedos have already been measured from Spitzer for Varuna and other large TNOs (Stansberry et al. 2004). Their albedos are even higher than 0.15, which strengthens the claim that $p = 0.04$ is too low. This conclusion might be a bias effect because the objects for which the albedo has been measured are all large. Perhaps the smaller TNOs could have lower albedos. Nevertheless there are data from small binary objects (discussed below) that can be interpreted as indicating higher albedos than the canonical 0.04 value.

Table 1. Compilation of geometric albedo measurements

Object	Geometric albedo (p_v)	Reference	Object	Geometric albedo (p_v)	Reference
2002TX300	>0.08	(a)	Quaoar	0.10	(b)
1995SM55	>0.04	(c)	1996TO66	>0.04	(c)
Chaos	>0.04	(c)	1999TC36	0.05	(c)
Huya	>0.08	(c)	Ixion	>0.15	(c)
2000AW197	0.09	(d)	1993SC	0.02	(e)
Varuna	0.06	(f)	1996TL66	>0.11	(e)
	0.04	(g)		0.03?	(e)

(a) Ortiz et al. 2004, (b) Brown & Trujillo 2004, (c) Altenhoff et al. 2004 (d) Margot et al. 2002 (e) Thomas et al. 2000 (f) Jewitt et al. 2002 (g) Lellouch et al. 2002

Time series photometry has shown remarkable variability for these objects. 27% of the bodies show variability above 0.15mag, 15% of the bodies have photometric ranges above 0.4mag and even 9% of the TNOs show variability above 0.6mag (Sheppard & Jewitt 2003, Ortiz et al. 2003). This variability is more pronounced than the variability of the Main Belt Asteroids of similar sizes. The photometric variations are caused by the rotation of a body that either has an irregular shape or surface markings or a combination of both. However, surface markings usually produce small magnitude variations. Color variations at the maxima or minima of the light curves could be indicative of variability induced by surface markings rather than irregular shape. Double peaked light curves can on the other hand indicate irregular shape rather than surface markings, but an appropriate albedo distribution can mimic any irregular shape light curve. If one assumes that the photometric ranges are solely due to irregular shape (which is strictly valid only for the large amplitude variation bodies) one can determine that the TNOs are more irregular than their main belt asteroids counterparts and perhaps this could be explained because the TNOs could be internally weak or rubble pile structures easily deformed by their spin. Extreme variations can perhaps only be explained in terms of contact or very close binaries (Sheppard & Jewitt 2004). The study of the rotational periods and photometric ranges within different TNO families might also reveal some trends, but currently the light curve data is still scarce.

Some interesting conclusions have been obtained for one of the largest TNOs by means of its lightcurve. Varuna shows a double peaked lightcurve of period 6.34h, with no color variations at the maxima or minima which indicates that the variability is due to a very elongated shape. Such an elongation is remarkable for an object this big (initially supposed to be 900km in diameter). Jewitt et al. 2000 concluded that this object could be almost strengthless if it were a rubble pile and be distorted as a response to rotation. Using equilibrium shapes of strengthless bodies they could constrain the density to be nearly 1000kg/m^3, which is also low for a big object even if the composition were mostly ice and rock. Jewitt et al. conclude that this low density could be the result of high porosity. Therefore, Varuna might be a high porosity rubble-pile structure. The very recent finding that the albedo of Varuna is 0.17 by Stansberry et al. (2004), indicates that the diameter is around 600km and therefore the constraint on the density should now be somewhat higher.

For Quaoar, which is even larger than Varuna, the longer rotation period and the smaller photometric range (Ortiz et al. 2003b) do not allow drawing important conclusions regarding density and structure. The study of the rotation rates of the largest TNOs is especially important because these bodies are not collisional products and therefore they must retain memory of the primordial angular momentum of the solar nebula at the time of their formation. The smaller TNOs are thought to be collisional products, and therefore they cannot give this primordial information.

A key question to the determination of masses and densities is the detection of binaries. This goal has been achieved (e.g. Veillet et al. 2002) and more than 8 binaries are currently known by means of HST observations and adaptive optics. The mass can be assed for binary objects provided that the rotation period and semimajor axis can be determined. The density determination needs the size estimates, but because the sizes cannot be determined (unless the primary was very big), an albedo has to be assumed and therefore the densities determined in such a manner are affected by large uncertainties. Thus, if one assumes an albedo of, say, 0.04 the diameter can be computed and the density too. For an albedo of 0.04 the densities obtained for some binary TNOs are in the order of 500 kg/m^3 (e.g. Osip et al. 2003) which is clearly below the expected values. For more reasonable 1000kg/m^3 densities, the albedos are approximately 0.09, consistent with the radiometrically derived albedos around 0.08.

The percentage of binaries is higher than in the asteroid belt, the binaries are more separated and the masses of the two components are usually similar, in contrast to the asteroid belt binaries. Noll et al. 2002 point out that there are at least 5% of binaries with angular separations of 0.15 arcsec and even 15% of the TNOs can be binaries with separations of 0.1arcsec. To account for the formation of a large fraction of binaries with components of similar sizes, different scenarios have been proposed (e.g. Weidenschilling 2002, Goldreich et al. 2002, Petit and Mousis 2004, Funato et al. 2004) but all of them require a large initial mass for the Trans-Neptunian disk, in agreement with the accretion models and the surface density of the solar nebula. So now there is little doubt that the initial Trans-Neptunian Belt was far more massive than it is today.

References

Altenhoff, W. J., Bertoldi, F., Menten, K. M. 2004, A&A, 415, 771

Barucci, M. A., Fulchignoni, M., Birlan, M., Doressoundiram, A., Romon, J., Boehnhardt, H. 2001, A&A, 371, 1150

Boehnhardt, H., Tozzi, G. P., Birkle, K., Hainaut, O., Sekiguchi, T., Vair, M., Watanabe, J., Rupprecht, G. 2001, A&A, 378, 653

Boehnhardt, H. et al. 2003, Earth Moon and Planets, 92, 145

Brown, R. H., Cruikshank, D. P., Pendleton, Y. 1999, ApJ, 519, L101

Brown, M. E. 2001, AJ, 121, 2804

Brown, M. E., Trujillo, C. A. 2004, AJ, 127, 2413

Brunini, A., Melita, M. D. 2002, Icarus, 160, 32

Cruikshank, D. P., Dalle Ore, C. M. 2003, Earth Moon & Planets, 92, 315

de Bergh, C., Boehnhardt, H., Barucci, M. A., Lazzarin, M., Fornasier, S., Romon-Martin, J., Tozzi, G. P., Doressoundiram, A., Dotto, E. 2004, A&A, 416, 791

Doressoundiram, A. 2003, Earth Moon & Planets, 92, 131

Duncan, M., Quinn, T., Tremaine, S. 1988, ApJ, 328, L69

Edgeworth, K. E. 1943, JBAA, 53, 181

Fernandez, J. A. 1980, MNRAS, 192, 481

Fornasier, S., Doressoundiram, A., Tozzi, G. P., Barucci, M. A., Boehnhardt, H., de Bergh, C., Delsanti, A., Davies, J., Dotto, E. 2004, A&A, 421, 353

Franklin, F. A., Lewis, N. K., Soper, P. R., Holman, M. J. 2004, AJ, 128, 1391

Funato, Y., Makino, J., Hut, P., Kokubo, E., Kinoshita, D. 2004, Nature, 427, 518

Gladman, B., Holman, M., Grav, T., Kavelaars, J., Nicholson, P., Aksnes, K., Petit, J.-M. 2002, Icarus, 157, 269

Goldreich, P., Lithwick, Y., Sari, R. 2002, Nature, 420, 643

Gomes, R. S. 2003, Icarus, 161, 404

Hughes, D. W. 2003, MNRAS, 345, 981

Ida, S., Larwood, J., Burkert, A. 2000, ApJ, 528, 351

Jewitt, D., Luu, J. 1998, AJ, 115, 1667

Jewitt, D., Luu, J., Trujillo, C. 1998, AJ,115, 2125

Jewitt, D. C., Sheppard, S. S. 2002, AJ, 123, 2110

G. P. Kuiper 1951, in *Astrophysics: A Topical Symposium*, ed. J. A. Hynek, New York: McGraw-Hill.

Larsen, J. A., Gleason, A. E., Danzl, N. M., Descour, A. S., McMillan, R. S., Gehrels, T., Jedicke, R., Montani, J. L., Scotti, J. V. 2001, AJ, 121, 562

Lazzarin, M., Barucci, M. A., Boehnhardt, H., Tozzi, G. P., de Bergh, C., Dotto, E. 2003, AJ, 125, 1554

Levison, H. F., Morbidelli, A. 2003, Nature, 426, 419

Malhotra, R. 1995, AJ, 110, 420

Morbidelli, A., Levison, H. F. 2004, AJ, 128, 2564

Noll, K. S., Stephens, D. C., Grundy, W. M., Millis, R. L., Spencer, J., Buie, M. W., Tegler, S. C., Romanishin, W., Cruikshank, D. P. 2002, AJ, 124, 3424

Ortiz, J. L., Gutierrez, P. J., Casanova, V., Sota, A. 2003, A&A, 407, 1149

Ortiz, J. L., Gutierrez, P. J., Sota, A., Casanova, V., Teixeira, V. R. 2003, A&A, 409, L13

Ortiz, J. L. et al. 2004, A&A, 420, 383

Osip, D. J., Kern, S. D., Elliot, J. L. 2003, Earth Moon & Planets 92, 409

Peixinho, N., Boehnhardt, H., Belskaya, I., Doressoundiram, A., Barucci, M. A., Delsanti, A. 2004, Icarus 170, 153

Peixinho, N., Doressoundiram, A., Delsanti, A., Boehnhardt, H., Barucci, M. A., Belskaya, I. 2003, A&A, 410, L29

Petit, J.-M., Mousis, O. 2004, Icarus 168, 409

Schaefer, B. E., Rabinowitz, D. L. 2002, Icarus 160, 52

Sheppard, S. S., Jewitt, D. C., Trujillo, C. A., Brown, M. J. I., Ashley, M. C. B.2000, AJ, 120, 2687

Sheppard, S. S., Jewitt, D. C. 2002, AJ, 124, 1757

Sheppard, S. S., Jewitt, D. C. 2003, Earth Moon & Planets 92, 207

Sheppard, S. S., Jewitt, D. 2004, AJ, 127, 3023

Stansberry, J. A., Cruikshank, D. P., Grundy, W. M., Emery, J. P., Osip, D. J., Fernandez, Y. R., vanCleve, J., Trilling, D. E. 2004, AAS Division for Planetary Sciences Meeting, 36, 43.01

Tegler, S. C., Romanishin, W. 1998, Nature, 392, 49

Thomas, N., Eggers, S., Ip, W.-H., Lichtenberg, G., Fitzsimmons, A., Jorda, L., Keller, H. U., Williams, I. P., Hahn, G., Rauer, H. 2000, ApJ, 534, 446

Trujillo, C. A., Luu, J. X., Bosh, A. S., Elliot, J. L. 2001, AJ, 122, 2740

Trujillo, C. A., Jewitt, D. C., Luu, J. X. 2001, AJ, 122, 457

Trujillo, C. A., Brown, M. E. 2002, ApJ, 566, L125

Veillet, C., Parker, J. W., Griffin, I., Marsden, B., Doressoundiram, A., Buie, M., Tholen, D. J., Connelley, M., Holman, M. J. 2002, Nature, 416, 711

Weidenschilling, S. J. 2002, Icarus, 160, 212

PLANET DETECTION WITH LARGE TELESCOPES AND INTERFEROMETRY

Andreas Quirrenbach
Sterrewacht Leiden
Postbus 9513, NL-2300 RA Leiden, The Netherlands
quirrenb@strw.leidenuniv.nl

Abstract Direct detection of extrasolar planets with coronography and interferometry from ground and space opens the possibility for spectroscopic studies of their surfaces and atmospheres. Astrometric interferometry can provide complementary information on planet statistics and on orbital dynamics in multiple systems. Progress expected over the next decade in these areas is discussed.

Keywords: Extrasolar planets, Interferometry, Extremely large telescopes, Adaptive optics, Coronography

1. Introduction and general goals

The discovery of a planet orbiting the star 51 Pegasi (Mayor & Queloz 1995) has opened a completely new field of astronomy: the study of extrasolar planetary systems. More than 100 planets outside our own Solar System are known to date, and new discoveries are announced almost every month. These developments have revolutionized our view of our own place in the Universe. We know now that other planetary systems can have a structure that is completely different from that of the Solar System. Moreover, the existential question whether other habitable worlds exist can for the first time in human history be addressed in a scientific way.

Further progress in our understanding of planetary systems will depend strongly on the development of new techniques, which can provide detailed information on the planets themselves and on their orbits. Direct detection of exoplanets with coronography and interferometric techniques will enable spectroscopic analyses of their surfaces and atmospheres. Measuring radius, temperature, albedo, atmospheric composition and structure, as well as characterizing surface mineralogy and establishing the presence of surface features, will extend the discipline of physical planetology from the Solar System to other worlds. Continued radial-velocity monitoring and precise astrometric

J. C. del Toro Iniesta (eds.), The Many Scales in the Universe, 235–251.

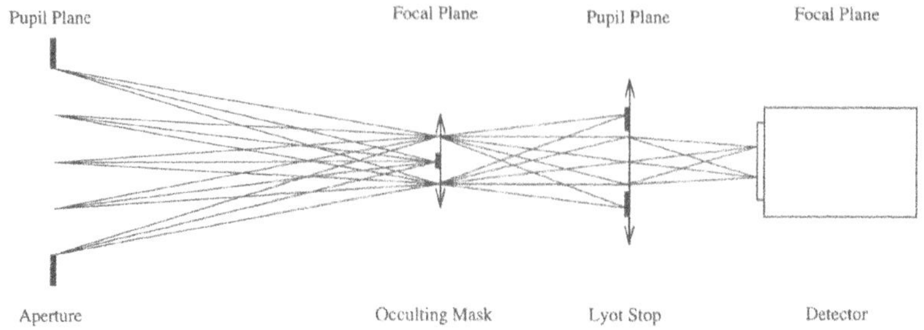

Figure 1. Optical schematic of a Lyot coronograph. Four optical planes are represented: the telescope aperture, the first image plane, the second pupil plane, and the final image plane. Optical rays have been traced for five different points in the entrance pupil, and for three different field positions. Adopted from Malbet (1996).

observations will yield complementary data on dynamical masses and orbital parameters, interactions and orbital resonances in multiple systems; this new field of planetary system dynamics will provide crucial information needed to test theories of planet formation and migration.

An important goal for the next decade is pushing down the detection limits of the various techniques towards lighter and smaller planets. Finding and characterizing Earth analogs would be an important step towards completing the Copernican Revolution, and is thus considered as a strong driver for the design of large ground-based telescopes as well as space missions.

2. Coronography with large telescopes

The direct detection of extrasolar planets requires achieving extremely high contrast ($10^6 \ldots 10^{10}$, depending on the planet parameters and observing wavelength) over very small angles ($0\farcs01 \ldots 1''$).

Foundations of Coronography

In a Lyot coronograph, a mask is placed in the focal plane to reject the light from the bright on-axis star. To prevent the light diffracted at this mask from entering the detector, a stop is placed in a second pupil plane, which is formed by suitable re-imaging optics (see Fig. 1). It is the combination of the occulting mask and the Lyot stop that leads to a high rejection rate, and thus enables the detection of faint objects close to a much brighter star.

The mode of operation of a Lyot coronograph can most easily be understood in the pupil plane (Fig. 2, for more details see Kuchner & Traub 2002). A look at the left column of this figure shows clearly how the choice of the diameter of the Lyot stop $D_{\rm Lyot}$ affects the performance of the coronograph. The undesired

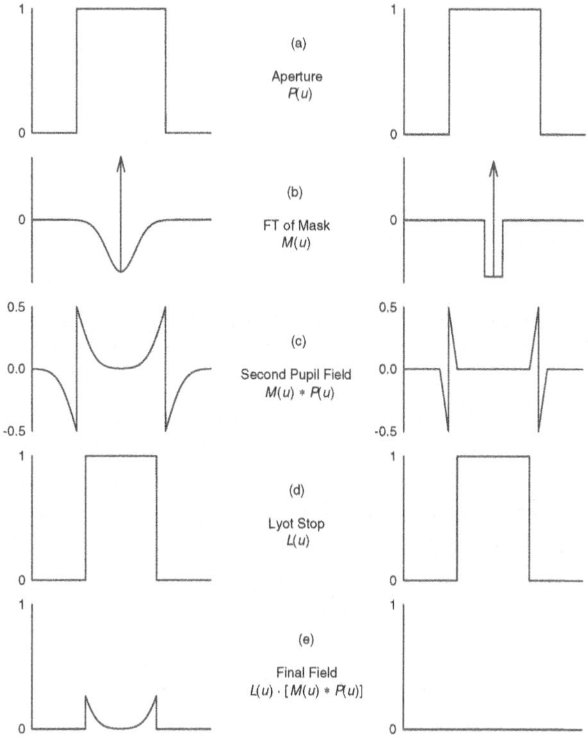

Figure 2. Principle of operation of a coronograph, shown in the pupil plane. An incoming plane wave encounters the primary aperture (a), then a mask in the first image plane (b). The mask intensity transfer function multiplies the image intensity; this corresponds to a convolution with the Fourier transform of the mask amplitude transfer function (c). The light is then passed through a Lyot stop in a second pupil plane (d). The final amplitude is shown in (e). The left column shows a Lyot coronograph with a Gaussian mask, the right-hand column one with a band-limited mask. Adopted from Kuchner & Traub (2002).

residual power from the on-axis star can be reduced by decreasing $D_{\rm Lyot}$, but this also reduces the throughput and resolution of the instrument for off-axis light from the faint companion. One could also make the focal plane spot larger, which reduces the stellar light diffracted near the edges of the aperture, but that means that a larger region around the star is not accessible anymore. There is thus a trade-off between the rejection of the starlight, the throughput and resolution away from the star, and the minimum separation from the star. This limitation can be overcome through the use of occulting masks with more complicated design (Fig. 2, right column).

For ground-based telescopes, however, the speckle halo due to phase errors induced by atmospheric turbulence is usually the main limitation of observations with high dynamic range. The most effective way of distinguishing between speckles and a true companion is provided by differential techniques, such as two-color difference imaging (Racine et al. 1999). If the planet has a deep absorption band not present in the stellar spectrum (e.g., a methane band), one can observe in a filter centered at this band and in a filter in the nearby continuum simultaneously; since the speckle pattern varies only slightly with wavelength, the planet can be identified in the difference image. This approach can be generalized to multi-band imaging or integral-field spectroscopy, in which a data cube containing a spectrum for each sky position is measured. The starlight reflected by planets is highly polarized; imaging polarimetry is therefore another good differential approach to planet detection (Stam et al. 2004).

The precision of speckle subtraction by these differential methods is ultimately limited by photon noise. A high-performance adaptive optics (AO) system is thus needed to reduce the light in the stellar speckle halo, and to sharpen the point spread function (psf) of the planet. The gain of an AO system that generates a psf with Strehl ratio S due to the speckle suppression is $\propto 1/\sqrt{1-S}$, the gain due to the sharpening of the planet image is $\propto S$; the performance of the AO system is thus characterized by $S/\sqrt{1-S}$.

The Potential of 30 m to 100 m Telescopes

An instrument combining a state-of-the-art adaptive optics system with an integral-field spectrograph or imaging polarimeter on an 8 m class telescope could detect the reflected light of "old" Jupiters or the thermal emission from young massive planets under favorable conditions (Feldt et al. 2003). Future larger telescopes will provide substantially better capabilities. For planet detection, increasing the telescope diameter D gives two separate advantages: first, the number of photons collected from the planet increases with D^2; second, the area subtended by the point spread function decreases with D^{-2}. Both advantages combine to a scaling of the observing time with D^{-4}. The goal of planet detection is thus a strong driver for telescope size.

A number of studies are currently underway for telescope concepts with diameters considerably larger than anything in existence today, up to the 100 m OWL concept (Brunetto et al. 2004). Assuming that any large future telescope will be equipped with highly sophisticated adaptive optics and thus approach diffraction-limited performance, the above scaling law suggests that the detection of Earth-like planets around stars at $5 \ldots 10$ pc distance may be marginally possible with 30 m telescopes, and should certainly be within reach of a 100 m aperture. Even spectroscopy of Earth twins could be considered, but one also has to keep in mind that formidable technological challenges have to be overcome. In addition to atmospheric seeing, wind-shake will contribute to the wavefront error; this puts strong constraints on the construction of very large telescope structures and on the adaptive optics design. The primary and secondary mirrors will consist of many small segments; diffraction at the segment edges will create a static array of hexagonal mini-psfs, which also has to be taken into account properly.

Selecting a good site will be critical for getting the most out of large telescopes, and the criteria that determine the expected performance of adaptive optics are quite involved (e.g., Quirrenbach 2002b). The intriguing suggestion that Dome C in Antarctica could provide seeing far superior to that at all known "traditional" mountain tops (Lawrence et al. 2004) should therefore lead to serious consideration of that site for planet-finding (and other) telescopes. In addition, telescopes operating in the cold Antarctic conditions provide a much lower background for observations in the thermal infrared.

3. Extrasolar planets with the VLTI

Whereas the construction of telescopes with diameters well in excess of 10 m will take a fair number of years, interferometers offering a number of attractive capabilities for planet searches are becoming available now. In this section I describe three techniques that are being implemented at ESO's Very Large Telescope Interferometer (VLTI).

Astrometry with PRIMA

Astrometric observations with an interferometer are based on measurements of the delay $D = D_{\rm int} + (\lambda/2\pi)\phi$, where $D_{\rm int}$ is the internal delay between the two interferometer arms measured by a metrology system, and ϕ the observed fringe phase. D is related to the baseline $\vec{B}$ by $D = \vec{B} \cdot \hat{s} = B \cos\theta$, where $\hat{s}$ is a unit vector in the direction towards the star, and θ the angle between $\vec{B}$ and $\hat{s}$. Each data point is thus a one-dimensional measurement of the position of the star. The second coordinate can be obtained from another measurement, either with a different pair of telescopes, or at a different time, when Earth rotation has changed the baseline orientation. The photon noise limit for the precision σ of an astrometric measurement is given by the expression $\sigma = \frac{1}{\rm SNR} \cdot \frac{\lambda}{2\pi B}$. Since high signal-to-noise ratios can be obtained for bright stars, σ can be orders of magnitude smaller than the resolution λ/B of the interferometer.

To reach this photon noise limit, it is of course necessary to control all other statistical and systematic errors. For ground-based instruments atmospheric angular anisoplanatism sets a fundamental limit of roughly $10\,\mu$as even over arcs as small as $10''$ between the target and astrometric reference (Shao & Colavita, 1992; Quirrenbach, 2000). To reach this limit, a metrology system that can monitor the internal path length with ~ 5 nm precision and the three-dimensional baseline vector with $\sim 50\,\mu$m precision is needed. The PRIMA

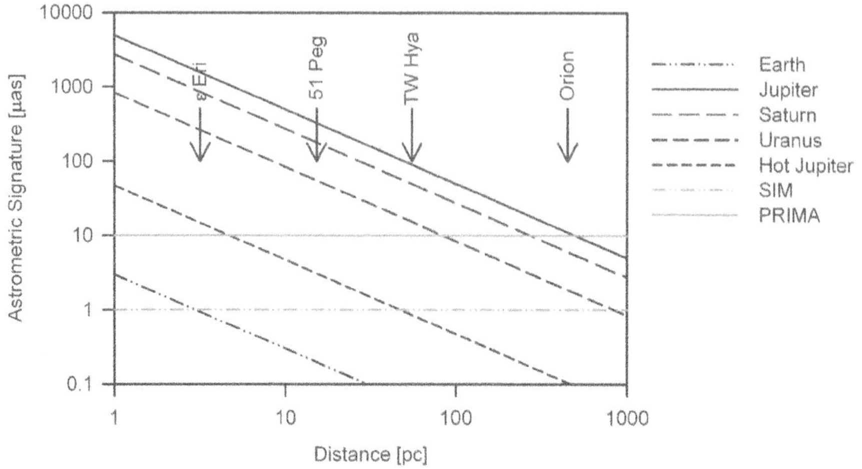

Figure 3. Astrometric signature (semi-amplitude) for five sample planets orbiting a Solar-mass star, as a function of distance. Anticipated detection limits for ground-based (VLTI PRIMA) and space-based (Space Interferometry Mission) instruments are also shown. Adopted from Quirrenbach (2003).

facility at the VLTI will provide this capability (Quirrenbach et al. 2004); a similar system is foreseen at the Keck interferometer. The most important goal of these projects are dynamical studies of extrasolar planets.

The principle of planet detection with astrometry is similar to that behind the Doppler technique: one infers the presence of a planet from the motion of its parent star around the common center of gravity. From simple geometry and Kepler's Laws it follows immediately that the astrometric signal θ of a planet with mass m_p orbiting a star with mass m_* at a distance d in a circular orbit of radius a is given by

$$\theta = \frac{m_p}{m_*} \cdot \frac{a}{d} = m \left(\frac{G}{4\pi^2}\right)^{1/3} \cdot \frac{m_p}{m_*^{2/3}} \cdot \frac{P^{2/3}}{d} \tag{1}$$

$$= 3\,\mu\text{as} \cdot \frac{m_p}{M_\oplus} \cdot \left(\frac{m_*}{M_\odot}\right)^{-2/3} \cdot \left(\frac{P}{\text{yr}}\right)^{2/3} \cdot \left(\frac{d}{\text{pc}}\right)^{-1} .$$

This signature is shown in Fig. 3 for five sample planets (analogues to Earth, Jupiter, Saturn, Uranus, and a "Hot Jupiter" with $m_p = 1\,M_{\text{jup}}$ and $P = 4\,$days) orbiting a $1\,M_\odot$ star. From this figure, the main strengths and difficulties of astrometric planet detection are readily apparent:

- The astrometric signature θ is small compared to the precision of "traditional" astrometric techniques ($\lesssim 1$ mas).

- The difficulty of detecting different types of planets varies greatly, with θ ranging from $\lesssim 1\,\mu$as to ~ 1 mas.

- The sensitivity of astrometry for a given type of planet drops linearly with d (unlike the radial-velocity technique).

- The detection bias of astrometry with orbital radius is opposite to that of the radial-velocity method, favoring planets at larger separations from their parent stars.

It should be pointed out that for circular orbits the observed astrometric signal is an ellipse with semi-major axis θ independent of the orbital inclination; the mass of the planet can therefore be derived directly from Eq. (1) if the mass of the parent star is known. The situation is a bit more complicated for non-circular orbits, but even in that case the orbital inclination can be determined from the astrometric data with techniques analogous to those used for fitting orbits of visual binaries (e.g., Binnendijk 1960).

The specific strengths of the astrometric method enable it to answer a number of questions that cannot be addressed by any other planet detection method. Among the most prominent goals of astrometric planet surveys are the following:

- Mass determination for planets detected in radial velocity surveys (without the $\sin i$ factor). The RV method gives only a lower limit to the mass, because the inclination of the orbit with respect to the line-of-sight remains unknown. Astrometry can resolve this ambiguity, because it measures two components of the orbital motion, from which the inclination can be derived.

- Confirmation of hints for long-period planets in RV surveys. Many of the stars with detected short-period planets also show long-term trends in the velocity residuals (Fischer et al. 2001). These are indicative of additional long-period planets, whose presence can be confirmed astrometrically.

- Inventory of planets around stars of all masses. The RV technique works well only for stars with a sufficient number of narrow spectral lines, i.e., fairly old stars with $m_* \lesssim 1.2\,M_\odot$. Astrometry can detect planets around more massive stars and complete a census of gas and ice giants around stars of all masses.

- Detection of gas giants around pre-main-sequence stars, signatures of planet formation. Astrometry can detect giant planets around young stars, and thus probe the time of planet formation and migration. Observations of pre-main-sequence stars of different ages can provide a critical test of the formation mechanism of gas giants. Whereas gas accretion on

$\sim 10\,M_{\oplus}$ cores requires ~ 10 Myr, formation by disk instabilities would proceed rapidly and thus produce an astrometric signature even at very young stellar ages (Boss 1998).

- Detection of multiple systems with masses decreasing from the inside out. Whereas the astrometric signal increases linearly with the semi-major axis a of the planetary orbit, the RV signal scales with $1/\sqrt{a}$. This leads to opposite detection biases for the two methods. Systems in which the masses increase with a (e.g., υ And, Butler et al. 1999) are easily detected by the RV technique because the planets' signatures are of similar amplitudes. Conversely, systems with masses decreasing with a are more easily detected astrometrically.

- Determine whether multiple systems are coplanar or not. Many of the known extrasolar planets have highly eccentric orbits. A plausible origin of these eccentricities is strong gravitational interaction between two or several massive planets (Lin & Ida 1997, Papaloizou & Terquem 2001). This could also lead to orbits that are not aligned with the equatorial plane of the star, and to non-coplanar orbits in multiple systems.

- Search for massive terrestrial planets orbiting low-mass stars in the Solar neighborhood. With a $10\,\mu$as precision goal, PRIMA at the VLTI will be able to look for rocky planets down to a limit of a few Earth masses around nearby M stars.

Differential Phase and Differential Closure Phase

Differential phase observations with a near-IR interferometer offer a way to obtain spectra of extrasolar planets. The method makes use of the wavelength dependence of the interferometer phase of the planet/star system, which depends both on the interferometer geometry and on the brightness ratio between the planet and the star (Quirrenbach 2000). The differential phase is strongly affected by instrumental and atmospheric dispersion effects. Difficulties in calibrating these effects might prevent the application of the differential phase method to systems with a very high contrast, such as extrasolar planets. A promising alternative is the use of spectrally resolved closure phases, which are immune to many of the systematic and random errors affecting the single-baseline phases.

Figure 4 shows the response of the AMBER instrument at the VLTI to a realistic model of the 51 Peg system, taking into account a theoretical spectrum of the planet as well as the geometry of the VLTI. Joergens & Quirrenbach (2004) have presented a strategy to determine the geometry of the planetary system and the spectrum of the extrasolar planet from such closure phase observations in two steps. First, there is a close relation between the nulls in the

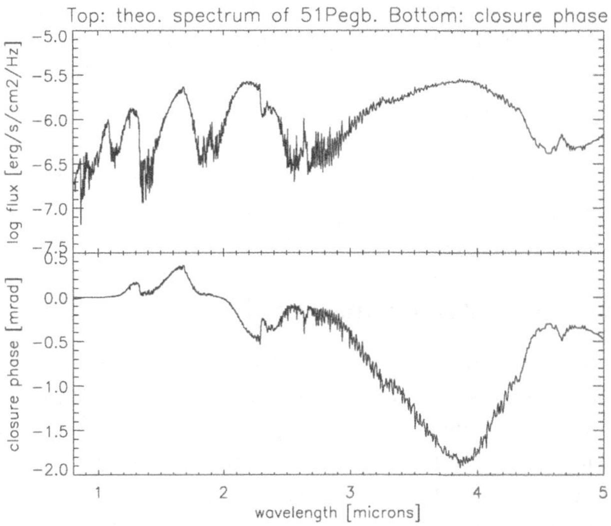

Figure 4. **Spectral signal of planet in the closure phase.** *Top panel:* Theoretical spectrum of the giant irradiated planet orbiting the solar-like star 51 Peg (Sudarsky, Burrows & Hubeny 2003). Clearly visible are CO and H_2O absorption bands in the near-IR. *Bottom panel:* Closure phases in milliradian based on the theoretical spectrum of 51 Peg B as well as of the host star and a simulated observation with the near-IR instrument AMBER at the VLTI using the three telescopes UT1, UT3 and UT4 (baseline lengths102 m, 62 m, and 130 m). It is evident that the spectrally resolved closure phases contain a wealth of spectral information of the planet.

closure phase and the nulls in the corresponding single-baseline phases: every second null of a single-baseline phase is also a null in the closure phase. This means that the nulls in the closure phase do not depend on the spectrum but only on the geometry, so that the geometry of the system can be determined by measuring the nulls in the closure phase at three or more different hour angles. In the second step, the known geometry can then be used to extract the planet spectrum directly from the closure phases.

Orientation of Stellar Rotation Axes

When a rotating star is observed with an interferometer oriented perpendicular to the stellar rotation axis, a phase difference between the red wings and the blue wings of stellar absorption lines is observed. No such gradient occurs for an interferometric baseline oriented parallel to the rotation axis. Measuring the position angle of the phase gradient thus allows determining the orientation of the stellar axis on the sky. More detailed modeling of the interferometric signal can also provide the inclination of the stellar rotation axis (Domiciano de Souza et al. 2004). Combining the information from interferometric spectroscopy (yielding the orientation of the stellar axis) with that from an astrometric planet detection (yielding the orientation of the planetary orbit) therefore enables one to determine whether the planet orbits in the equatorial plane of the star, or to measure the inclination of its orbit if it does not. This is not a trivial piece of information at all: if the orbital evolution of planetary systems is dominated by few-body scattering processes, one might expect to find orbits that are not aligned with the stellar angular momentum (Lin & Ida 1997, Papaloizou & Terquem 2001).

4. The space interferometry mission

While many interesting astrometric projects on extrasolar planets can be carried out from the ground, overcoming the fundamental limitations imposed by atmospheric anisoplanatism requires going to space. Orbits that leave the immediate vicinity of the Earch (e.g., L2 orbits or drift-away orbits) provide the added bonus of a quiet environment with stable heat flux and low vibration levels. NASA's Space Interferometry Mission (SIM), to be launched in 2010, will exploit these advantages to perform a diverse astrometric observing program (Unwin & Turyshev 2002). SIM is essentially a single-baseline interferometer with 30 cm telescopes on a baseline of length 10 m. SIM is a pointed mission, i.e., targets can be observed whenever there is a scientific need (subject only to scheduling and Solar exclusion angle constraints), and the integration time can be matched to the desired signal-to-noise ratio. The limiting magnitude of SIM with "reasonable" observing times (of order one hour per visit) is thus $m_V \approx 19 \ldots 20$. The main scientific driver of SIM is observing extrasolar planets, but a wide range projects addressing Galactic and extragalactic astrophysics will also be carried out (e.g., Quirrenbach 2002a).

For any fixed spacecraft orientation, SIM will be able to access stars in a field with a diameter of $\sim 15°$. Within each such "tile", a few stars will be selected before the mission to define an astrometric reference grid. A basic observing block will consist of observations of the target object(s) interleaved with observations of the grid stars in the tile; the measured delay differences will thus yield one-dimensional positions of the target(s) relative to the grid stars. The second coordinate can be obtained by rotating the spacecraft around the line of sight by an angle close to $90°$. During the course of the mission, the grid stars will be visited regularly, about four to five times per year. By observing overlapping tiles, a full-sky reference grid can be constructed in the same manner as overlapping plates have been used to assemble all-sky astrometric catalogs. The inclusion of quasars in the grid will ensure that this reference system will represent an inertial frame. The expected performance of SIM in this "global astrometry" mode is $4\,\mu$as precision on the derived astrometric parameters (position, parallax) at the end of the nominal 5-year mission.

Many terms in the error budget of SIM measurements depend on the angle between the target and the astrometric references. It is thus possible to improve the accuracy by choosing references close ($\lesssim 1°$) to the object of interest. For this "narrow-angle" mode, an accuracy of $1\,\mu$as for each measurement is envisaged. (Note that this number has to be divided by $\sqrt{N}$ with $N = 30 \ldots 100$ for the number N of visits to obtain the "mission accuracy" of the derived standard astrometric parameters. This is the fair comparison with the "global" mode and with the figures usually quoted for other missions.) These narrow-angle measurements will not be in an inertial reference frame, but only with respect to the selected reference stars. The narrow-angle precision will thus not apply to absolute parallaxes and proper motions, but this is the relevant number for determining the astrometric wobble due to unseen companions.

SIM enjoys two major advantages over PRIMA: an order-of-magnitude better precision, and more relaxed constraints on the selection of possible targets. (The latter is because being above the atmosphere, SIM does not suffer from atmospheric anisoplanatism, which dictates the choice of reference stars within a few tens of arcseconds to the target in ground-based observations.) These characteristics enable a high-precision survey of $\sim$ 200 nearby stars reaching down to $1 \ldots 3\,M_{\oplus}$ (depending on stellar mass and distance), and a less-sensitive survey of $\sim$ 2000 stars establishing better statistics on massive planets in the Solar neighborhood. In addition, SIM will observe a sample of pre-main-sequence stars to investigate the epoch of planet formation.

5. Darwin and the terrestrial planet finder

Over the past few years, ESA and NASA have formulated rather similar goals for a mission — known as "Darwin" in Europe and as "Terrestrial Planet Finder" in the US — aimed at a detailed study of Earth-like planets orbiting nearby stars. Such a mission must detect radiation from any Earth-like planets in the habitable zones surrounding $\sim$ 150 solar type (spectral types F, G, and K) stars, characterize the orbital and physical properties of all detected planets to assess their habitability, and characterize the atmospheres and search for potential biomarkers among the brightest candidates for Earth-like planets.

From this requirement it follows that Darwin / TPF must be able to separate planets in the habitable zone from their parent stars. This translates to an angular resolution of $\sim$ 35 mas, corresponding to 0.35 AU at 10 pc, or 0.7 AU at 20 pc. The most important challenge is the very large contrast between the planet and the parent star, $\sim 1 : 10^{10}$ in the visible, and $\sim 1 : 10^{6}$ near $10\,\mu$m.

Spectroscopy of Planetary Atmospheres

Life on Earth has a profound impact on the composition of the atmosphere, producing abundant oxygen, and methane in extreme thermodynamic disequi-

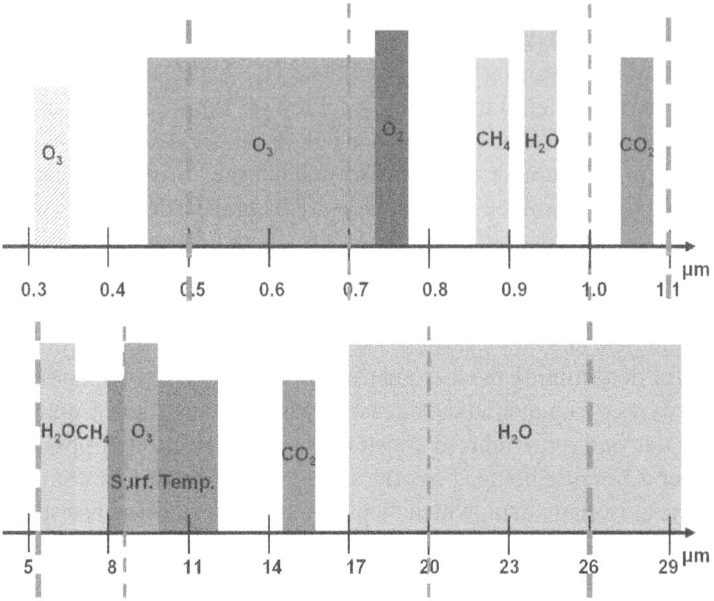

Figure 5. Important diagnostic bands in the visible (top) and mid-infrared (bottom) wavelength range. The dashed vertical lines indicate the wavelength coverage needed for a basic (water, oxygen) and extended analysis of the planetary atmosphere.

librium. The Earth fly-by of the Galileo spacecraft in 1990 demonstrated beautifully that these signatures of life can be detected by remote sensing; in addition the spacecraft detected "a widely distributed surface pigment with a sharp absorption edge in the red part of the visible spectrum" (Sagan et al. 1993); this is the "vegetation red edge" characteristic of land plants. It is plausible to assume that any form of abundant life on an extrasolar planet might transform the planetary atmosphere in a similar way, and thus produce a signature that can be detected spectroscopically (Fig. 5, see also Des Marais et al. 2002).

The mid-infrared spectral region ($\sim 6 \ldots 20\,\mu$m) contains a number of important spectral features that can be used to diagnose the presence of H_2O, CO_2, and O_3, which serves as a proxy for O_2 (Angel et al. 1986). Methane also has a strong absorption band in this spectral region, but it would be extremely hard to detect in an Earth-like atmosphere. In the past, however, before the rise of oxygen, biogenic CH_4 may have been more abundant in the Earth's atmosphere by a factor $100 \ldots 1000$ (Catling et al. 2001); at this level it would also be detectable and serve as an indicator of bacterial life. One cautionary remark is in order, however: the detection and interpretation of infrared absorption bands could be complicated by the presence of cirrus clouds, which tend to reduce the depth of these features (Smith et al. 1993).

There are of course also many molecular absorption bands in the visible and near-IR spectral ranges. Using the oxygen A band at 7600 Å as a tracer of life was indeed proposed much earlier than the mid-IR signatures discussed above (e.g. Owen 1980). Ozone has a broad band near 6000 Å, and a very strong band in the UV that is responsible for the short-wavelength cutoff in the transmission of our atmosphere. Water also has very strong absorption bands; the gaps between these bands define the astronomical J, H, K, and L bands in the near-IR. In addition, the CH_4 and CO_2 bands in the 2 μm spectral region could serve as a diagnostic of elevated levels of these gases.

Mining the wealth of information potentially available in the visible / near-IR has its own complications, however. A wide range of plausible values of the albedo translates into a corresponding uncertainty in the planetary brightness. There are large differences between the colors and spectra reflected from oceans, snow, deserts, and forested areas. Clouds also cause very marked changes of albedo, UV-visible color, and depth of the oxygen bands (Burrows et al. 1999). It is of course possible to take advantage of these effects with photometric measurements, from which the rotation period, presence of marked surface features, fractional cloud cover, and perhaps seasonal variations of these quantities could be derived (Ford et al. 2001). Again, caution has to be exercised in the interpretation of the data, as for example aerosols can reduce the contrast and therefore suppress diurnal variations (King et al. 1999).

Whether information is gathered in the mid-IR or in the visible/near-IR, careful modeling of the planetary atmosphere is required to rule out an abiotic origin of O_2, CH_4, or any other presumed indicator of life (e.g. Schindler & Kasting 2000). One has to keep in mind that the biosphere, atmosphere, hydrosphere, and lithosphere interact in many complicated ways. Even reconstructing the processes that lead to the rise of oxygen in our own atmosphere remains a challenging task, in spite of the wealth of available data (Kasting 2001). It will certainly be much more difficult to interpret the glimpses of atmospheric chemistry that we may be obtain by remote sensing of extrasolar terrestrial planets.

On balance, there are good arguments in favor of the visible, and equally good reasons to prefer the mid-infrared wavelength range for Darwin / TPF. For the time being, NASA has decided to implement a visible-light mission first, whereas ESA is pursuing the mid-infrared option.

Mid-Infrared Interferometry

To achieve the required angular resolution in the mid-infrared, the Darwin / TPF mission requires an interferometer with baselines of at least 50 . . . 100 m to separate the planets from their host stars. It appears that mounting each telescope (and probably the beam combiner) on a separate spacecraft will be much easier than building a single large structure, and therefore a "free-flyer" approach has been adopted.

In an infrared interferometer the starlight can be suppressed by introducing an achromatic π phase shift in one arm of the interferometer, so that there is destructive interference for light arriving on-axis (Bracewell 1978). The most important characteristic of such a nulling interferometer is the quality of this destructive output, called the null depth N. In other words, the observed starlight is reduced by a factor N compared to a constructive interferometer. Denoting the residual phase (i.e., the difference between the actual phase and the "best compromise" for all wavelengths within the bandpass and the two polarization states) by $\Delta\phi$, the rms phase difference due to dispersion mismatch, averaged over the bandpass, by $\Delta\phi_\lambda$, the phase difference between the two polarization states by $\Delta\phi_{s-p}$, the relative rotation angle between the two beams by α, and the normalized intensity mismatch by $\delta I/I \approx \delta \ln I$, the null depth is given by (Serabyn 2000)

$$N = \frac{1}{4}\left[(\Delta\phi)^2 + (\Delta\phi_\lambda)^2 + \frac{1}{4}(\Delta\phi_{s-p})^2 + \alpha^2 + (\delta \ln I)^2\right] . \qquad (2)$$

It is apparent from this equation that very precise control of wavefront defects and precise matching of the interferometer arms is required to achieve a null depth of $\sim 10^{-5}$, as required for the spectroscopy of Earth-like planets.

In a single-baseline interferometer one would also have to know the absolute null depth very precisely – otherwise a planet could not be distinguished from a slightly larger contribution from the stellar photons leaking through the "destructive" output of the interferometer. This is almost impossible to achieve in practice, therefore proposed interferometric implementations of Darwin / TPF are based on at least three telescopes forming separate nulling interferometers, whose outputs can be combined with time-variable phases, which produces a modulated signal proportional in strength to the brightness of the planet (e.g., Kaltenegger et al. 2004). A Darwin mission consisting of three 3 m telescopes and capable of performing the survey of ~ 150 stars described above could be ready for launch around 2015.

Visible-Light Coronography

The visible wavelength range has an important technical advantage over the mid-IR: the required spatial resolution can be achieved with a ~ 20 times smaller aperture diameter. It is therefore not necessary to use an interferometer for the direct detection of planets at these shorter wavelengths. Nevertheless, building a telescope that can give a 10^{10} dynamic range over a separation of ~ 35 mas is a daunting task. A space telescope equipped with a coronograph and deformable mirror for suppressing scattered light may be a viable solution, in particular if an optimized coronograph design is used (see Fig. 2, right column). It should be pointed out that a circular pupil is not necessarily the optimum geometry. An elliptical pupil with an axial ratio of roughly 2 : 1 (e.g., $3 \times 6\,\mathrm{m}$) would give sufficient resolution along the major axis, and it could be launched much more easily than a telescope with a 6 m circular aperture. More exotic pupils may also offer significant advantages. Nisenson & Papaliolios (2001) present simulations of an apodized square aperture, which gives a very clean point spread function, except in the two directions aligned with the edges. Non-apodized pupils that give extremely dark point spread functions along one axis can also be designed (Kasdin et al. 2003). A non-circular telescope would have to be rotated around the line-of-sight for observations aimed at planet detection; the spectroscopic follow up could be done with the optimum orientation. For any of the suggested geometries, manufacturing an optical system with the required exquisite control of scattered light remains the largest technological hurdle.

6. Conclusions

Progress in interferometry, adaptive optics, instrumentation, and telescope construction over the next decade will create many new opportunities for studies of extrasolar planets. Extremely large telescopes on the ground, visible-light coronography and mid-infrared interferometry from space will all push the sensitivity and contrast of direct planet detection down to the level needed for studies of Earth-like planets (see also Angel 2003). A natural precursor and by-product of these efforts will be the capability of getting detailed information on the physical properties of Jovian planets. Interferometric closure-phase measurements could also contribute in this area.

Interferometric astrometry, first from the ground, then from space, will meanwhile provide a census of planets around nearby stars, and planet statistics as a function of stellar mass and age. Complete sets of orbital and stellar rotation parameters in multiple systems will provide critical tests of planet formation and migration theories. The second decade of exo-planetary science promises to become as exciting as the first.

References

Angel, R. 2003, in *Towards other Earths – DARWIN/TPF and the search for extrasolar terrestrial planets.* Eds. Fridlund, M., & Henning, T., ESA, SP-539, 221

Angel, J.P.R., Cheng, A.Y.S., & Woolf, N.J. 1986, Nature, 322, 341

Binnendijk, L. 1960, Philadelphia: University of Pennsylvania Press.

Boss, A.P. 1998, Nature, 393, 141

Bracewell, R.N. 1978, Nature, 274, 780

Brunetto, E., Dierickx, P., Gilmozzi, R., Le Louarn, M., Koch, F., et al. 2004, in *Second Bäckaskog workshop on extremely large telescopes.* Eds. Ardeberg, A.L., & Andersen, T., SPIE, Vol. 5382, 159

Burrows, J.P., Weber, M., Buchwitz, M., Rozanov, V., Ladstätter-Weissenmayer, A., et al. 1999, JAtmSci, 56, 151

Butler, R.P., Marcy, G.W., Fischer, D.A., Brown, T.M., Contos, A.R., et al. 1999, ApJ, 526, 916

Catling, D.C., Zahnle, K.J., & McKay, C. 2001, Science, 293, 839

Des Marais, D.J., Harwit, M.O., Jucks, K.W., Kasting, J.F., Lin, D.N.C., et al. 2002, Astrobiology, 2, 153

Domiciano de Souza, A., Zorec, J., Jankov, S., Vakili, F., Abe, L., & Janot-Pacheco, E. 2004, A&A, 418, 781

Feldt, M., Hippler, S., Henning, T., Gratton, R., Turatto, M., Waters, R., & Quirrenbach, A. 2003, in *Scientific frontiers in research on extrasolar planets.* Eds. Deming, D., & Seager, S., ASP Conference SeriesVol. 294, 569

Fischer, D.A., Marcy, G.W., Butler, R.P., Vogt, S.S., Frink, S., & Apps, K. 2001, ApJ, 551, 1107

Ford, E.B., Seager, S., & Turner, E.L. 2001, Nature, 412, 885

Joergens, V., & Quirrenbach, A. 2004, in *New frontiers in stellar interferometry.* Ed. Traub, W.A., SPIE, Vol. 5491, 551

Kaltenegger, L., Karlsson, A., & Neger, T. 2004, AdSpR, 34, 618

Kasdin, N.J., Vanderbei, R.J., Spergel, D.N., & Littman, M.G. 2003, ApJ, 582, 1147

Kasting, J.F. 2001, Science, 293, 819

King, M.D., Kaufman, Y.J., Tanré, D., & Nakajima, T. 1999, BullAmMetSoc 80, 2229

Kuchner, M.J., & Traub, W.A. 2002, ApJ 570, 900

Lawrence, J.S., Ashley, M.C.B., Tokovinin, A., & Travouillon, T. 2004, Nature, 431, 278

Lin, D.N.C., & Ida, S. 1997, ApJ, 477, 781

Malbet, F. 1996, A&AS, 115, 161

Mayor, M., & Queloz, D. 1995, Nature, 378, 355

Nisenson, P., & Papaliolios, C. 2001, ApJ, 548, L201

Owen, T. 1980, in *Strategies for the search for life in the Universe.* Ed. Papagiannis, M.D., 177

Papaloizou, J.C.B., & Terquem, C. 2001, MNRAS, 325, 221

Quirrenbach, A. 2000, in *From extrasolar planets to cosmology: the VLT opening symposium.* Eds. Bergeron, J., & Renzini, A., 462

Quirrenbach, A. 2002a, in *From optical to millimetric interferometry: Scientific and technological challenges.* Eds. Surdej, J., Swings, J.P., Caro, D., & Detal, A., Université de Liège, 51

Quirrenbach, A. 2002b, in *Astronomical site evaluation in the visible and radio range.* Eds. Vernin, J., Benkhaldoun, Z., & Muñoz-Tuñón, C., ASP Conference Series, Vol. 266, 516

Quirrenbach, A. 2003, in *Towards other Earths – DARWIN/TPF and the search for extrasolar terrestrial planets.* Eds. Fridlund, M., & Henning, T., ESA, SP-539, 19

Quirrenbach, A., Henning, T., Queloz, D., Albrecht, S., Bakker, E., et al. 2004, in *New frontiers in stellar interferometry.* Ed. Traub, W.A., SPIE, Vol. 5491, 424

Racine, R., Walker, G.A.H., Nadeau, D., Doyon, R., & Marois, C. 1999, PASP, 111, 587

Sagan, C., Thompson, W.R., Carlson, R., Gurnett, D., & Hord, C. 1993, Nature, 365, 715

Schindler, T.L., & Kasting, J.F. 2000, Icarus, 145, 262

Serabyn, E. 2000, in *Interferometry in optical astronomy.* Eds. Léna, P.J., & Quirrenbach, A., SPIE, Vol. 4006, 328

Shao, M., & Colavita, M.M. 1992, A&A, 262, 353

Smith, W.L., Ma, X.L., Ackerman, S.A., Revercomb, H.E., & Knuteson, R.O. 1993, JAtmSci, 50, 1708

Stam, D.M., Hovenier, J.W., & Waters, L.B.F.M. 2004, A&A, 428, 663

Sudarsky, D., Burrows, A., & Hubeny, I. 2003, ApJ, 588, 1121

Unwin, S., & Turyshev S. 2002, *Science with the Space Interferometry Mission.* Jet Propulsion Laboratory

AVO FIRST SCIENCE

Discovery of type 2 quasars

Mark G. Allen
Observatoire de Strasbourg UMR 7550
allen@astro.u-strasbg.fr

Paolo Padovani
ST-ECF, European Southern Observatory

Piero Rosati
European Southern Observatory

Nicholas A. Walton
Institute of Astronomy, Cambridge

Abstract We present the first science for the Astrophysical Virtual Observatory (AVO) project. The exploitation of astronomical data beyond classical identification limits with interoperable tools for statistical identification of sources is now a reality. We show how the AVO prototype has been used to identify high redshift type 2 AGN in the two GOODS fields. Using an empirical estimator for the X-ray luminosity, we find many high power candidates which qualify as active galactic nuclei (AGN). Thirty-one of these sources have high estimated powers ($> 10^{44}$ erg/s) and therefore qualify as optically obscured quasars, the so-called QSO 2. This significantly increases the number of such known objects in the GOODS fields, and provides new estimates for the QSO-2 number density. This work shows how the first Virtual Observatory interoperability gains are leading to scientific discovery.

Keywords: Quasars, AGN, Virtual Observatories, Databases

1. Introduction

The properties of active galactic nuclei (AGN) can largely be explained by the physics of the black hole, accretion disk, jet, and obscuring torus, along

J. C. del Toro Iniesta (eds.), The Many Scales in the Universe, 253–260.

with the observational effect of the viewing angle. The unified model for AGN (Urry & Padovani 1995) proposes that Type 1 AGN have an unimpeded view of the central nucleus and accretion disk whereas in Type 2 objects the view is obscured by the torus. While many examples of local, and therefore relatively low-power, type 2 AGN are known (Seyfert 2s), their high-power counterparts, that is optically obscured, radio-quiet type 2 QSO have been difficult to find. This is largely because QSO 2 objects are heavily reddened and therefore fall through the "standard" (optical) methods of quasar selection. However, hard X-rays generated in the accretion disk are able to penetrate the torus, and indicate the presence of a hidden AGN (Norman et al. 2002). In addition to their importance for AGN models, type 2 sources are also expected comprise a significant fraction of the X-ray background and are therefore also cosmologically relevant.

In this paper we describe how the Astrophysical Virtual Observatory (AVO) prototype has been used to identify high redshift type 2 QSOs in the two Great Observatories Origins Deep Survey (GOODS) fields. This represents one of the first scientific results for Virtual Observatories. Full details are available in Padovani et al. (2004).

2. Astrophysical virtual observatory

Astronomy is at a turning point, data volumes and rates are increasing to unprecedented levels, posing significant challenges for data processing, delivery and analysis. Moreover, contemporary science questions demand use of the full electromagnetic spectrum, requiring that data from different telescopes, analysis tools, on-line services and archived information are readily compatible.

The Virtual Observatory (VO) initiatives seek to build the framework for such interoperable systems. The vision is that astronomy resources (data archives, catalogue services, analysis tools, published information, theoretical models) be accessible and usable as if they were all installed on your local desktop computer.

VO initiatives are now at a stage where prototype tools can be utilised to produce scientific results. Real gains have been made in the areas of accessing and describing remote data sets, manipulating image and catalogue data, and performing remote calculations. These tools are enabled by the VO infrastructure and interoperability standards that are being developed cooperatively by all the VO projects under the auspices of the IVOA (http://www.ivoa.net). The Astrophysical Virtual Observatory Project (AVO: http://www.euro-vo.org) is a "phase A" program (jointly funded by the European Commission under FP5-Fifth Framework Programme) to study the scientific requirements and technologies necessary to build a VO for European astronomy.

The AVO project is driven by its strategy of regular annual scientific demonstrations of VO technology. For this purpose progressively more complex demonstrators are being constructed. The current AVO prototype is based on CDS Aladin tool. This allows efficient interactive visualization and manipulation of image and catalogue data. Links and access to other VO services including remote data archives and image servers are provided via the GLU registry of services (http://simbad.u-strasbg.fr/glu/glu.htx) and also by plug-in style connections between software components. Tabular data is handled in VOTable format, on-line images and spectral data are accessed by new VO protocols, (Simple Spectral Access, and Simple Image Access) and other enhanced access mechanisms such as the Aladin image server tree-view function. The description and interoperability of images and tabular data are aided by metadata such as UCDs and IDHA data-model.

The *First Science* event for the AVO project was held in January 2004 where we demonstrated the use of VO tools for two scientific scenarios. This demonstration was truly multi-wavelength, using heterogeneous and complex data covering the whole electromagnetic spectrum. The science scenarios demonstrated the use of MERLIN and VLA (radio), ISO (spectra and images), 2MASS (infrared), UNSO, ESO 2.2m/WFI and VLT/FORS (spectra), and HST/ACS (optical), XMM and Chandra (X-ray) data and catalogues. The galactic scenario was a multi-wavelength investigation of young stellar objects (YSOs). The extragalactic science scenario, the subject of this conference article, demonstrated the discovery significant numbers of type 2 QSOs in the GOODS survey fields, and led to the publication of the first AVO science paper (Padovani et al. 2004).

3. Finding type 2 AGN in the GOODS fields

The GOODS (HDFN and CDFS) fields are the most data-rich, deep survey areas on the sky. Of most interest for finding type 2 AGN are the X-ray (Chandra) catalogues that cover a large region around both fields (Alexander et al. 2003), and the Hubble Space Telescope ACS $BViz$-band imaging data (Giavalisco et al. 2004). The prototype allows access to these data via a "data tree" mechanism that provides a means of browsing, and selecting data from very large remote data sets. Catalog data can be selected, imported and interoperated with image data from any VO compliant source. Figure 1 shows a snapshot of the prototype in action. Here we see the data tree on the left showing all the data available for the CDFS field. Browsing over the tree automatically generates field of view outlines of the image data, and conversely browsing over a point in the image highlights all data available at the point. This hierarchical data tree mechanism allows one to easily handle large sets such as the GOODS data set, or the relevant part of an enormous archive, all alongside any small local data set.

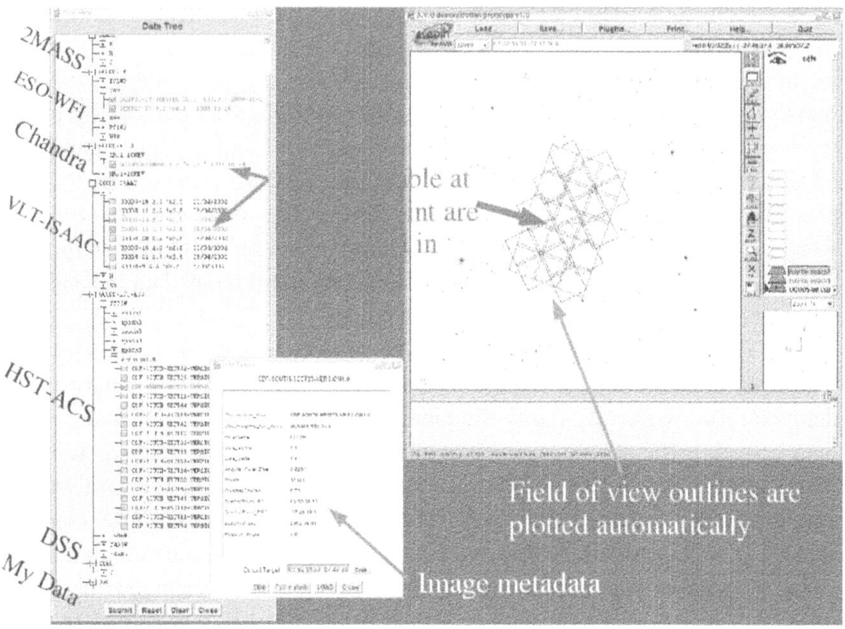

Figure 1. The AVO prototype in action.

The two key physical properties that we use to identify type 2 AGN candidates are that they be obscured, and that they have sufficiently high power to be classed as an AGN and not a starburst. To find these candidates, we use a relatively simple method based on the X-ray and optical fluxes. Firstly we select candidate AGN from the X-ray data. Ideally, to identify type 2 AGN via their X-ray emission, one would need X-ray spectra to select sources with flat spectral indices, indicative of absorption. Typical count rates of X-ray sources in the GOODS fields are however relatively low for spectral fitting, therefore we select sources based on their X-ray hardness ratio, a measure of the fraction of hard photons relative to soft photons, $HR = (H-S)/(H+S)$, where H is the hard X-ray counts ($2.0-8.0$ keV) and S is the soft X-ray counts ($0.5-2.0$ keV). Out of the 829 X-ray sources we find 294 such absorbed sources.

To estimate the X-ray power for candidate type 2 sources we then use the correlation described by Fiore et al. (2003), between the $f(2-10\ \mathrm{keV})/f(R)$ ratio and the hard X-ray power. This correlation allows estimation of the hard

X-ray power even when the objects are so faint that no spectrum, and hence no redshift are available. The basis of this correlation is the fact that the $f(2 - 10\ \text{keV})/f(R)$ ratio is roughly equivalent to the ratio between the nuclear X-ray power and the host galaxy R band luminosity. Since the host galaxy R band luminosities (unlike the X-ray power) show only a modest amount of scatter, this flux ratio is a good indicator of X-ray power.

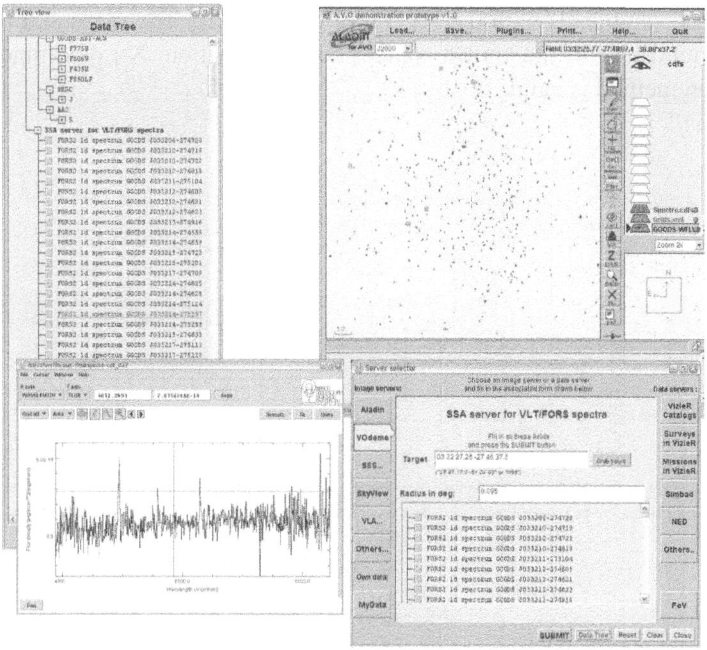

Figure 2. Spectral Access in the AVO prototype. Spectra available via the Simple Spectral Access protocol may be discoved by a query around a sky coordinate, and then browsed in the data tree. The spectra slit positions and angles are drawn on the image, and the Specview plugin to the prototype allows visualization and analysis of the spectra. The figure displays VLT-FORS spectra in the CDFS.

To apply this correlation, we must first match the absorbed X-ray candidates with their optical counterparts. The cross match facility in the AVO prototype tool allows us to do this. The "best match" mode provides the closest positional match pairs within a threshold radius and performs the necessary table join. Subsequent filtering allows further match constraints to be applied so that only sources with matches within the total X-ray plus optical positional error are accepted. The "all matches" and "no matches" modes of the cross matcher also allows checking for multiple matches, cases of object confusion, and unmatched sources. With these criteria all of the CDFS X-ray candidates have optical counterparts, and only a small number of X-ray candidates in the

HDFN are unmatched. Modeling of the number of expected matches shows the false match rate to be between 8 and 15%.

The X-ray luminosities for these sources, calculated with the Fiore relation as described above, are all above 10^{42} erg/s and thus all qualify as AGN. Some of these candidates are already known via spectroscopic studies of the GOODS fields by Szokoly et al. (2004) and Barger et al. (2003). Access to the Szokoly et al. data via the VO Simple Spectral Access protocol allows us to easily identify these candidates with the AVO tool (see Fig. 2). The measured redshifts and luminosities of the known sources confirms that our X-ray luminosity estimation method is valid.

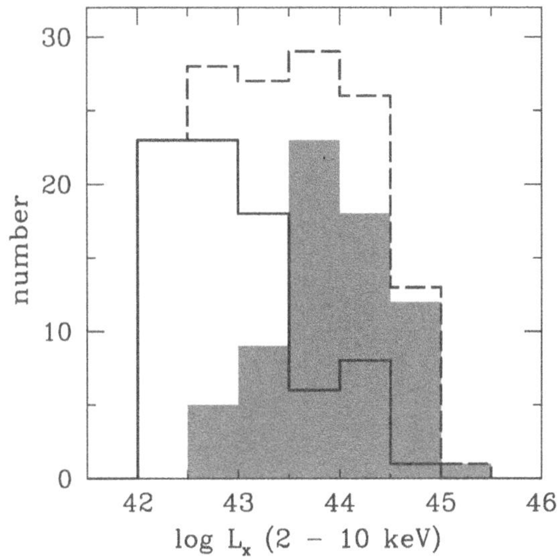

Figure 3. X-ray power distribution of type 2 AGN candidates (filled), previously known type 2 AGN(solid line), and the total (dashed line).

The result of these selection criteria, after removing the known sources, yields a total of 68 new type 2 AGN candidates. Of these, 31 have X-ray luminosities in excess of 10^{42} erg/s so qualify as QSO 2 candidates. Figure 3 shows the luminosity distribution of the new candidates, along with the known type 2 AGN, and the total distribution. The new candidates fill a gap in the high luminosity end of this distribution. This is due to our method going much deeper than previous spectroscopic studies allowing us to probe type 2 AGN at higher redshifts. Based on their derived X-ray powers, our candidates have redshifts, $z \sim 3$, and $z \sim 4$ for the QSO 2s.

This work brings to 40 the number of QSO 2s in the GOODS fields, an improvement of a factor $\sim$ 4 compared to the only nine such sources previously known. As expected, being still unidentified, our sources are very faint: their median ACS i magnitude is ~ 26.3, which corresponds to $R \sim 26.8$, compared to $R \sim 22$ for typical CDFS sources with redshift determinations. Therefore spectroscopic identification is not possible for most of these objects, even with the largest telescope currently available. In this way VO techniques are allowing us to sample a region of redshift-power parameter space which is much harder to reach with *classical* methods.

In terms of surface density, we find >330 type 2 QSOs per square degree down to a flux limit of 10^{-15} erg cm^{-2} s^{-1} in the 0.5-8 keV band. This is larger than current estimates and some predictions (Gandhi et al. 2004) and suggests that the QSO 2 surface density has been underestimated at faint flux limits.

4. Next AVO science developments

The AVO is promoting science with VO tools through two further developments: the next (Jan 2005) science demonstration, and the EURO-VO Science Reference Mission (SRM).

The Jan 2005 demonstrations will see the roll-out of the first version of the EURO-VO portal through which astronomers will gain access to a wide range of data and computation capabilities. In particular we will show use of distributed workflows, registry harvesting, and wrapping of sophisticated astronomical applications as Web Services. This demonstration will also mark the transition of AVO into EURO-VO.

For the SRM, the AVO team and Science Working Group will define the key scientific goals of a full-fledged EURO-VO. It will consist of a number of science cases covering a broad range of astrophysics with related scientific and technical requirements.

5. Summary

This work shows that prototype VO tools have evolved beyond the demonstration level to become respectable research tools. The main advantage provided by these prototype tools is that they allow efficient, interoperable use of image, spectral and catalog data in a very flexible way. Components such as the data tree provide a way to easily access the relevant pixels in large data sets, and the ability to check catalogue measurements and positions with the original data. The simple positional cross matcher in the AVO prototype was vital for manipulating information from different sources. The advances described here are however, only the first interoperability gains. As VO initiatives move from "phase A" studies such as AVO, into building robust VO services we expect significant gains in the areas of VO registries, VOQL a query language, and the data model for describing astronomy data on all levels. The working groups of the IVOA are making rapid progress in all these areas via open forums (http://www.ivoa.net/forum) to which *any* interested parties may contribute.

References

Alexander, D. M., Bauer, F. E., Brandt, W. N., et al. 2003, AJ, 126, 539
Barger, A. J., Cowie, L. L., Capak, P., et al. 2003, AJ, 126, 632
Fiore, F., Brusa, M., Cocchia, F., et al. 2003, A&A, 409, 79
Gandhi, P., Crawford, C. S., Fabian, A. C., & Johnstone, R. M. 2004, MNRAS, 348, 529
Giavalisco, M., Ferguson, H. C., Koekemoer, A. M., et al. 2004, ApJ, 600, L93
Norman, C., Hasinger, G., Giacconi, R., et al. 2002, ApJ, 571, 218
Padovani, P., Allen, M. G., Rosati, P., & Walton, N. A. 2004, A&A, 424, 545
Szokoly, G. P., Bergeron, J., Hasinger, G., et al. 2004, ApJS, in press [astro-ph/0312324]
Urry, C. M., & Padovani, P. 1995, PASP, 107, 803

EURO50: A EUROPEAN 50 M ADAPTIVE OPTICS EXTREMELY LARGE TELESCOPE

Arne Ardeberg and Torben Andersen
Lund Observatory
Sweden
arne@astro.lu.se; torben@astro.lu.se

Abstract Science drivers, technical challenges and solutions are discussed for a 50 m Adaptive-Optics (AO) Extremely Large Telescope (ELT). Among programmes requiring ELT data, special attention is given to the evolution from proto-star shells, via proto-planetary discs to planet formation. Earth-like planets and life signatures are discussed. Kuiper-Belt objects (KBOs) are taken up. Star formation and spatially and temporally resolved stellar observations are discussed. Use and advantage of stellar clusters as probes of galactic evolution is discussed, also for distant galaxies. First-generation supernovae are taken up as tracers of the first stellar generations but also as back-ground objects for mapping of the inter-galactic medium. A proposed European ELT, the Euro50, is described. The Gregory-type optical system, with elliptical primary, f/0.85, and secondary mirrors is discussed as are focus arrangements. The compact, light-weight mechanical structure is discussed together with mirror alignment. Wind-tunnel and computational fluid dynamics tests of an enclosure concept are described. Special attention is given to the adaptive and live optics concepts. Multi-conjugate AO and control of the deformable secondary mirror are discussed together with laser guide stars. The over-all control system is described. Integrated modelling is discussed with emphasis on the importance on wind-load control, an item of crucial importance for an ELT with diffraction-limited operation.

Keywords: Proto-planetary discs, Earth-like planets, Stellar evolution, Galactic evolution, First stars, Adaptive Optics, Extremely Large Telescopes, Optical systems, Deformable mirrors - Mechanical structures, Wind load - Mirror alignment, Enclosure, Deformable mirrors, Laser guide stars, Integrated modelling

1. Introduction

Lately, the observable universe has expanded, as has our planetary system. New details have emerged and new objects have been identified. Many explanations exist for the flow of new insight. Still, there can be no doubt regarding the major role of new telescopes. Among them, the largest ones for visual and adjacent wavelengths have had a strong influence on the progress. Since 1990,

J. C. del Toro Iniesta (eds.), The Many Scales in the Universe, 261–294.

several Very Large Telescopes (VLTs) got into operation. An additional number of VLTs are in progress. These telescopes, with primary mirrors of 8 to 10 m, have been highly successful. With them, a stream of break-through results has been obtained. The VLTs provide ample proof of the power of a combination of a large light-collecting surface and high image quality, backed by new technology. Raising progress, the new advance has created new challenges to our understanding of the universe. Many targets are beyond VLT reach. requiring larger apertures and higher spatial resolution. The growing amount of scientific challenges has, together with new developments of technology, defined a strong demand for new telescopes with primary mirror diameters larger than 20 m, the extremely large telescopes (ELTs).

The required light collection and spectral and temporal resolution demand primary mirrors of several hundred square metres or more. The required spatial resolution needs diffraction-limited image quality. Fortunately, the science demands are matched by development in systems design, optical design, optical manufacture, segment alignment systems, adaptive optics, schemes for natural and artificial reference objects, integrated modelling and analysis. Also, instrumentation and detectors are in vigorous development. New site demands are met by new testing and evaluation methods. ELT development is strong, as shown by recent progress (Ardeberg and Andersen, 2004; Oschmann, 2004).

2. From a Swedish to a European project

ELT development started early at Lund Observatory (Ardeberg et al., 1992). Emphasising light collection and image quality, the Live Optics concept was developed (Ardeberg et al., 1996). A full adaptive optics system was designed and a 50 m telescope aperture was adopted (Andersen et al., 1999). An adaptive optics ELT, the Euro50, was refined, while the project got a wider European support (Andersen et al., 2002, 2003, 2004a, 2004b). Since 2002, six institutes from five European countries support Euro50. They are Lund Observatory (chair), Lund University, Sweden, Instituto de Astrofísica de Canarias (IAC), La Laguna, Spain, Department of Physics, National University of Galway, Ireland, Optical Science Laboratory, University College London, and National Physical Laboratory, United Kingdom, and Tuorla Observatory, Turku University, Finland. Currently, new development is taking place in favour of a common European ELT. Under the auspices of Opticon, the European coordination organization for astronomy, a programme for such a telescope is under discussion with the European Commission (EC) (Gilmore, 2004). In 2004, EC has, in its Frame Programme 6 (FP6) granted support for a design study of a European ELT. All Euro50 teams take part in this study.

3. A European ELT – Why?

Projects requiring ELT data include from solar-system objects to the first stars and from tiny objects to large structures. Of prime importance are high spatial, spectral and temporal resolution as well as a large light-collecting surface. Formulating a unique ELT science case is an impossible task. It depends on the type of ELT envisioned and scientific interests. Still, the science programmes dominating the ELT discussion may be summarized in some representative groups. One group refers to proto-stars, proto-stellar shells, planetary discs and systems, Earth-like planets and life signatures. Another group deals with star formation and stellar evolution. A connecting group concerns formation, structure and evolution of galaxies. As a final group, we have the first stars.

From proto-stars to circum-stellar discs

Circum-stellar envelopes transform into discs in around 106 years (Strom, 1995). This is commonly accepted, in spite of the limited data behind the conclusion. The quality of observations of circum-stellar discs has improved significantly. In 2004, around twenty proto-planetary discs are well recognized. Still, their small angular sizes and faintness have allowed few and limited studies only. A high frequency of circum-stellar discs around forming stars is indicated by K-band silhouette disc images (Elston et al., 2003). Still, detailed disc data are relatively rare. In-depth studies of proto-planetary discs have not been fully successful in more than a few cases. A large diversity seems indicated.

An asymmetric and multi-warped proto-planetary disc orbits the A5V star Pic (Smith and Terrile, 1984; Wahhaj et al., 2003; Weinberger et al., 2003). Another system orbits the M1Ve star AU Mic (Kalas et al., 2004). While Pic is 100 times as luminous as AU Mic, the latter has a life time 500 times that of Pic. The extent of the AU Mic disc seems a tenth of that of the Pic disc. An asymmetric proto-planetary disc orbits the B9.5Ve star HD 141569 (Mouillet et al., 2001). Its surface density distribution indicates a circum-primary disc in an eccentric binary system (Augereau and Papaloizou, 2004). Much higher, spatial and spectral, resolution is needed for studies of nascent planets.

A few other proto-planetary systems were observed in some detail. All but one orbit stars more massive and hotter than the Sun. Results for proto-planetary discs support tentative conclusions only. Adequate understanding of planet formation requires a planetary-disc sample much larger, less biased and much better spatially and spectroscopically resolved. Today, we cannot claim to understand even basic parts of the process from circum-stellar envelopes to orbiting planets (Mouillet, 2004). An adequate sample of proto-planetary systems must include central stars of different temperature, mass and heavy-element abundance. They must be in different evolutionary phases and interstellar environments and represent both isolated and interacting objects. We require imaging, photometry and polarimetry with high spatial and/or spectral resolution, masses, linear sizes, morphology, chemistry and dynamics. Proto-planetary discs are challenging objects with the power-law nature of mass distribution and small angular extensions. An ELT is needed.

From circum-stellar discs to planets

In circum-stellar discs, gas is accelerated and expelled and the dust grains left unshielded. Collisions and radiation pressure rapidly destroy them. Planetesimals are formed by dust, interact and create proto-planetary embryos (Rafikov, 2003a,b). Gravitationally scattering planetesimals, the embryos convert to proto-planets. Planetesimals collide producing dust grains. The processes are fundamental. However, a study needs high spatial and spectral resolution and work with high contrast ratios (Mouillet et al., 2003). An ELT is required.

T Tauri stars are excellent for studies of forming stars, circum-stellar shells and planets. They evolve to solar-mass stars. T Tauri stars are, however, found at distances beyond 100 pc only. Thus, Earth-like planets appear very close to their parent stars and are faint. In circum-stellar discs, grains with sizes up to some millimetres should emit significantly from μm to some mm. The emission recorded is little more than that expected from the stars alone (André and Montmerle, 1994; Osterloh and Beckwith, 1995; Beckwith and Sargent, 1996). Surveys for signatures of molecular CO tell the same story. Neither Zuckerman et al. (1995) nor Duvert et al. (2000) found significant signs of molecular CO emission lines. We must conclude that circum-stellar discs are nearly void of gas and dust after 5x106 years. From NIR and mm wavelength data, Lada (2003) confirmed this and discussed a time as restricted as 3x106 years for the window open to planet formation. Owing to Bary et al. (2002, 2003), the non-observed circum-stellar discs may be simply explained. The vanishing emission from dust and CO may be caused by planets building up. The dust in the discs may be largely consumed and bound in planetesimals, embryos and proto-planets. Still, large amounts of gas can be available and be captured for further planet growth. This would make the disc depletion less puzzling than so far understood.

First giant planets

Long, consensus held that formation of the first giant planets required around 107 years, from circum-stellar disc creation (Lissauer, 1993; Pollack et al., 1996; Hollenbach et al., 2000; Natta et al., 2000; Grady et al., 2000). Still, Boss (2000) proposed that gas giant planets, due to gravitational instabilities, form 103 years after disc stabilization. Boss (2001, 2002) and Boss et al. (2002) provided more evidence for rapid planet formation (Bary et al., 2003).

Still, in the generally adopted picture of giant-planet formation, circum-stellar discs dissipate too rapidly to allow giant-planet formation. With disc instabilities a prominent mechanism forming gas giant planets, also stars with low heavy-element abundance should have such planets (Boss, 2001). Our sample of central stars of planetary systems does not support this, although no definite contradiction exists. Still, observations show that gas giant planets are abundant enough not to be anomalous.

For formation of massive planets in circum-stellar discs, theorists identify a mass limit around ten Jupiter masses (Bate et al., 2003). However, the list of observed extra-solar planets includes some planets with masses well above this limit (Chabrier, 2003a,b). Planet formation swifter than 5x106 years may exist but needs good timing. An efficient series of processes might produce gas giant planets fast enough to comply with gas and dust dissipation. Still, in terms of planet-production theory, starting from gas, continuing via rocks and planetesimals to proto-planets massive enough to attract adequate amounts of gas, time seems a problem. Early formation of planets may be detected through surveys for ring-shaped gaps in circum-stellar discs. The small angular extensions require high spatial resolution. Waiting for telescopes with sufficient image quality, the disc gaps may, partly, be investigated via spectral energy distribution anomalies (Rice et al., 2003).

Simulated interaction of giant planets and surrounding discs with magneto-hydrodynamic turbulence (Steinacker and Papaloizou, 2002; Papaloizou and Nelson, 2003; Nelson and Papaloizou, 2003; Winters et al., 2003; Pickett et al., 2003; Papaloizou et al., 2004; Nelson and Papaloizou, 2004) showed disc turbulence with significant fluctuations to be important. The turbulent scenario differs from that of laminar disc behaviour, especially for low-mass planets and migration. Haghighipour and Boss (2003a,b) may solve the time-shortage problem. From small-solid motions in the vicinity of local density maxima, considering the gas drag force, they find solids of suitable mass and size to migrate to local density maxima in 103 years. Here, collisions and coagulation may accelerate the formation of planetesimals. Nevertheless, there can be little doubt that the early phases of planetary systems will remain a vital challenge for theorists and observers alike.

A growing number of known planetary systems

In October 2004, 117 planetary systems are confirmed (Schneider, 2004), one with four planets. There is little doubt that planetary systems are frequent. Several identification methods exist, especially if planets orbit nearby stars (Trimble and Aschwanden, 2004). Potentially the most obvious method, direct imaging has, for smaller planets, to wait for an ELT. Monitoring of radial-velocity, light level and proper motion are of high interest as well as microlensing and, for forming planetary systems, disc morphology.

Transit systems

From light-variations, in the OGLE survey (OGLE, 2004a,b), four planet systems were identified. Part of many transit candidates, they were confirmed, showing radial-velocity and light variations in phase (Konacki et al., 2003a; Bouchy et al., 2004). OGLE-TR-113b and OGLE-TR-132b, form, with OGLE-TR-56b, a group with shorter periods than any other planets, a giant-planet class, the 'Very Hot Jupiter-Type Planets' added to 'Hot Jupiter-Type Planets' (Basri, 2004). The OGLE survey found no 'Hot Jupiter-Type Planet', though radial-velocity statistics imply the OGLE sensitivity to 'Hot Jupiter-Type Planets' to be 10–20 times that to 'Very Hot Jupiter-Type Planets'. Also, radial-velocity surveys should have found several 'Very Hot Jupiter-Type Planets' (Basri, 2004). HD 209458 is the only transit system, for which it has been possible to exploit the transit data to some of their potential (Brown et al., 2001; Burrows et al., 2002; Charbonneau, 2003; Vidal-Madjar et al., 2003; Burrows et al., 2003). Multi-colour light-curve and radial-velocity monitoring combined with spectrophotometry provide a wealth of information.

An improved sample of planetary systems

Our planetary-system sample suffers from bias from selection and observation. The bias towards high-mass and small orbits is observational. Also, the bias of central stars and their spectral types is clear. The interesting small-mass stars were studied by Guenther and Wuchterl (2003) using an ESO VLT and repeated radial-velocity observations of 26 stars and brown dwarfs lighter than 0.1 solar mass. For only one object, radial-velocity variations gave significant indications of a companion with planetary mass.

How frequent are planetary systems and formation? What favours formation? How do the nature and stability depend on the stars, their temperature, mass, metallicity, birth place and interaction with other stars? Which are the relations between planets and stars and the numbers of major planets, smaller planets and masses and sizes of planets? How many planets are solid and gaseous? How do these qualities relate to planet size and star distance? How are semi-major axis, orbital period and eccentricity of larger and smaller planets and system size distributed? How evolved are outer system regions? How frequent and stable are Earth-like planets? Are comet families frequent? How does our planetary system fit the system distribution? Is it a typical representative or is it atypical?

One way to handle these questions is through a survey. The volume must be large enough to contain enough planetary systems for statistics and an adequate amount of Earth-like planets. Still, it must be small enough to make the survey efficient. From Hipparcos and Tycho data, we get 2500 stars within 20 pc. With a standard luminosity function, we find a total of 5000 star systems, to the brown dwarf limit. The only observed data regard hot and very hot Jupiter-type planets (Marcy et al., 2000). Relatively large incidences of planetary systems and Earth-like planets are assumed (Rebolo and Zinnecker, 2004). Already a 10

At 500–2000 nm, the spatial resolution of a diffraction limited 50 m ELT is 0.003—0.012 arcsec. At 20 pc, this corresponds to 0.06-0.24 AU. Even for cool stars, these are realistic detection data. For a planetary-system survey, including a search for Earth-like planets, a 50 m ELT can handle a 20 pc sample, the coverage depending on the temperature of the central star. Also for faint stars, data for radial velocitity and photometry are required. At 20 pc, an M0V star has apparent magnitudes at 550 nm, 700 nm and 900 nm, in the VRI system, of V = 9.6, R = 8.3 and I = 7.4. An M5V star has V = 13.0, R = 11.2 and I = 9.5. In terms of photon collection, VLTs handle planet-system stars, securing data for major planets, down to V = 11.0. Thus, a 50 m ELT can handle stars down to V = 14.5. Thus, within 20 pc, a 50 m ELT can deliver data for main-sequence stars to M5. The ELT survey should be coordinated with corresponding surveys other telescopes. We mention especially the Gaia (Marcy et al., 2004) and Darwin (ESA, 2003) missions.

Earth-like planets

For Earth-like planets, contrast ratios limit imaging and bio-signature surveys. The ratios vary with stellar luminosity, temperature and distance, planetary diameter, albedo, surface, atmosphere, dust and clouds. Planets scatter stellar light and emit thermal radiation. Brightness ratios star-planet vary with relative distance and wavelength. We can vary stellar and planetary parameters, keep planetary temperature variations in bio limits, deriving contrast ratios. In the visual, we get 108 to 3×1010, in the NIR 5×107 to 1010. We want direct images of such planets. The limiting distance for this imaging is uncertain due to lack of proper knowledge concerning both contrast ratios and our ability to handle them successfully. To be able to investigate the Earth-like planets in detail and to study their life-supporting properties, we need spectroscopy, a further challenge. For Earth-like planets, adequate spectroscopy with a 50 m ELT should be possible at least for a smaller, and closer, sample of the systems resulting from the survey.

The solar system

Is the solar system typical or atypical? How did it form and evolve? The Kuiper Belt (KB) is important but our understanding of it is limited. VLTs

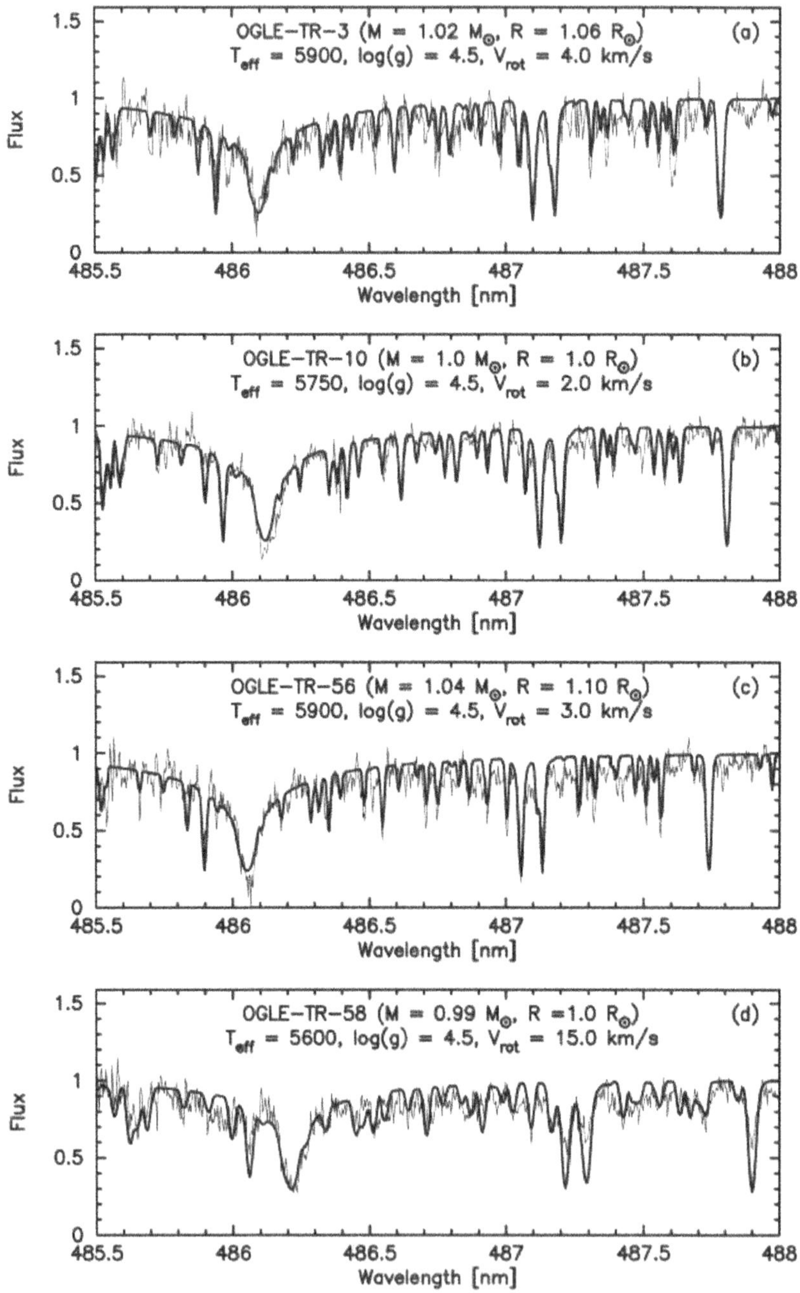

Figure 1. Co-added spectrograms around H line for four OGLE transit candidates with best-fitting synthetic spectra superposed as smooth lines. Derived stellar parameters are indicated (Konacki et al., 2003b).

and space telescopes give neither the spatial resolution nor the light collection required. An ELT allows us to survey binary KB objects (KBOs), monitor radial velocities and light curves and measure diameters (Snel et al., 2004). We get masses, densities and collision data, giving insight into KBO evolution. We get data on KBO shapes and rotations, approaching fundamental questions concerning the KB. Like for Pluto, we can study large-KBO atmospheres. Distant as they are, these planets have very slow angular motions. Thus, for occultation observations, we will be forced to employ rather faint stars, needing an ELT. An ELT target is the subsurface and surface activity of Io and Triton, coupled to atmospheric ejections and patterns. Time series in a range of wavelength will bring new understanding of processes important as such and also with regard to biospheres. Surface mapping is a target even for asteroids as is time-series high-resolution observations of comets.

Star formation

Star formation is normally initiated by high-mass stars covered by clouds. Our studies of the birth places, their structure, chemistry and kinematics, and birth processes are restricted. Sometimes, winds from neighbour stars rip the clouds apart (Nürnberger and Petr-Gotzens, 2002; Nürnberger et al., 2002; Nürnberger and Stanke, 2003; Nürnberger, 2003), revealing initial structures of massive stars, as in NGC 3603 IRS 9. Here, the brightest protostars were exposed. At NIR and MIR wavelengths, a pioneering study of a massive nascent star could be made. IRS 9A has a luminosity of around 105 solar luminosities and its circum-stellar envelope a mass more than a tenth of a solar mass. For further progress on such sources, narrow-band NIR imaging is required as is spectroscopy for age data. Possible multiplicity is a further challenge. High spatial NIR and MIR spatial resolution is needed combined with efficient photon collection.

The most massive stars

For the most massive stars, formation via accretion of gas and dust has been regarded as impossible, although questioned (Yorke and Sonnhalter, 2002). These stars should result from merging of intermediate-mass stars (Bonnell et al., 1998). With AO and other VLT NIR imaging, other observations added, Chini et al. (2004) studied early phases of massive-star formation. In M17, using deep IR imaging, they studied a heavy molecular cloud and found an opaque silhouette against a nebular background associated with a reflection nebula. They showed the system to comply perfectly with predictions for a high-mass forming star surrounded by a huge accretion disc. The object was found to have a bipolar mass outflow. The protostar seems to have collected around 20 solar masses, and the accretion process is still active. The circum-stellar disc holds at least 100 solar masses of gas for continued stellar growth. The result strongly indicates that also massive stars can form through accretion processes similar to those building up low-mass stars. Deeper insight calls for ELT observations with high spatial resolution and high contrast ratios at NIR wavelengths combined with dual-linear polarimetry (Vakili, 2003).

Initial mass function

An important parameter is the initial mass function (IMF) (Chabrier, 2003b). An extended single-object IMF, down to brown dwarfs, agrees with estimated L- and T-dwarf densities. It gives a brown dwarf disc density close to that of stars. Young stellar clusters share the disc field population. A significant brown-dwarf population seems excluded in globular clusters and the spheroidal system. The characteristic star-formation mass falls with time. Disc and spheroid IMF mass-to-light ratio gives values 1.5 times smaller than a Salpeter function. A Jeans-mass spectrum replaces the single Jeans mass.

Rotation and shape of stars

Fast stellar rotation and stellar stability are normally not compatible. Many related problems have been investigated by van Belle et al. (2001), Vauclair and Théado (2003), Théado and Vauclair (2003a,b) and Reiners et al. (2003). Whatever the case, the results for the form of the B3Ve star Achernar (Domiciano de Souza et al., 2003) are highly puzzling. One year after its documentation, in June 2003, no adequate explanation has been forwarded of its highly flattened shape coupled to an apparently stable total behaviour of the star. We must ask if we really do understand stars.

Resolving stellar discs

High-resolution imaging of stars is important for studies of physics, outer layers, energy production, structure and oscillations. Moving, variable spots are valuable for stellar physics and component interaction but also for variations interpreted as due to planets. With HST, limited imaging has been possible in a few exceptional cases (Gilliland and Dupree, 1996; Karovska et al., 1997). Also with VLTs equipped with adaptive optics systems, resolved stellar surface imaging is possible for a few extended stars only. With an ELT, we can image many stars in detail. The spatial resolution and photon collection allow smaller features and narrow wavelength bands. We can combine resolved imaging with narrow pass bands centred on spectral lines, adding continuum bands. High surface resolution can be obtained with adequate fluxes in image sub-elements. From comparison of pass band images, we can derive element distribution in the upper stellar layers. ELT surface maps can be combined with high-resolution spectroscopy for studies of chemistry and surface activity. Line profiles, line asymmetries and wavelength shifts can be studied from stellar disc centres out to the limb, revealing small-scale movements and the aspect angle, supporting analysis of line formation and the structure of the outer layers.

Time-resolved observations of stars

With ELT photon collection efficiency, millisecond-resolved photometry can be made for white dwarfs and their micro-variability studied. This is needed for understanding of white-dwarf atmospheres and evolution of granular and other surface features. High-speed photometry can probe compact objects. White dwarfs, neutron stars and black holes with accretion discs can be monitored for studies of processes resulting from gas flowing onto their surfaces. These flows feed events marked by very high energy. The extreme compactness, small extensions and high magnetic fields set the scene for events beyond our grasp. Targets are oscillations and flares. For compact objects without clouds, surface high-frequency fluctuations, radial and non-radial oscillations are fundamental for studies of the internal structure. All these observations will give highly valuable information on objects for which our knowledge and understanding are poor. High-speed photometry and high-speed spectroscopic monitoring are excellent tools for probing of the nature of neutron stars. With ELT photometry and spectroscopy, we can also derive masses of neutron stars and black holes. Using astrometry, we can add further data.

Evolution of galaxies

Few major events in astronomy are as poorly understood as early galaxy formation. Even the subsequent evolution of galaxies is only partly understood. We observe interacting galaxies but have vague ideas how mutual actions influence their structure and evolution. From recent observations of massive galaxies around 10 Gyr old (Glazebrook et al., 2004; Cimatti et al., 2004), there are strong reasons to conclude that the commonly accepted picture of hierarchical merging of galaxies is in problems.

Our understanding of galactic evolution is very limited also for our closest neighbours in the Local Cluster of galaxies (van den Bergh, 2000). Obviously, our understanding is poor even regarding M31. While M31 and the Galaxy are similarly classified, they show important dissimilarities regarding the halo populations (Durrell et al., 2001; Ferguson et al., 2002; Brown et al., 2003). For the globular clusters of the two galaxies, we see clear similarities but also striking differences, as observed by van den Bergh. The complexity is obvious in an over-all evolutionary perspective. However, at the same time, there are important general similarities between M31 and the Galaxy (Stephens et al., 2003).

A representative sample of galaxies

Understanding of galactic evolution requires a representative sample, covering at least all major types of galaxies. Such a sample is not even marginally available in the Local Group of galaxies. To obtain an adequate sample of galaxies, we must reach and resolve the members of the most nearby, Virgo and Fornax, clusters of galaxies, located at 16-20 Mpc. In practice, such observations outside the limits of the Local Cluster of galaxies are definitely impossible even with ten-metre telescopes with full adaptive optics.

Difficult to study

It is difficult to study stellar populations in galaxies. Their discs are often unfavourably oriented. Different disc populations, in different evolutionary phases, overlap and are hard to disentangle. Different populations mix in colour-magnitude diagrams, luminosity functions and abundance functions and cause spurious results. Also for favourable disc orientations, super-positioning of structures at different depths gives rise to erroneous conclusions. The interstellar dust in galaxies is irregularly distributed, adding problems for population photometry. In conclusion, it is difficult to find isolated, well-defined and homogeneous stellar populations and determine their evolutionary status with precision.

Stellar Clusters and Evolution of Galaxies

Stellar clusters are close to ideal for evolutionary studies. In practice, cluster members can be seen as codistant and coeval and to share the same initial abundance of heavy elements. Mostly, cluster stars can be taken as affected by constant effects of foreground interstellar extinction. Normally, internal cluster extinction can be seen as negligible or common to all member stars. Stellar clusters have long been used as probes for studies of evolution in the Galaxy. Still, they are even better probes in external galaxies, especially distant ones. Limited in dimensions, clusters are easy to identify and disentangle from field stars. Their number densities make their statistics robust against field-star influence. They can be studied unaffected by the orientation of parent galaxies and they can be traced and studied over large distances and in most circumstances.

Stellar clusters are fundamental for evolutionary parameters. An important example is the initial mass function (IMF). Recent work on the IMF by Chabrier (2003b) confirms the high usefulness of clusters for evolutionary studies. Of special importance for our purpose is the fact that it seems well confirmed that young stellar clusters represent the same stellar population as the stars in the surrounding disc fields of their parent galaxies.

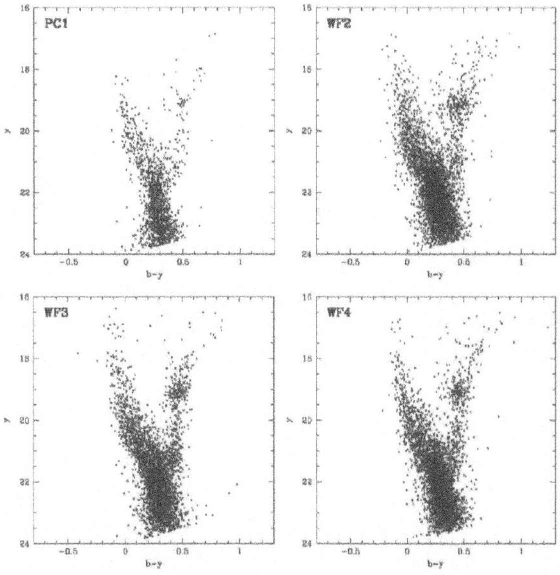

Figure 2. CMDs for four fields in the LMC Core, obtained in y and b of the Strömgren uvby system (Linde and Ardeberg, 2004). Data were obtained with the Hubble Space Telescope and its Wide Field Planetary Camera.

Image crowding and evolutionary studies

While of high advantage, the small dimensions of clusters imply a high volume density of member stars. Thus, effects of image crowding are more pronounced for cluster members than for field stars and increase with distance. With image quality set by atmospheric turbulence, image crowding severely limits stellar clusters as evolutionary targets, even for most parts of the Galaxy. For external galaxies, even in our immediate vicinity, the effects of image crowding have, in practice, been prohibitive with respect to evolutionary studies. Also with the best image quality currently available, image crowding is severe for the use of clusters as probes of evolution in external galaxies. With an image quality of around 0.1 arcsec, the Hubble Space Telescope (HST) permits evolutionary studies with stellar clusters in the most nearby external galaxies but not beyond, not even in the Local Cluster of galaxies. VLTs with AO systems will allow corresponding studies of evolution in most or all galaxies in the Local Cluster but not at larger distances.

Clusters, evolution and ELTs

Compared to a ten-metre AO VLT, a 50 m ELT, as the Euro50 (Andersen et al., 2003, 2004a,b), offers five times higher spatial resolution. Simulations show that a 50 m ELT can resolve the members of star clusters and study evolutionary parameters in detail for galaxies in the Virgo and Fornax clusters (Ardeberg and Linde, 2004a,b,c). Out to these clusters of galaxies, we can measure and employ data for the main-sequence turn-off point (TOP), the horizontal branch (HB), the asymptotic giant branch (AGB) and the red giant branch (RGB). For young clusters, evolutionary parameters, including TOP data, can be obtained even for galaxies beyond the Virgo and Fornax clusters. All the time, photometry with an AO ELT requires special consideration (Ardeberg, 2004a,b).

Frayn (2003) derived maximum distances possible for TOP photometry versus population age and limiting magnitude. For a comparison between our results and those of Frayn, it should be noted that while we refer to stellar clusters, Frayn derives his data for field populations. Our sample is better defined than FraynŠs but more vulnerable to image crowding. Taking this into account, our results and those of Frayn are in good agreement.

With a 50 m ELT and stellar clusters, we can derive evolutionary data from CMDs, with TOP-based ages, for member galaxies of both the Virgo and Fornax clusters. For older populations, data can be obtained via HB, AGB, RGB and LF photometry. This offers new possibilities for investigations of the evolution of galaxies as a function of type, the dependence on galactic environment and with respect to interactions with other galaxies. If we choose integral colour data for the clusters, we can study evolutionary parameters, albeit in lower resolution, also for galaxies at distances as high as and beyond 500 Mpc.

First stars and supernovae

Observation of the formation of the first stellar generations is a challenge. We cannot, with any precision, tell when or how it took place. Our lack of insight regarding first star formation includes the Galaxy (Shigeyama and Tsujimoto, 2003). A reasonable guess is that initial star formation occurred corresponding to redshifts $z = 20$ or maybe earlier (Yoshida et al., 2003). Subsequently, pioneer stellar generations produced large numbers of supernovae (SN). These SN should be visible from around $z = 10$, or earlier (Bromm et al., 2003). Concerning early star formation and its consequences, SN are excellent tracers and tools.

Type Ia SN, produced in binary systems, have bright, well defined light curves (Riess et al., 1999a,b; Valentini et al., 2003) and appear in most types of galaxies. They are perfect probes of the earliest star formation (Madgwick et al., 2003). SN of types Ib-Ic and II, from single-star core collapses, are excellent for studies of early phases of production of heavy elements. We note that all types of safe classification, from light curves or spectroscopy, requires around a factor of ten more light than detection. With the extreme distances of target SN, this implies that they are beyond VLT reach.

The time from formation of a future SN progenitor to its final explosion is uncertain (Dahlén and Fransson, 1999). Even so, the SN can provide very important information on the evolution of the earliest generations of stars (Livio and Riess, 2003). In addition, they can contribute essential data on large structures (Beasley et al., 2003). Relatively recent SN of type Ia have demonstrated significant differences in terms of peak luminosity. These variations may be caused by scatter in the abundance of heavy elements of the progenitors (Timmes et al., 2003). Stanek et al. (2003) and Kawabata et al. (2003) have found evidence for connections between hypernovae and gamma-ray bursts.

SN, early large-scale evolution and inter-galactic medium

SN are excellent probes of the early evolution of galaxies and large-scale events of star birth. This is not least true concerning pioneer production of dust (Nozawa and Kozasa, 2003). At a larger scale, data for early SN can contribute very much to reliable mapping of the evolving universe. Of special interest is the expansion at these epochs. We can add studies of SN at less extreme distances (Valentini al., 2003; Della Valle and Panagia, 2003) and derive an improved set of data for studies of acceleration and deceleration.

Studies of large-scale evolution (Perlmutter et al., 1999; Goldhaber et al., 2001) require solid distance calibration. Highly improved bright standard-source sets for calibrations of distance over very large volumes of space are well within reach, if we can tie together observational data of a range of SN including from first-generation to more recent representatives (Nugent et al., 2002) plus data from gravitational lensing and other light-source standards useful for smaller distances. A determined effort in this sense will have an excellent influence on the reliability of the determination of cosmological constants (Knop et al., 2003; Tonry et al, 2003; Bolton and Burles, 2003; Wang et al, 2003).

New light-standard sets, including the most distant SN, serve studies of the intergalactic medium. The improved, in depth and density, grid of background objects is attractive for mapping of the intergalactic medium over much larger volumes than now possible. At the same time, also the space resolution of the grid will be improved. This will add with respect to both quality and quantity to the data now available (Pichon et al., 2003).

4. A European ELT - How

Highlights

The Euro50 is an optical and infrared telescope with an equivalent primary mirror diameter of about 50 metres. The telescope has an enclosure to protect it against adverse weather and to provide a shelter for wind during observations. Figure 3 shows the telescope in its enclosure. The telescope has a segmented, aspherical primary mirror. Adaptive optics is integrated into the telescope optics and the secondary mirror is deformable with over 3000 actuators. In addition, there is a deformable mirror near the Gregorian focus to provide dual-conjugate adaptive optics. Observations are possible in the seeing limited regime or with single or dual-conjugated adaptive optics. The Observatory at Roque de los Muchachos, Canary Islands, is a possible site for the Euro50.

Optics

The optical system is of the Gregorian type with an elliptical primary mirror and a concave, elliptical secondary providing an aplanatic layout (Goncharov, 2003). See Figure 4. The f-ratio of the primary is f/0.85. The primary mirror has 618 hexagonal segments that each is about two metres from side to side. There are 104 different types of segments. The secondary mirror is deformable and has a hexagonal shape measuring 4 metres from side to side. The elevation axis of the telescope is located behind the primary mirror, providing ample space for relay optics, adaptive optics and instrumentation inside the primary mirror cell. Observations are possible in the primary focus (using a special corrector (Andersen et al., 2003), in the Gregorian focus, or in either of the two Nasmyth foci with dual conjugate adaptive optics or seeing-limited using a focal reducer. The achievable field in the Gregorian focus is as large as 8 arcminutes. However, it is unlikely that this will be exploited often because the atmosphere limits the field to about Ø0.5 arcminute for single conjugate and about Ø1 arcminute for dual conjugate adaptive optics when all error sources are taken into account.

A comparison was made between an asperical and a spherical primary mirror design for the Euro50 and it was found that they are about equally expensive. The aspherical design calls for fewer optical elements, thereby increasing throughput and easing the automatic alignment task.

Mechanics

The mechanical design can be seen in the right-hand part of Figure 3. The telescope structure is largely of steel with a mass of about 3000 tonnes and a height of about 90 meters. The lowest locked rotor resonance frequency is

Figure 3. The Euro50 in its enclosure. On the graphics to the right, a part of the enclosure has been cut away to depict telescope details.

2.7 Hz. Many structural members are identical, thereby providing cost benefit due to mass fabrication. However, considering the large amount of steel and weldings, production cost largely will be defined by steel cost and welding time so the potential for significant cost reductions by mass fabrication is limited.

Where the elevation axis goes through the primary mirror cell, there is a special structure (Şstrong-beamŤ) that, at the same time, provides a stiff elevation axis and a tunnel for placement of relay optics and instrumentation. There is also another tunnel perpendicular to the elevation and tube axes and intersecting with the first tunnel, and there is a tunnel with the shape of an elevator shaft protruding downwards from the Gregorian focus on the tube axis. This tunnel is used for the laser guide star wavefront sensors that, due to the changing distance to the laser guide stars in the Mesosphere, must be moved axially during observations to re-focus.

The segments of the primary mirror are suspended on whiffle trees providing three virtual supports with actuators capable of adjusting tip, tilt and piston of the primary mirror segments for alignment of the segments during operation. Position feedback is established from edge sensors measuring offsets between adjacent segments at the corners.

The secondary mirror unit is supported by a tripod similar to those used for radio telescopes. See Figure 5. To provide higher stiffness, carbon fibre reinforced tripod legs are used. The legs are thin to avoid excessive light loss and to ensure that no segment is completely obscured. This would be detrimental to the primary mirror segment alignment camera function.

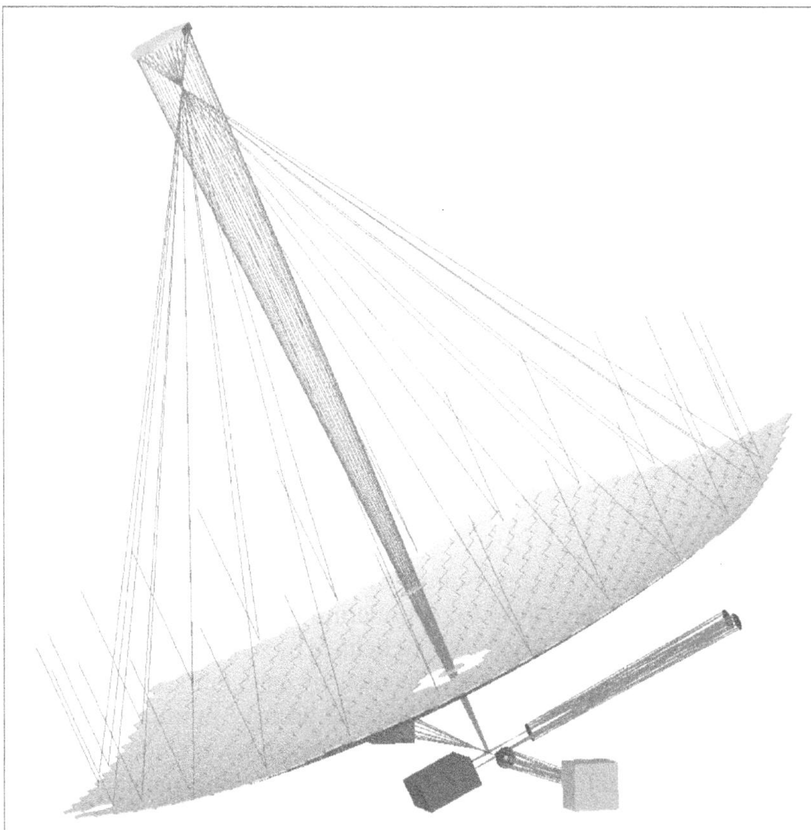

Figure 4. Optical design of the Euro50.

Figure 5. Suspension of secondary mirror unit on air bellows with force actuators for 5 degree-of-position control.

The telescope has hydrostatic bearings providing a smooth movement with no stick-slip in spite of the large telescope mass. The drives are of the direct-drive type without gears, again providing better tracking performance than drives with a high gear ratio. Smooth tracking is highly important for the Euro50 (and any ELT) because of the high image quality and, although tip-tilt can be removed by the adaptive optics, residuals should be kept at a low level.

Enclosure

The Euro50 enclosure is comparable in size to some of the largest buildings existing. The design shown in Figure 3 is based on a feasibility study (Andersen & Christensen, 2000). The structural design of the enclosure is complex due to the high wind loads. Clamping mechanisms for the main hatches are required to provide an integral structure for the survival wind case when the hatches are closed. The complete enclosure rotates on large wheel bogies. There is an elaborate thermal control system with large ventilation gates on the sides to provide air flushing during observations, and with a large cooling plant active during the day to ensure maximum thermal stabilization already at the beginning of the observing runs.

Wind tunnel studies and calculations using computational fluid dynamics calculations (see Figure 13) have been carried out (Riewaldt et al., 2004) to determine the survival wind loads and the efficiency of the sheltering effect of the building during observations.

Adaptive Optics

Adaptive optics (AO) is an integrated part of the telescope system (Owner-Petersen et al., 2003). It will initially be implemented in the K-band and later also for the visible. The secondary mirror is deformable and its form can be adjusted with a bandwidth of about 500 Hz. Figure 6 shows the adaptive optics system below the primary mirror. Light from the secondary mirror goes through a dual-wedge atmospheric dispersion compensator (Owner-Petersen & Goncharov, 2004) to a beam splitting plane mirror. The laser guide star light passes through the beam splitter and the light from the science object is reflected. The beam splitting mirror can direct light towards a focal reducer for non-AO observations (seeing-limited), towards a science instrument for single conjugate AO (SCAO) observations or to a collimating relay mirror that sends the light to a second deformable mirror and back to another relay mirror before focusing the light onto a science instrument for dual conjugate AO (DCAO). Alternatively, for DCAO, the light may be directed to clamshell relay mirrors for transmission to the second Nasmyth focus. Ground-layer adaptive optics (GLAO) using highly off axis natural guide stars is also possible using the SCAO scheme.

The Euro50 multi-conjugate adaptive optics has two deformable mirrors. Obviously, three deformable mirrors will potentially give a better image quality at the price of more reflections and higher cost. The curve in Figure 7 shows the Strehl ratio achievable over the field for a typical situation with 2 and 3 deformable mirrors, respectively. As can be seen, the advantage of using three deformable mirrors is not overwhelming, thereby justifying the use of two deformable mirrors.

Adaptive optics is an integral part of the telescope, and the form and the position of the secondary mirror can be controlled under fast, real-time computer control. The secondary mirror is suspended in a lightweight mirror cell made of a carbon fibre reinforced composite floating on air bellows to take gravity forces but with very low suspension stiffness. The position of the mirror unit is maintained by large force actuators that control rigid-body motion in five degrees of freedom based upon an inertial reference unit with laser gyros and accelerometers. The form of the deformable mirror is controlled using force actuators. The mirror is thin (2.2 mm) and made of a carbon fibre composite. The secondary mirror control algorithms are presently being studied in detail and a prototype actuator is being tested. Figure 8 shows the actuator. It has a suction cup to connect to the back of the deformable mirror. With some design

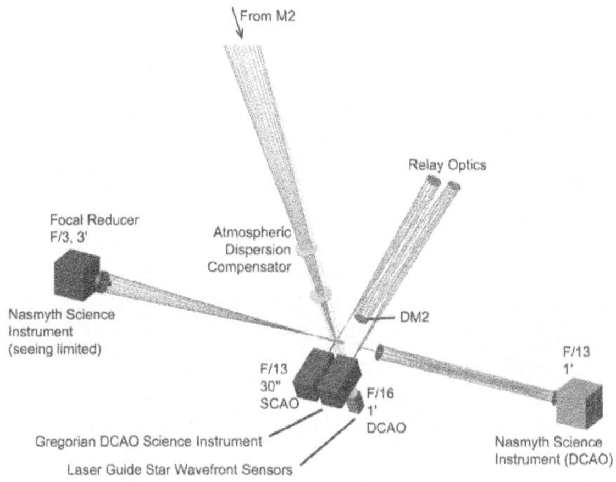

Figure 6. Adaptive optics of the Euro50.

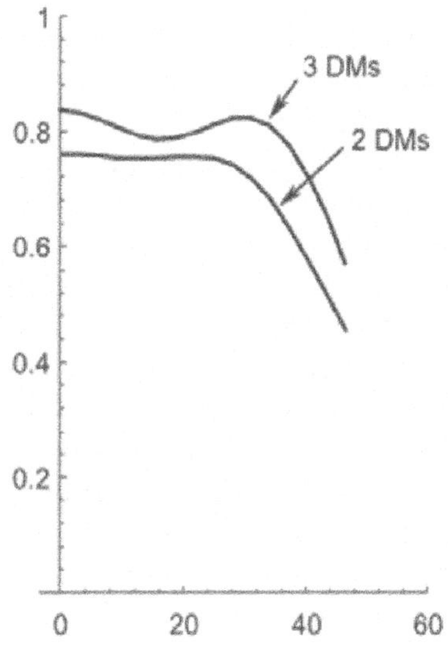

Figure 7. Strehl ratio of the field radius (arcseconds) using two and three mirrors, respectively.

effort, an actuator spacing as small as 15-20 mm seems achievable. For the initial AO in the K-band, a spacing of 68 mm is required.

For exoplanet studies it is not difficult to find a bright guide star for adaptive optics but for many other observations there are no natural guide stars available. For this reason, laser guide stars are needed. A proposed laser guide star scheme is shown in Figure 9. More studies are needed, in particular related to

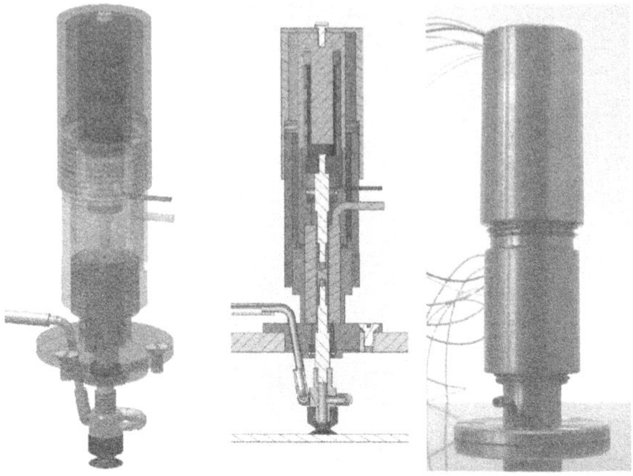

Figure 8. Prototype of a deformable mirror actuator.

provision of suitable lasers. The exact number of laser guide stars required is still under discussion. The results obtained so far suggest that 13 laser guide stars are needed for SCAO and 37 for DCAO.

Laser guide stars have inherent difficulties, for instance related to the cone effect and the perspective elongation. The cone effect may be overcome using a Hanning window (Andersen et al., 2002), and perspective elongation using an electronically adjustable zoom-lens that tracks light pulses through the mesosphere (Beckers et al., 2003a & 2003b).

The Euro50 has wavefront sensors for both natural guide stars and laser guide stars. There are as many laser guide star wavefront sensors as there are laser guide stars. For DCAO, it is not possible to relay laser guide star light all the way through the optical train to the final focus. The laser guide stars are too much out of focus to allow a satisfactory transmission through the second deformable mirror and its relay optics. Instead, an artificial laser point source is inserted in the Gregorian focus after the first deformable mirror, and its light is transmitted via the second deformable mirror to the final focus where it is analyzed by a wavefront sensor. A control loop is closed around the first deformable mirror using the laser guide stars, and another loop around the second deformable mirror using the artificial laser point source. Combining information from the wavefront sensor in the Gregorian focus with that from the wavefront sensor in the final focus makes it possible to predict what a laser guide star wavefront sensor in the final focus would have seen if it were feasible. We call this concept the virtual wavefront sensor principle (Andersen et al, 2003), seemingly a necessity for successful implementation of MCAO/DCAO on ELTs.

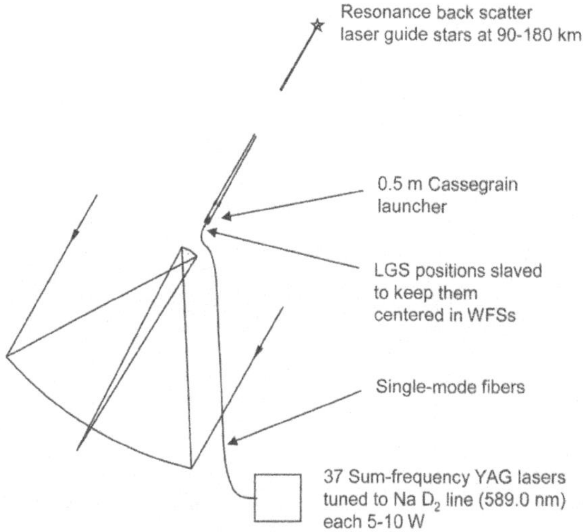

Figure 9. Laser guide star concept for the Euro50.

Live Optics

Adaptive optics not only corrects for atmospheric turbulence effects but also for telescope aberrations, for instance due to wind. Similarly, the primary mirror alignment system not only corrects for structural telescope misalignments but also for low-order atmospheric aberrations. Together, the ensemble of the primary mirror segment alignment system, the adaptive optics, the secondary mirror control system and the main axes servos form an intricate control system called live optics (see Figure 10). Live optics performs the wavefront control task that is crucial for the success of any ELT.

Figure 11 shows the situation in more detail. To the left, many of the sensors of the complete system have been depicted and to the right, the actuators are shown. It is the task of the telescope designers to define the highly multivariable control system to ensure that the wavefront control objectives are met. The different parts of the control system will each cover different types of spatial frequencies, temporal frequencies, and wavefront magnitudes as shown in the example of Figure 12. Much more information on this can be found in (Andersen et al., 2003).

Integrated modeling

The live optics is a highly complex system involving the telescope optics, the mechanical structure, the servomechanisms, the primary mirror alignment system, the control system and adaptive optics. Verification of the correct func-

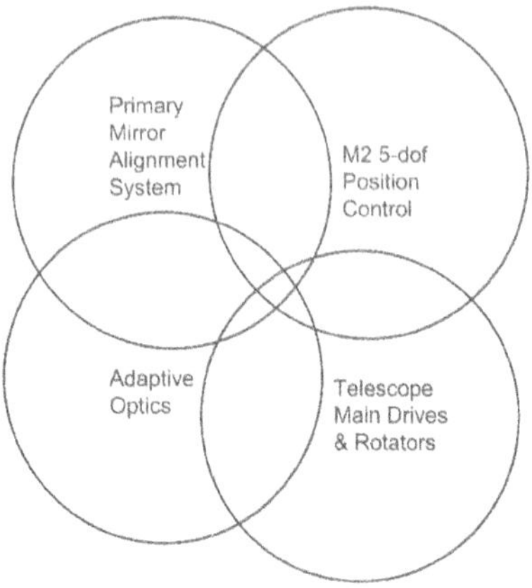

Figure 10. Graphical illustration of the different parts of the live optics and how they overlap.

tion of live optics calls for cross-disciplinary models that, due to their complex nature, typically must be solved numerically by simulation.

For the Euro50, an integrated model involving the telescope structure, the live optics control system, primary mirror segment control, and adaptive optics has been formulated (Andersen et al., 2004b). The number of degrees of freedom is around 16000 without adaptive optics and 1 million with AO. The structural model originates from the finite element model created in Ansys with about 70000 nodes. A truncated modal representation is applied. The optics model is based upon ray tracing using a sensitivity matrix for computation of deviations from a stationary operation point. The segment alignment system has edge sensor feedback and uses a singular value decomposition approach for control of the segments. The individual primary mirror segments were not included in the finite element model but have been taken into account as 618 rigid bodies using the general equations of motion. The adaptive optics model includes wavefront sensor, control algorithms, and deformable mirror control. All optical effects, including atmosphere and telescope aberrations, are combined in the telescope exit pupil.

It is important to have reliable estimates of disturbances from wind, sensor noise, and optical effects in the atmosphere. Wind load on the primary mirror poses a particularly difficult problem. To study this further, a wind tunnel test and a computational fluid dynamics calculation were carried out (Riewaldt et al., 2004). Figure 13 shows the boundary layer wind tunnel test setup and the mesh used for the CFD calculation. There was reasonably good agreement between the results from the two approaches. Calculations with the total integrated model with the telescope under wind load were performed. As an example, Figure 14 and Figure 15 show the results from such a simulation.

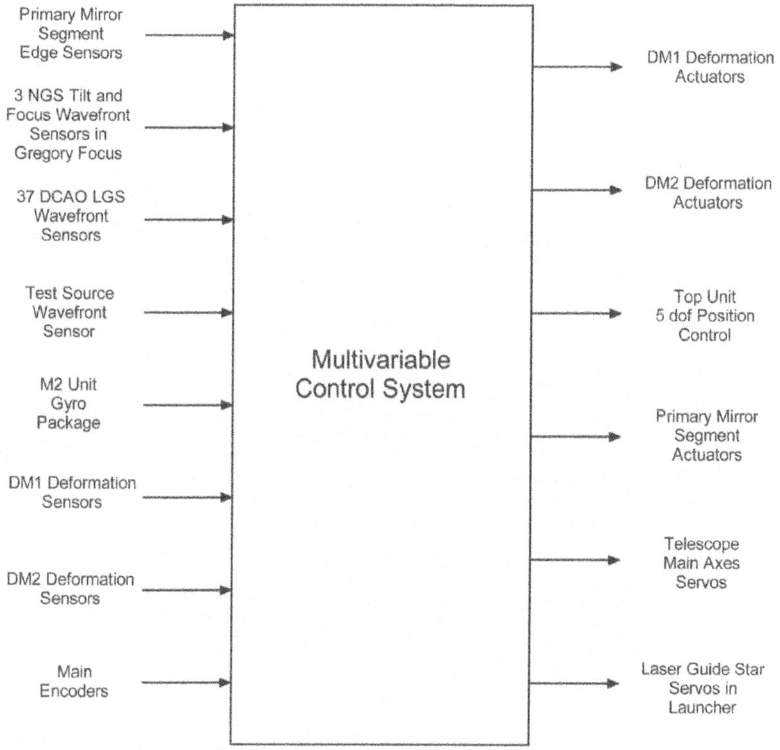

Figure 11. General layout of the live optics control system for wavefront control.

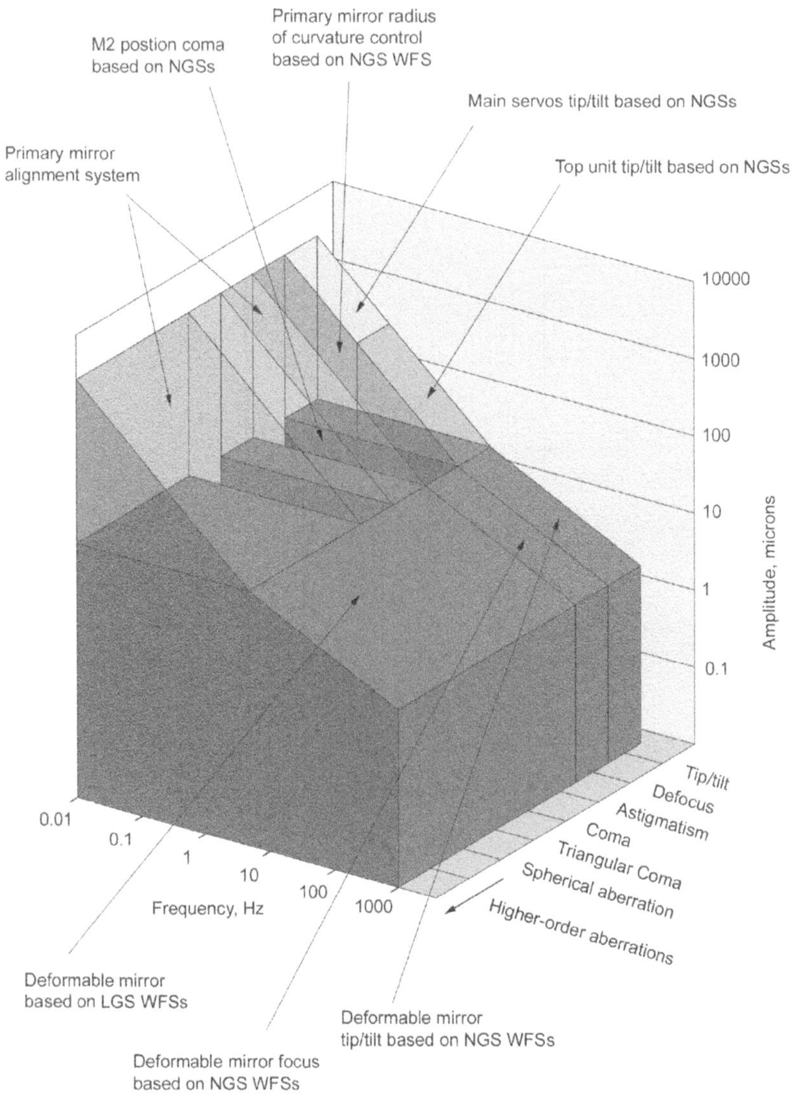

Figure 12. Example showing some operation ranges for different subsystems of the live optics over temporal frequencies, spatial frequencies and wavefront magnitudes.

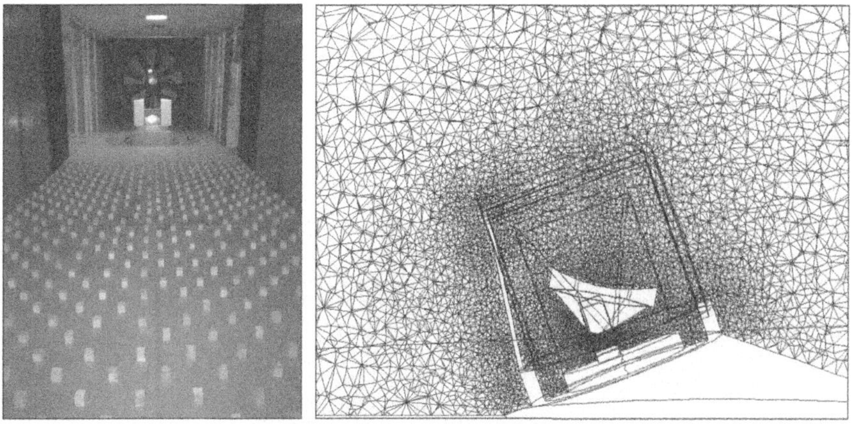

Figure 13. Wind tunnel setup in a boundary layer wind tunnel (left) and mesh for a CFD calculation (right).

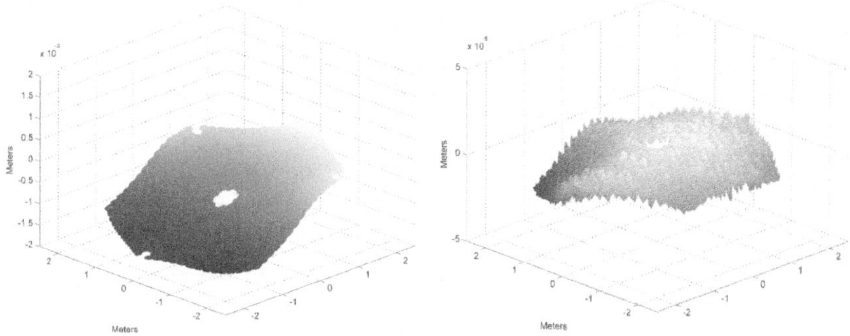

Figure 14. Example of results from runs with the integrated model showing optical path differences at the exit pupil of the telescope. Left: Wind disturbance with no correction by the primary mirror segment alignment system, the secondary mirror rigid body motion control, and adaptive optics control of the secondary mirror. Right: The same load case, however with all components of the live optics system, including single conjugate adaptive optics, engaged. There are residuals of the order of 20-30 microns for this run. Studies are on the way to further optimize the design.

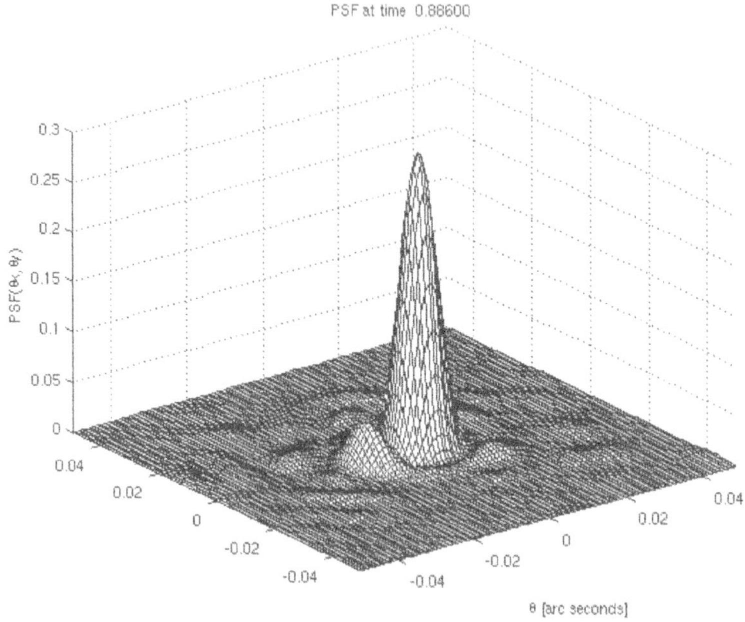

Figure 15. Point spread function corresponding to the optical path difference plot in the right-hand part of Figure 14. The scale has been normalized to provide a value of 1 for the diffraction-limited case.

Simulations have shown that there will be considerable residual aberrations from wind shaking of the telescope to be corrected by the adaptive optics system. In fact, the residual aberrations from the telescope are larger than the aberrations from the atmosphere. It is therefore important that the adaptive optics system be designed not only with the atmosphere in mind but it must also be adapted to the nature of the telescope aberrations.

References

Andersen, T., Ardeberg, A., Beckers, J., Flicker, R., Jessen, N.-C., Gontcharov, A., Mannery, E., Owner-Petersen, M., 1999: The Proposed Swedish 50 m Extremely Large Telescope, in Bäckaskog Workshop on Extremely Large Telescopes, ESO Conference and Workshop Proceedings No. 57, pp. 72-82.

Andersen, T., Ardeberg, A., Beckers, J., Gontcharov, A., Owner-Petersen, M., Riewaldt, H., 2004a: Euro50, in 2nd Bäckaskog Workshop on Extremely Large Telescopes, Proc. SPIE vol. 5382, pp. 169-182.

Andersen, T., Ardeberg, A., Beckers, J., Gontcharov, A., Owner-Petersen, M., Riewaldt, H., Snel, R., Walker, D., 2002: The Euro50 Extremely Large Telescope, in Future Giant Telescopes, Proc. SPIE vol. 4840, pp. 214-225.

Andersen, T., Ardeberg, A., Owner-Petersen, M., 2003: Euro50 Ű A 50 m Adaptive Optics Telescope, Lund Observatory, ISBN 91-631-4317-8.

Andersen, T., Ardeberg, A., Riewaldt, H., Lastiwka, M., Quinlan, N., McNamara, K., Wang, X., Enmark, A., Owner-Petersen, M., Shearer, A., Fan, C., Moraru, D., 2004b: Status of the Euro50 Project, in Ground-based Telescopes, Proc. SPIE vol. 5489, pp. 407-416.

Andersen, T. and Christensen, P. H., "Is there an Upper Limit to the Size of Enclosures?", Proc. SPIE, Vol. 4004, 2000.

Andersen, T., Enmark, A., Moraru, D., Owner-Petersen, M., Riewaldt, H., Browne, M., Fan, C., and Shearer, A., "A Parallel Integrated Model of the Euro50", Proc. SPIE, Vol. 5497, 2004.

André, P., Montmerle, T., 1994, Astrophys. J. 420, 837.

Ardeberg, A., 2004a, Science Drivers of a 50 m AO Telescope, 2nd Bäckaskog Workshop on Extremely Large Telescopes, Proc. SPIE vol. 5382, pp. 37-46.

Ardeberg, A., 2004b, Diffraction-Limited Light Collection: Promises and Challenges of a 50 m ELT, Ground-based Telescopes, Proc. SPIE vol. 5489, pp. 23-34.

Ardeberg, A., Andersen, T., 2004: 2nd Bäckaskog Workshop on Extremely Large Telescopes, Proc. SPIE vol. 5382.

Ardeberg, A., Andersen, T., Lindberg, B., Owner-Petersen, M., Korhonen, T., Søndergård, P., 1992: Breaking the 8 m Barrier, One Approach for a 25 m Class Optical Telescope, ESO Conference on Progress in Telescopes and Instrumentation Technologies, ESO Conference and Workshop Proceedings No. 42, pp. 75-78.

Ardeberg, A., Andersen, T., Owner-Petersen, M., Jessen, N.-C., 1996: A 25 m Live Optics Telescope, in Telescopes of Today and Tomorrow, Proc. SPIE vol. 2871, pp. 585-596.

Ardeberg, A., Linde, P., 2004a, Clusters as Probes of Distance, Evolution and Chemistry of Galaxies, 2nd Bäckaskog Workshop on Extremely Large Telescopes, Proc. SPIE vol. 5382, pp. 47-56.

Ardeberg, A., Linde, P., 2004b, Evolution of Distant Galaxies from Cluster Photometry, Ground-based Telescopes, Proc. SPIE vol. 5489, pp. 72-83.

Ardeberg, A., Linde, P., 2004c, Evolution of Galaxies from ELT Photometry of Stellar Clusters, these proceedings.

Augereau, J. C., Papaloizou, , J. C. B., 2004, Astron. Astrophys. 414, 1153.Barlow, M. J., Crawford, I. A., Diego, F., Dryburgh, M., Fish, A. C., Howarth, I. D., Spyromilio, J., Walker, D. D., 1995, Mon. Not. Roy. Astron. Soc. 272, 333.

Bary, J. S. Weintraub, D. A., Kastner, J. H., 2002, Astrophys. J. 576, L 73.

Bary, J. S. Weintraub, D. A., Kastner, J. H., 2003, Astrophys. J. 586, 1136.

Basri, G., 2004, Nature 430, 24.

Bate, M. R., Lubow, S. H., Ogilvie, G. I., Miller, K. A., 2003, Mon. Notes Roy. Astron. Soc. 341, 213.

Beasley, M. A., Kawata, D., Pearce, F. R., Forbes, D. A., Gibson, B. K., 2003, Astrophys. J. 596, L 187.

Beckers, J., Owner-Petersen, M., and Andersen, T., "Sodium Beacon Wavefront Sensing for the Euro50 in the Presence of Perspective Elongation", Proc. SPIE, Vol. 5382, 2003a.

Beckers, J., Owner-Petersen, M., and Andersen, T., "Correction of perspective elongation of sodium beacons in the 50-meter Euro50 telescope", Proc. SPIE, Vol. 5169, 2003b.

van Belle, G. T., Ciardi, D. R., Thompson, R. R., Akeson, R. L., Lada, E. A., 2001, Astrophys. J. 559, 1155.

van den Bergh, S., 2000, The galaxies of the Local Group, Cambridge Univ. Press, Cambridge UK, New York.

Bolton, A. S., Burles, S., 2003, Astrophys. J. 592, 17.

Bonnell, I., Bate, M., Zinnecker, H., 1998, Mon. Not. Roy. Astron. Soc. 298, 93.

Boss, A. P., 2000, Astrophys. J. 536, L 101.

Boss, A. P., 2001, Astrophys. J. 563, 367.

Boss, A. P., 2002, Astrophys. J. 567, L 153.

Boss, A. P., Wetherhill, G. W., Haghighipour, N., 2002, Icarus 156, 291.

Bouchy, F., Pont, F., Santos, N. C., Melo, C, Mayor, M., Queloz, D., Udry, S., 2004, Astron. Astrophys. 421, L 13.

Bromm, V., Yoshida, N., Hernquist, L., 2003, Astrophys. J. 596, L 135.

Brown, T. M., Charbonneau, D., Gilliland, R. L., Noyes, R. W., Burrows, A., 2001, Astrophys. J. 552, 699.

Brown, T. M., Fergusson, H. C:, Smith, E., Kimble, R. A., Sweigart, A. V., Renzini, A., Rich, R. M., VandenBerg, D. A., 2003, Astrophys. J. 592, L 17.

Burrows, A., Ram, R. S., Bernath, P., Sharp, C. M., Milsom, J. A., 2002, Astrophys. J. 577, 986.

Burrows, A., Sudarsky, D., Hubbard, W. B., 2003, Astrophys. J. 594, 545.

Chabrier, G., 2003a, Astrophys. J. 586, L 133.

Chabrier, G., 2003b, Publ. Astron. Soc. Pacific 115, 763.

Charbonneau, D., 2003, Nature 422, 124.

Chini, R., Hoffmeister, V., Kimeswenger, S., Nielbock, M., Nürnberger, D., Schmidtobreick, L., Sterzik, M., 2004, Nature 429, 155.

Cimatti, A., Daddi E., Renzini, A., Cassata, P., Vanzella, E, Pozzetti, L., Cristiani, S., Fontana, A., Rodighiero, G., Mignoli, M, Zamorani, G., 2004, Natue 430, 184.

Dahlén, T., Fransson, C., 1999, Astron. Astrophys. 350, 349.

Della Valle, M., Panagia, N., 2003, Astrophys. J. 587, L 71.

Domiciano de Souza, A., Kervella, P, Jankov, S, Abe, L, Vakili, F, di Folco, E, Paresce, F, 2003, Astron. Astrophys. 407, L47.

Durrell, P. R., Harris, W. E., Pritchet, C. J., 2001, Astron. J. 121, 2557.

Duvert, G., Guilloteau, S., Ménard, F., Simon, M., Dutrey, A., 2000, Astron. Astrophys. 355, 165.

Elston, R. J., Lada, E. A., Alves, J., Ferreira, B., Lada, C. J., Levine, J., Muench, A. A., Raines, S.N., Rashkind, N., Roman, C., 2003, Nearly Edge on Silhouette Disks From the FLAMINGOS Molecular Cloud Survey, AAS, May 2003, 28.09.

ESA, 2003: http://www.esa.int/esaSC/120382_index_0_m.html

Ferguson, A. M. N., Irwin, M. J., Ibata, R. A., Lewis, G. F., Tanvir, N. R., 2002, Astron. J. 124, 1452.

Frayn, C., 2003, PhD Thesis, University of Cambridge.

Gilliland, R., Dupree, A., 1996, AAS 187.

Gilmore, G., 2004, OPTICON, a (small) part of European Astronomy, 2nd Bäckaskog Workshop on Extremely Large Telescopes, Proc. SPIE vol. 5382, pp. 138-141.

Glazebrook, K., Abraham, R. G., McCarthy, P. J., Savaglio S., Chen, H.-W., Crampton, D., Murowinski, R., Jørgensen, I., Roth, K, Hook, I., Marzke, R. O., Carlberg, R G., 2004, Nature 430, 181.

Goldhaber, G., Groom, D. E., Kim, A., Aldering, G., Astier, P., Conley, A., Deustua, S. E., Ellis, R., Fabbro, S., Fruchter, A. S., Goobar, A., Hook, I., Irwin, M., Kim, M., Knop, R. A., Lidman, C., McMahon, R., Nugent, P. E., Pain, R., Panagia, N., Pennypacker, C. R., Perlmutter, S., Ruiz-Lapuente, P., Schaefer, B., Walton, N. A., York, T., 2001, Astrophys. J. 558, 359.

Goncharov, A.. Extremely Large Telescopes: Optical Design and Wavefront Correction. 2003. Lund Observatory.

Grady, C A., Sitko, M. L., Russell, R. W., Lynch, D. K., Hanner, M. S., Pérez, M. R., Bjorkman, K. S., de Winter, D., 2000, in Protostars and Planets IV, Mannings,W., Boss, A. P., Russell, S. S. (eds.), The University of Arizona Press, Tucson, p. 613.

Guenther, E. W., Wuchterl, G., 2003, Astron. Astrophys. 401, 677.

Haghighipour, N., Boss, A. P., 2003a, Astrophys. J. 583, 996.

Haghighipour, N., Boss, A. P., 2003b, Astrophys. J. 586, 1442.

Hollenbach, D. J., Yorke, H.W., Johnstone, D., 2000, Disk Dispersal Around Young Stars, in Protostars and Planets IV, Mannings,W., Boss, A. P., Russell, S. S. (eds.), The University of Arizona Press, Tucson, p. 401.

Kalas, P., Liu, M. C., Matthews, B., 2004, Science 303, 1990.

Karovska, M., Hack, W., Raymond, J., Guinan, E., 1997, Astrophys. J. 482, L 175.

Kawabata, K. S., Deng, J., Wang, L., Mazzali, P., Nomoto, K., Maeda, K., Tominaga, N., Umeda, H., Iye, M., Kosugi, G., Ohyama, Y., Sasaki, T., Höflich, P., Wheeler, J.C., Jeffery, D. J., Aoki, K., Kashikawa, N., Takata, T., Kawai, N., Sakamoto, T., Urata, Y., Yoshida, A., Tamagawa, T., Torii, K., Aoki, W., Kobayashi, N., Komiyama, Y., Mizumoto, Y., Noumaru, J., Ogasawara, R., Sekiguchi, K., Shirasaki, Y., Totani, T., Watanabe, J., Yamada, T., 2003, Astrophys. J. 593, L 19.

Knop, R. A., Aldering, G., Amanullah, R., Astier, P., Blanc, G., Burns, M. S., Conley, A., Deustua, S. E., Doi, M., Ellis, R., Fabbro, S., Folatelli, G., Fruchter, A. S., Garavini, G., Garmond, S., Garton, K., Gibbons, R., Goldhaber, G., Goobar, A., Groom, D. E., Hardin, D., Hook, I., Howell, D. A., Kim, A. G., Lee, B. C., Lidman, C., Mendez, J., Nobili, S., Nugent, P. E., Pain, R., Panagia, N., Pennypacker, C. R., Perlmutter, S., Quimby, R., Raux, J., Regnault, N., Ruiz-Lapuente, P., Sainton, G., Schaefer, B., Schahmaneche, K., Smith, E., Wood-Vasey, A. L., Yasuda, N., 2003, Astrophys. J. 598, 102.

Konacki, M., Torres, G., Jha, S., Sasselov, D. D., 2003a, Nature 421, 507.

Konacki, M., Torres, G., Sasselov, D. D., Jha, S., 2003b, Astrophys. J. 597, 1076.

Lada, E. A., 2003, Evolution of Circumstellar Disks in Young Stellar Clusters, AAS, May 2003, 24.06.

Linde, P., Ardeberg, A., 2004, in manuscript.

Lissauer, J. J., 1993, Planet Formation, in Ann. Rev. Astron. Astrophys. Vol. 31, Burbidge; G., Layzer, D., Sandage, A. (eds,), Annual Rev. Inc., Palo Alto.

Livio, M., Riess, A. G., 2003, Astrophys. J. 594, L 93.

Madgwick, D. S., Hewett, P. C., Mortlock, D. J., Wang, L., 2003, Astrophys. J. 599, L 33.

Marcy, G. W., Butler, R. P., Vogt, S. S., 2000, Astrophys. J. 536, L 43.

Marcy, G.; Butler, R. P., Fischer, D. A., Vogt, S. S., 2004: Exoplanets: New Parameter Space, in Astrometry Conference 2004, Astrometry in the Age of the Next Generation of Large Telescopes, 18-20 October, 2004, Lowell Observatory, in print. Mouillet, D., 2004, Science 303, 1982.

Mouillet, D., Augereau, J. C., Chauvin, G., Lagrange, A.-M., Ménard, F., 2003, Protoplanetary Disks and High Dynamic Observations, in Astronomy with High Contrast Imaging: From Planetary Systems to Active Galactic Nuclei, Aime, C., Soummer, R. (eds.), EAS Publ. Ser. Vol. 8, EDP Sciences, Les Ulis, France, p. 37.

Mouillet, D., Lagrange, A. M., Augereau, J. C., Ménard, F., 2001, Astron. Astrophys. 372, L 61.

Natta, A., Grinin, V. P., Mannings, V., 2000, Properties and Evolution of Disks around Pre-Main-Sequence Stars of Intermediate Mass, in Protostars and Planets IV, Mannings,W., Boss, A. P., Russell, S. S. (eds.), The University of Arizona Press, Tucson, p. 559.

Nelson, R. P., Papaloizou, J. C. B., 2003, Mon. Not. Roy. Astron. Soc. 339, 993.

Nelson, R. P., Papaloizou, J. C. B., 2004, Mon. Not. Roy. Astron. Soc. 350, 849.

Nozawa, T., Kozasa, T., 2003, Astrophys. J. 598, 785.

Nugent, P., Kim, A., Perlmutter, S., 2002, Publ. Astron. Soc. Pacific 114, 803.

Nürnberger, D. E. A., 2003, Astron. Astrophys. 404, 255.

Nürnberger, D. E. A., Bronfman, L., Yorke, H. W., Zinnecker, H., 2002, Astron. Astrophys. 394, 253.

Nürnberger, D. E. A., Petr-Gotzens, M. G., 2002, Astron. Astrophys. 382, 537.

Nürnberger, D. E. A., Stanke, Th., 2003, Astron. Astrophys. 400, 223.

OGLE, 2004a, http://sirius.astrouw.edu.pl/ ogle/

OGLE, 2004b, http://bulge.princeton.edu/ ogle/

Oschmann, Jr., J. M., 2004: Ground-based Telescopes, Proc. SPIE vol. 5489.

Osterloh, M., Beckwith, S. V. W., 1995, Astrophys. J. 439, 288.

Owner-Petersen, M. and Goncharov, A., "Some Consequences of Atmospheric Dispersion for ELTs", Proc. SPIE, Vol. 5489, 2004.

Owner-Petersen, M., Andersen, T., Goncharov, A., and Beckers, J., "Control Strategy for the Adaptive Optics of the Euro50", Proc. SPIE, Vol. 4840, 2003, pp. 427-435.

Papaloizou, J. C. B., Nelson, R. P., 2003, Mon. Not. Roy. Astron. Soc. 339, 983.

Papaloizou, J. C. B., Nelson, R. P., Snellgrove, M. D., 2004, Mon. Not. Roy. Astron. Soc. 350, 829.

Perlmutter, S., Aldering, G., Goldhaber, G., Knop. R. A., Nugent, P., Castro, P. G., Deustua, S., Fabbro, S., Goobar, A., Groom, D. E., Hook, I. M., Kim, A. G., Kim, M. Y., Lee, J. C., Nunes, N. J., Pain, R., Pennypacker, C. R., Quimby, R., Lidman, C, Ellis, R. S., Irwin, M., McMahon, R. G., Ruiz-Lapuente, P., Walton, N., Schaefer, B., Boyle, B. J., Filippenko, A. V., Matheson, T., Fruchter, A. S., Panagia, N., Newberg, H. J. M., Couch, W. J., 1999, Astrophys. J. 517, 565.

Pichon, C., Scannapieco, E., Aracil, B., Petitjean, P., Aubert, D., Bergeron, J., Colombi, S., 2003, Astrophys. J. 597, L 97.

Pickett, B. K., Mejía, A. C., Durisen, R. H., Cassen, P. M., Berry, D. K., Link, R. P., 2003, Astrophys. J. 590, 1060.

Pollack, J. B., Hubickyj, O., Bodenheimer, P., Lissauer, J. J., Podolak, M., Greenzweig, Y., 1996, Icarus 124, 62.

Rafikov, R. R., 2003a, Astron. J. 125, 906.

Rafikov, R. R., 2003b, Astron. J. 125, 942.

Rebolo, R., Zinnecker, H., 2004, ELT: Stars and Planets Summary, Opticon-ELT Science Case Meeting, Florence, November 2004.

Reiners, A., Schmitt, J. H. M. M., 2003, Astron. Astrophys. 398, 647.

Rice, W. K. M., Wood, K., Armitage, P. J., Whitney, B. A., Bjorkman, J. E., 2003, Mon. Not. Roy. Astron. Soc. 342, 79.

Riess, A. G., Filippenko, A. V., Li, W., Schmidt, B. P., 1999a, Astron. J. 118, 2668.

Riess, A. G., Filippenko, A. V., Li, W., Treffers, R. R., Schmidt, B. P., Qiu, Y., Hu, J., Armstrong, M., Faranda, C., Thouvenot, E., Buil, C., 1999b, Astron. J. 118, 2675.

Riewaldt, H., Lastiwka, K., Quinlan, N., McNamara, K., Wang, X., Andersen, T., and Shearer, A., "Wind on the Euro50 Enclosure", Proc. SPIE, Vol. 5495, 2004.

Schneider, J., 2004, http://www.obspm.fr/encycl/catalog.html

Shigeyama, T., Tsujimoto, T., 2003, Star Formation History in Galaxies Inferred from Stellar Elemental Abundance Patterns, in Proc. IAU 8th Asian-Pacific Regional Meeting, vol. I, Ikeutchi, S., Hearnshaw, J., Hanawa, T., ASP Conf. Proc. vol. 289, pp. 263-266.

Smith, B. A., Terrile, R. J., 1984, Science 226, 1421.

Snel, R., Ardeberg, A., Flicker, R., 2004, Some aspects of science with a 50 m AO telescope, in : 2nd Bäckaskog Workshop on Extremely Large Telescopes, Proc. SPIE vol. 5382, pp. 57-66.

Stanek, K. Z., Matheson, T., Garnavich, P. M., Martini, P., Berlind, P., Caldwell, N., Challis, P., Brown, W. R., Schild, R., Krisciunas, K., Calkins, M. L., Lee, J. C., Hathi, N., Jansen, R. A., Windhorst, R., Echevarria, L., Eisenstein, D. J., Pindor, B., Olszewski, E. W., Harding, P., Holland, S. T., Bersier, D., 2003, Astrophys. J. 591, L 17.

Steinacker, A., Papaloizou, J. C. B., 2002, Astrophys. J. 571, 413.

Stephens, A. W., Frogel, J. A., DePoy, D. L., Freedman, W., Gallart, C., Jablonka, P., Renzini, A., Rich, R. M., Davies, R., 2003, Astron. J. 125, 2473.

Strom, S. E., 1995, Rev. Mex. Astron. Astrofís. Conf. Ser. 1, 317.

Théado, S., Vauclair, S., 2003a, Astrophys. J. 587, 784.

Théado, S., Vauclair, S., 2003b, Astrophys. J. 587, 795.

Timmes, F. X., Brown, E. F., Truran, J. W., 2003, Astrophys. J. 590, L 83.

Tonry, J. L., Schmidt, B. P., Barris, B., Candia, P., Challis, P., Clocchiatti, A., Coil, A. L., Filippenko, A. V., Garnavich, P., Hogan, C., Holland, S. T., Jha, S., Kirshner, R. P., Krisciunas, K., Leibundgut, B., Li, W., Matheson, T. Phillips, M. M., Riess, A. G., Schommer, R., Smith, R. C., Sollerman, J., Spyromilio, J., Stubbs, C. W., Suntzeff, N. B., 2003, Astrophys. J. 594, 1.

Trimble, V., Aschwanden, M. J., 2004, Publ. Astron. Soc. Pacific 116, 187.

Vakili, F., 2003, High Dynamic Imaging of Luminous Massive Stars, in Astronomy with High Contrast Imaging: From Planetary Systems to Active Galactic Nuclei, Aime, C., Soummer, R. (eds.), EAS Publ. Ser. Vol. 8, EDP Sciences, Les Ulis, France, p. 49.

Valentini, G., Di Carlo, E., Massi, F., Dolci, M., Arkharov, A. A., Larionov, V. M., Pastorello, A., Di Paola, A., Benetti, S., Cappellaro, E., Turatto, M., Pedichini, F., DŠAlessio, F., Caratti o Garatti, A., Li Causi, G., Speziali, R., Danziger, I. J., Tornambé, A., 2003, Astrophys J. 595, 779.

Vauclair, S., Théado, S., 2003, Astrophys. J. 587, 777.

Vidal-Madjar, A., Lecavelier des Etangs, A., Désert, J.-M., Ballester, G. E., Ferlet, R., Hébrard, G., Mayor, M., 2003, Nature 422, 143.

Wahhaj, Z., Koerner, D. W., Ressler, M. E., Werner, M. W., Backman, D. E., Sargent, E. L., 2003, Astrophys. J. 584, L 27.

Wang, Y., Freese, K., Gondolo, P., Lewis, M., 2003, Astrophys. J. 594, 25.

Weinberger, A. J., Becklin, E. E., Zuckerman, B., 2003, Astrophys. J. 584, L 33.

Winters, W., Balbus, S., Hawley, J., 2003, Astrophys. J. 589, 543.

Yorke, H., Sonnhalter, C., 2002, Astrophys. J. 569, 846.

Yoshida, N., Abel, T., Hernquist, L., Sugiyama, N., 2003, Astrophys. J. 592, 645.

Zuckerman, B., Forveille, T., Kastner, J. H., 1995, Nature 373, 494.

Author Index

The manufacturer's authorised representative in the EU is Springer Nature Customer Service Centre GmbH, Europaplatz 3, 69115 Heidelberg, Germany. If you have any concerns regarding our products, please contact ProductSafety@springernature.com

Printed and bound by CPI Group (UK) Ltd, Croydon, CR0 4YY
30/09/2026
02225436-0006